AF597918

METHODS IN MOLECULAR BIOLOGY™

Series Editor
John M. Walker
School of Life Sciences
University of Hertfordshire
Hatfield, Hertfordshire, AL10 9AB, UK

For further volumes:
http://www.springer.com/series/7651

Protein Acetylation

Methods and Protocols

Edited by

Sandra B. Hake

Department of Molecular Biology, Adolf-Butenandt Institute, München, Germany

Christian J. Janzen

Department of Cell & Developmental Biology, Biocenter University of Würzburg, Würzburg, Germany

Editors
Sandra B. Hake
Department of Molecular Biology
Adolf-Butenandt Institute
München, Germany

Christian J. Janzen
Department of Cell & Developmental Biology
Biocenter University of Würzburg
Würzburg, Germany

ISSN 1064-3745 ISSN 1940-6029 (electronic)
ISBN 978-1-62703-304-6 ISBN 978-1-62703-305-3 (eBook)
DOI 10.1007/978-1-62703-305-3
Springer New York Heidelberg Dordrecht London

Library of Congress Control Number: 2012955678

Printed on acid-free paper

Humana Press is a brand of Springer
Springer is part of Springer Science+Business Media (www.springer.com)

Preface

The addition of chemical tags to proteins has long been recognized as one important mechanism to regulate different functions of the target protein. Of particular interest is the enzymatic addition of one acetyl group to proteins, a reversible process that has been discovered more than 40 years ago, and which can occur co- and posttranslationally. Highly specific acetyltransferases catalyze the addition of an acetyl group to target sequences in proteins, and deacetylases specifically remove these chemical tags if necessary.

Since the initial discovery, many thousands of proteins have been identified to be acetylated, and immense research power has been attributed worldwide into experiments to solve the biological implications of each and every protein acetylation.

Two particular sites of protein acetylation have been described intensively: the N-terminal methionine residue of a nascent protein and lysine residues within a protein.

In this book a comprehensive collection of methods describing several different topics of protein acetylation is assembled. Starting with different methods used for initial identification of protein acetylation, such as mass spectrometry, column- and gel electrophoresis-based approaches, and ending with methods to computationally predict and to functionally study the biological response to protein acetylation, this book covers most technical aspects involved in the studies of protein acetylation.

We hope that the readers will be satisfied with this large collection of methods and will succeed in their own research on protein acetylation.

München, Germany — *Sandra B. Hake*
Würzburg, Germany — *Christian J. Janzen*

Contents

Contributors

MOHAMMED ARIF • *Transcription and Disease Laboratory, Molecular Biology and Genetics Unit, Jawaharlal Nehru Centre for Advanced Scientific Research, Bangalore, India*

THOMAS ARNESEN • *Department of Molecular Biology, University of Bergen, Bergen, Norway; Department of Surgery, Haukeland University Hospital, Bergen, Norway*

AMRITA BASU • *Broad Institute of MIT and Harvard, Cambridge, MA, USA*

RODERICK S.P. BENSON • *Imagen Biotech Ltd, Manchester, UK*

JEF D. BOEKE • *Department of Molecular Biology and Genetics, The High Throughput Biology Center, John Hopkins University School of Medicine, Baltimore, MD, USA*

JOANNA CHOWDRY • *Molecular Gastroenterology Research Group, Department of Oncology, University of Sheffield, Sheffield, UK*

BERNARD M. CORFE • *Molecular Gastroenterology Research Group, Department of Oncology, University of Sheffield, Sheffield, UK*

BO DAI • *Center for Regenerative Biology, University of Connecticut, Storrs, CT, USA; Institute for Stem Cell Biology and Regenerative Medicine, Stanford Cancer Center, Stanford University School of Medicine, Stanford, CA, USA*

CAROLINE A. EVANS • *Department of Chemical and Biological Engineering, ChELSI Institute, University of Sheffield, Sheffield, UK*

RUNE H. EVJENTH • *Department of Molecular Biology, University of Bergen, Bergen, Norway*

JULIA M. GAJER • *Institute of Pharmaceutical Sciences, University of Freiburg, Freiburg, Germany*

TIELONG GAO • *Department of Chemistry, Georgia State University, Atlanta, GA, USA*

BENJAMIN A. GARCIA • *Department of Molecular Biology, Princeton University, Princeton, NJ, USA; Department of Chemistry, Princeton University, Princeton, NJ, USA*

MICHAEL A. GEEVES • *School of Biosciences, University of Kent, Canterbury, UK*

KRIS GEVAERT • *Department of Medical Protein Research, VIB, Ghent, Belgium; Department of Biochemistry, Ghent University, Ghent, Belgium*

CHARLES GIARDINA • *Department of Molecular and Cell Biology, University of Connecticut, Storrs, CT, USA*

MARKUS GRAMMEL • *The Laboratory of Chemical Biology and Microbial Pathogenesis, The Rockefeller University, New York, NY, USA*

GARETH J. GRIFFITHS • *Imagen Biotech Ltd, Manchester, UK*

JOHN R. GRIFFITHS • *Paterson Institute for Cancer Research, University of Manchester, Manchester, UK*

HOWARD C. HANG • *The Laboratory of Chemical Biology and Microbial Pathogenesis, The Rockefeller University, New York, NY, USA*

ALEXANDER-THOMAS HAUSER • *Institute of Pharmaceutical Sciences, University of Freiburg, Freiburg, Germany*

Peter Højrup • *Department of Biochemistry and Molecular Biology, University of Southern Denmark, Odense, Denmark*

Cecil J. Howard • *Department of Biochemistry and The Ohio State Biochemistry Program, The Ohio State University, Columbus, OH, USA*

Matthew Johnson • *School of Biosciences, University of Kent, Canterbury, UK*

Manfred Jung • *Institute of Pharmaceutical Sciences, University of Freiburg, Freiburg, Germany; Freiburg Institute of Advanced Studies (FRIAS), University of Freiburg, Freiburg, Germany*

Dhanasekaran Karthigeyan • *Transcription and Disease Laboratory, Molecular Biology and Genetics Unit, Jawaharlal Nehru Centre for Advanced Scientific Research, Bangalore, India*

Josephine Kilner • *Department of Chemical and Biological Engineering, ChELSI Institute, University of Sheffield, Sheffield, UK*

G.V. Pavan Kumar • *Light Scattering Laboratory, Chemistry and Physics of Materials Unit, Jawaharlal Nehru Centre for Advanced Scientific Research, Bangalore, India*

Tapas K. Kundu • *Transcription and Disease Laboratory, Molecular Biology and Genetics Unit, Jawaharlal Nehru Centre for Advanced Scientific Research, Bangalore, India*

Yu-Yi Lin • *Department of Oncology, National Taiwan University Hospital, Institute of Biochemistry and Molecular Biology, College of Medicine, National Taiwan University, Taipei, Taiwan*

Jin-Ying Lu • *Department of Laboratory Medicine, National Taiwan University Hospital, Institute of Molecular Medicine, College of Medicine, National Taiwan University, Taipei, Taiwan*

Chandan Mandal • *Cancer and Cell Biology Division, Council of Scientific and Industrial Research (CSIR)-Indian Institute of Chemical Biology, Kolkata, India*

Chitra Mandal • *Cancer and Cell Biology Division, Council of Scientific and Industrial Research (CSIR)-Indian Institute of Chemical Biology, Kolkata, India*

Daniel P. Mulvihill • *School of Biosciences, University of Kent, Canterbury, UK*

Chandrabhas Narayana • *Light Scattering Laboratory, Chemistry and Physics of Materials Unit, Jawaharlal Nehru Centre for Advanced Scientific Research, Bangalore, India*

Jennifer J. Ottesen • *Department of Biochemistry and The Ohio State Biochemistry Program, The Ohio State University, Columbus, OH, USA*

Saw Yen Ow • *Department of Chemical and Biological Engineering, ChELSI Institute, University of Sheffield, Sheffield, UK*

Romeo Papazyan • *Department of Pharmacology and Molecular Sciences, Johns Hopkins University School of Medicine, Baltimore, MD, USA*

Michael G. Poirier • *Department of Physics, The Ohio State Biochemistry Program, The Ohio State University Medical Center, The Ohio State University, Columbus, OH, USA; Department of Biochemistry, The Ohio State Biochemistry Program, The Ohio State University Medical Center, The Ohio State University, Columbus, OH, USA; Department of Molecular Virology, Immunology, and Medical Genetics, The Ohio State University Medical Center, The Ohio State University, Columbus, OH, USA*

Theodore P. Rasmussen • *Department of Pharmaceutical Sciences, University of Connecticut, Storrs, CT, USA; Department of Molecular and Cell Biology, University of Connecticut, Storrs, CT, USA; University of Connecticut Stem Cell Institute, Farmington, CT, USA*

BART RUTTENS • *Department of Medical Protein Research, VIB, Ghent, Belgium; Department of Biochemistry, Ghent University, Ghent, Belgium*
JOHN C. SHIMKO • *Department of Biochemistry and The Ohio State Biochemistry Program, The Ohio State University, Columbus, OH, USA*
SOUMIK SIDDHANTA • *Light Scattering Laboratory, Chemistry and Physics of Materials Unit, Jawaharlal Nehru Centre for Advanced Scientific Research, Bangalore, India*
DUNCAN L. SMITH • *Paterson Institute for Cancer Research, University of Manchester, Manchester, UK*
SEAN D. TAVERNA • *Department of Pharmacology and Molecular Sciences, Johns Hopkins University School of Medicine, Baltimore, MD, USA*
RICHARD D. UNWIN • *Centre for Advanced Discovery and Experimental Therapeutics (CADET), Central Manchester University Hospitals NHS Foundation Trust and School of Medicine, University of Manchester, Manchester, UK*
PETRA VAN DAMME • *Department of Medical Protein Research, VIB, Ghent, Belgium; Department of Biochemistry, Ghent University, Ghent, Belgium*
ANTHONY D. WHETTON • *School of Cancer and Enabling Sciences, Wolfson Molecular Imaging Centre, Manchester Academic Health Science Centre, University of Manchester, Manchester, UK*
PHILLIP C. WRIGHT • *Department of Chemical and Biological Engineering, ChELSI Institute, University of Sheffield, Sheffield, UK*
WEI XU • *State Key Lab of Genetic Engineering, Institutes of Biomedical Sciences, Fudan University, Shanghai, China*
CHAO YANG • *Department of Chemistry, Georgia State University, Atlanta, GA, USA*
BARRY M. ZEE • *Department of Molecular Biology, Princeton University, Princeton, NJ, USA*
XUMIN ZHANG • *Department of Food Science, Aarhus University, Årslev, Denmark*
SHIMIN ZHAO • *State Key Lab of Genetic Engineering, Institutes of Biomedical Sciences, Fudan University, Shanghai, China*
YUJUN GEORGE ZHENG • *Department of Chemistry, Georgia State University, Atlanta, GA, USA*
HENG ZHU • *Departments of Pharmacology, The High Throughput Biology Center, John Hopkins University School of Medicine, Baltimore, MD, USA; Department Molecular Sciences, The High Throughput Biology Center, John Hopkins University School of Medicine, Baltimore, MD, USA*

Chapter 1

Validation of Protein Acetylation by Mass Spectrometry

Barry M. Zee and Benjamin A. Garcia

Abstract

Due to the key role of posttranslational modifications (PTMs) such as lysine acetylation in numerous signaling and regulatory pathways, the ability to identify novel PTMs and to quantify their abundances provides invaluable information for understanding these signaling networks. Currently, mass spectrometry (MS) arguably serves as the most high-throughput and unbiased platform for studying PTMs. Here we detail experimental and analytical procedures for the characterization of lysine acetylation on proteins in general and on histones in particular, which are among the most highly modified proteins in eukaryotic cells.

Key words: Mass spectrometry, Lysine acetylation, Posttranslational modification, Proteomics, Histones

1. Introduction

The identification of posttranslational modifications (PTMs) is often more technically challenging than the identification of the modified protein itself for many both biological and technical reasons. First, the complexity of modifications, ranging from small groups such as acetylation to highly branched moieties such as poly-ADP-ribosylation, vastly dwarves the sequence complexity found with the 20 canonical amino acids. Second, most modifications occur at sub-stoichiometric levels relative to the total protein pool, often rendering the detection of such modifications below the dynamic range of the method used. Third, most enzymatically catalyzed modifications are reversible, and thus are dynamic in nature and have limited transient lifetimes. Undeniably the most extensively studied class of modifications in molecular biology is serine/threonine/tyrosine phosphorylation, where such investigations are

Sandra B. Hake and Christian J. Janzen (eds.), *Protein Acetylation: Methods and Protocols*, Methods in Molecular Biology, vol. 981, DOI 10.1007/978-1-62703-305-3_1, © Springer Science+Business Media, LLC 2013

greatly facilitated by several methods including ^{32}P-radiolabeling, phosphorylation-specific antibodies, phosphospecific gel staining, and enrichment using various approaches such as immobilized metal affinity chromatography (1–4).

In contrast to phosphorylation, relatively fewer tools are available for the identification of lysine acetylation. Such methodological deficit is particularly disconcerting given recent investigations identifying approximately 3,500 acetylated residues in mammalian cells (5, 6) comparable to the approximately >10,000 phosphorylated residues identified also in mammalian cells (7). Additionally, the number of reports describing protein phosphorylation far surpasses those studies characterizing acetylated proteins. Mass spectrometry has emerged as the most rigorous platform for both the identification and quantification of lysine acetylation (8). Essentially using the same logic as Edman degradation using phenylthiohydantoin or dansyl-chloride, MS sequencing of a putative acetylated peptide relies on the fragmentation of a peptide into smaller peptide fragments sharing a common N- or C-terminus. The difference in mass between peptide fragments sharing either a common N- or C-terminus essentially corresponds to the residue mass of a single additional amino acid between the fragments. When that difference cannot be accounted by the residue mass of lysine alone but rather by the mass of lysine plus a nominal mass shift of 42 Da (or more accurately a mass of 42.010 Da), the possibility of an acetylated rather than unmodified lysine becomes highly likely. In this chapter, we detail the latest advancements in the MS analysis of lysine acetylation for proteins in general and for histones in particular, which require a more specialized workflow.

2. Materials

All reagents and samples should be handled with no contact to bare skin in order to minimize keratin contamination or to any detergents such as those used in glassware cleaning. Unless otherwise noted, all reagents can be stored at room temperature.

2.1. Preparation of Protein Samples for MS Analysis

1. Sequencing-grade trypsin, stored at –80°C (see Note 1).
2. 100 mM ammonium bicarbonate, pH 8, stored at 4°C (see Note 2).
3. 10 mM dithiothreitol (DTT) in 100 mM ammonium bicarbonate, prepared fresh prior to use.
4. 55 mM iodoacetamide in 100 mM ammonium bicarbonate, prepared fresh prior to use and stored in the dark.
5. Acetonitrile (see Note 3).

6. Glacial acetic acid.
7. Peptide desalting kits, such as ZipTip pipette tips (Millipore) or Proxeon StageTips C18 material (Thermo Scientific) (see Note 4).
8. Razor blades (only if proteins were resolved by SDS-PAGE).
9. Coomassie blue destain solution: 40% Volume methanol, 10% volume glacial acetic acid, 50% volume water.

2.2. Preparation of Histone Samples for MS Analysis

1. Sequencing-grade trypsin, stored at −80°C (see Note 1).
2. 100 mM ammonium bicarbonate, pH 8, stored at 4°C (see Note 2).
3. Anhydrous isopropanol (see Note 5).
4. Anhydrous propionic anhydride (see Note 5).
5. Ammonium hydroxide solution (28.0–30% NH_3 basis).

3. Methods

SDS-PAGE of protein samples is best reserved for complex solutions containing approximately 10 or more expected proteins. Using SDS-PAGE for relatively simple solutions will reduce the digestion efficiency and increase the risk of introducing acrylamide adducts that will complicate peptide sequencing. Highly complex protein samples containing approximately 100 or more expected proteins should be fractionated, for instance via reversed-phase liquid chromatography, prior to SDS-PAGE. Similar logic is applicable to histone preparations, depending on the efficiency of the purification method used. All steps, unless otherwise noted, occur in room temperature.

3.1. In-Gel Digestion of Protein Bands and Peptide Extraction from Gel Bands

1. Excise out desired protein bands (see Note 6) with clean razor blades, and dice into approximately 1 mm^3 pieces. Place gel pieces within a 1.5 mL tube. Ensure that razor blades are wiped clean with methanol prior to cutting another band to minimize cross contamination between samples.
2. If bands were stained with Coomassie blue, destain the band for at least 1 h to reduce the amount of residual Coomassie that could impede trypsin digestion efficiency and could eventually be introduced into the mass spectrometer. If bands are irreversibly stained, for instance by silver staining, proceed to next step.
3. Replace destain solution with 100 mM ammonium bicarbonate and shake or rotate vigorously for 5 min. Remove as much ammonium bicarbonate as possible and dehydrate gel slices using a vacuum centrifuge (see Note 7).

4. Add a sufficient volume of 10 mM DTT to cover gel and reduce samples for 1 h at 51°C. Afterwards, allow tubes to cool to room temperature and replace DTT with equal volume of 55 mM iodoacetamide. Incubate for 1 h in the dark.

5. Remove iodoacetamide and wash gel pieces with 100 mM ammonium bicarbonate by shaking/rotating for 5 min. After the wash, replace ammonium bicarbonate with acetonitrile to dehydrate the gel band and shake/rotate for 5 min. Remove acetonitrile and dry gel pieces completely.

6. Swell gels with minimal volume (i.e., enough to submerge the gel pieces) of 12.5 ng/μL sequencing-grade trypsin in 100 mM ammonium bicarbonate for approximately 30 min on ice. After gel pieces are fully swollen, incubate samples at room temperature for 6–12 h (see Note 8).

7. Collect solution that was not absorbed within gel into another tube, and dehydrate with minimal volume of 75% acetonitrile/5% acetic acid while shaking/rotating for 5–10 min. Afterwards, pool solution with the previously collected digest solution in the new tube and rehydrate gel pieces with minimal volume of 100 mM ammonium bicarbonate while shaking/rotating for 5–10 min. Pool this solution into the same new tube, and repeat the dehydration/rehydration step one additional time.

8. After pooling the 100 mM ammonium bicarbonate, wash for the second time, and dehydrate the gel with 100% acetonitrile while shaking/rotating for 5–10 min. Without collecting the acetonitrile, add 100 mM ammonium bicarbonate at 3× excess volume of the acetonitrile solution while shaking/rotating for 5–10 min. Pool the acetonitrile/ammonium bicarbonate solution into the same new tube as before. Repeat the dehydration/rehydration step one additional time. Remove the acetonitrile content in the pooled solution using a vacuum centrifuge prior to desalting using commercially available kits.

3.2. In-Solution Digestion of Protein Samples

1. Adjust pH of the solution to 7.5–9, and determine protein concentration using the most reliable assay available. Add DTT to final concentration of 10 mM and reduce for 1 h at 51°C. Add excess volume of iodoacetamide for final concentration of approximately 55 mM and alkylate for 45 min in the dark.

2. Dry the solution to near completeness, with residual volume <5 μL using a vacuum centrifuge, and add sequencing-grade trypsin at a 1:10–1:20 enzyme:substrate concentration. Allow to digest for approximately 5–6 h at 37°C. Quench reaction with addition of glacial acetic acid to approximately pH = 4. Samples are now ready for desalting using commercially available kits (see Note 4).

3.3. In-Solution Derivatization of Histone Samples (9)

1. Adjust pH of histone samples to approximately pH 8. Determine histone concentration using the most convenient assay available. If histones are dried, solubilize the proteins with 100 mM ammonium bicarbonate. Prepare a 3:1 isopropanol:propionic anhydride reagent by volume and add reagent to histone sample at approximately half sample volume for derivatization.
2. Quickly adjust pH of samples with addition of ammonium hydroxide to return to pH 8. If samples become too basic at pH≥10, adjust by addition of isopropanol:propionic anhydride reagent. Once pH is adjusted, incubate samples at 37°C for 15 min.
3. Reduce sample to original volume using a vacuum centrifuge, and repeat the derivatization one additional time. Afterwards, add 100 mM ammonium bicarbonate at 3–4× excess volume of the sample and digest with sequencing-grade trypsin at 1:20 enzyme:substrate concentration at 37°C for 5–6 h. Quench digestion with addition of acetic acid to pH 4.
4. Reduce sample volume, and after readjusting to pH 8, repeat derivatization method two times. After the second derivatization of the tryptic peptides, histone peptides are now ready for desalting using commercially available kits.

3.4. LC–MS Operation and Analysis of MS Data

Depending on the instrument available, there are various setups for peptide sequencing. This chapter assumes upfront sample separation with an online HPLC followed either by a linear ion trap quadrupole mass spectrometer or a hybrid linear ion trap-orbitrap mass spectrometer.

1. The optimal HPLC gradient should be determined for each protein sample. Gradient parameters include the buffer composition, rate of change of the buffers, and gradient duration. Generally, reversed-phase chromatography provides a versatile gradient for resolving most tryptic peptides.
2. For nonhistone proteins, a typical instrument method would comprise the acquisition of one full mass spectrum (MS) followed by 5–10 data-dependent tandem mass spectra (MS/MS) for the 5–10 most abundant peptide ions detected from the mass spectrum. Collisional induced dissociation (CID) of the precursor ion should be appropriate for most tryptic peptides (see Note 9).
3. After the MS run is completed, proceed to a bioinformatic search of putative acetylated peptides. Various software packages are available commercially, and other programs are developed in-house (see Note 10).

3.5. Validation of Putative Acetylated Lysines on Histone and Nonhistone Proteins

1. Confirm whether the precursor mass found in the MS is consistent with an acetyl moiety (42.010 Da) or the nearly isobaric trimethyl moiety (42.046 Da); if the samples were analyzed on a high-mass-accuracy mass spectrometer, such as the hybrid linear ion trap-orbitrap this should be possible (see Note 11). Shown in Fig. 1 are the total ion chromatograms for the doubly charged ions from three histone peptides: the histone H3 unmodified 9–17 peptide $_{pr}K_{pr}STGGK_{pr}APR$ (Fig. 1a), the H3K9me3 9–17 peptide $_{pr}K_{me3}STGGK_{pr}APR$ (Fig. 1b), and the H3K9ac 9–17 peptide $_{pr}K_{ac}STGGK_{pr}APR$ (Fig. 1c). Note that pr=propionyl group added to unmodified lysines. Differentiation of the H3K9me3 and H3K9ac peptides based on accurate mass can be accomplished, as the mass of the peptide in Fig. 1b is closer to the calculated mass of the H3K9me3 (1.89 ppm error for me3 compared to 35.9 ppm error for H3K9ac), while the mass of the peptide in Fig. 1c is closer to the calculated mass of the H3K9ac (0 ppm error for me3 compared to 34 ppm error for H3K9ac).
2. Inspect manually the MS/MS corresponding to the putative acetylated peptide. First confirm the presence of particular b (n-terminal)- and y (c-terminal)-ion fragments that arise only from the acetylated residue. For instance, in the 9–17 histone H3 peptide $_{pr}K_{ac}STGGK_{pr}APR$ (pr=propionyl group added to unmodified lysines), the presence of the expected y4–y8 fragment ions is more critical than the presence of the expected y1–y3 fragment ions for localizing the acetyl moiety to K9 rather than K14 (Fig. 2). Similar logic can be applied for finding the appropriate b-ion series. One should be concerned with not only the extent of b- and y-ion coverage but also the overall noise of the MS/MS. Maximum b- and y-ion coverage is readily observed for a very noisy MS/MS. An ideal MS/MS should contain relatively few ions that cannot be accounted by the expected CID fragmentation pattern (see Note 12).
3. Confirm that the acetylated histone peptide elutes *earlier* than the respective unmodified histone peptide if a reversed-phase gradient were used due to the absence of a propionyl group on the acetylated peptide and thus a decrease in relative hydrophobicity (see Note 13). Furthermore, a peptide containing an acetylated lysine should elute *later* than the same peptide containing a trimethylated lysine due to charge stabilization on the side-chain amine from the three methyl groups compared to charge removal on the amine from the acetyl group (see Note 14).

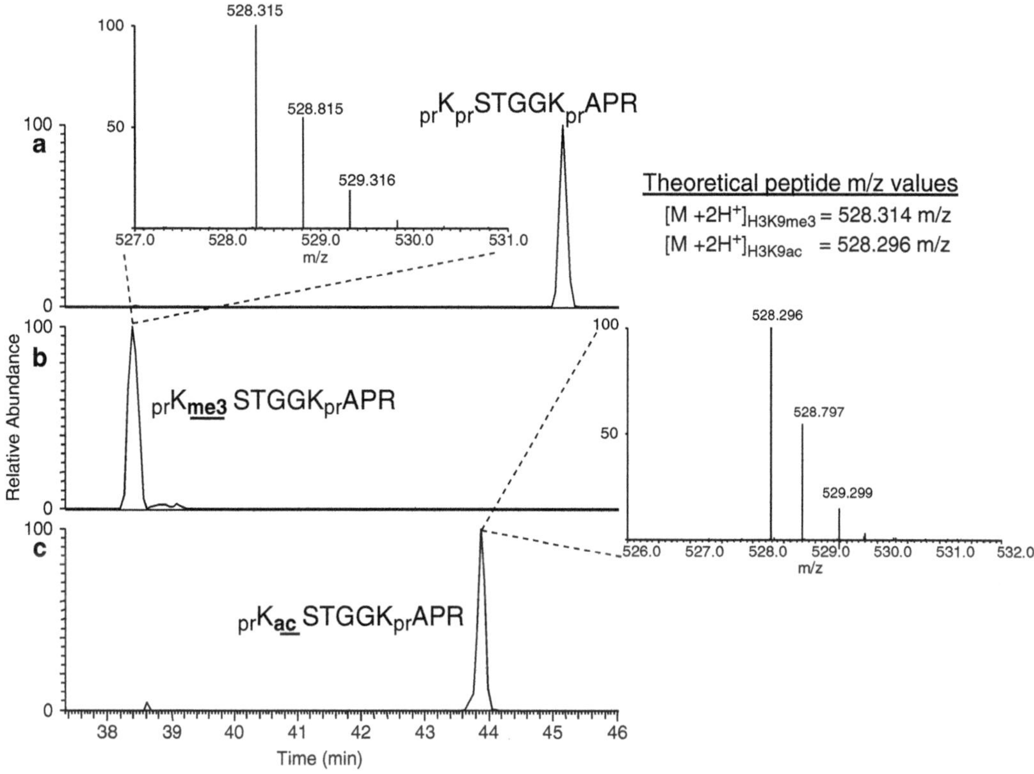

Fig. 1. Extracted ion chromatograms of the doubly charged 9–17 peptide (KSTGGKAPR) on histone H3 unmodified on K9 (**a**), trimethylated on K9 (**b**), and monoacetylated on K9 (**c**). Differences in retention time under reversed-phase chromatography facilitate PTM assignment of propionylated histone peptides. In general, the acetylated peptide elutes earlier than the unmodified peptide, which elutes earlier than the monomethylated peptide. Inserts depict the MS of the respective trimethylated and monoacetylated histone peptides acquired in the orbitrap. Note that the spacing of the isotopes (~0.5 *m*/*z*) indicates the histone peptide as doubly charged. Abbreviations: *pr* propionyl, *me3* trimethyl, *ac1* acetyl.

4. Notes

1. Despite the catalytic robustness of trypsin, the protease should not undergo multiple freeze–thaw cycles for maximal digestion efficiency. Do not substitute with tissue culture-grade trypsin due to the increased likelihood of autolysis and reduced substrate specificity resulting from the digested trypsin. Furthermore, one may substitute the ammonium bicarbonate buffer with the accompanied buffer provided by the vendor if the mass spectrometer detects a high abundance of undigested intact proteins. The use of ammonium bicarbonate is simply for convenience with later preparation steps.
2. Over time, the ammonium bicarbonate buffer may lose its buffering capacity. Thus, one should confirm the buffer has pH 8 prior to use.

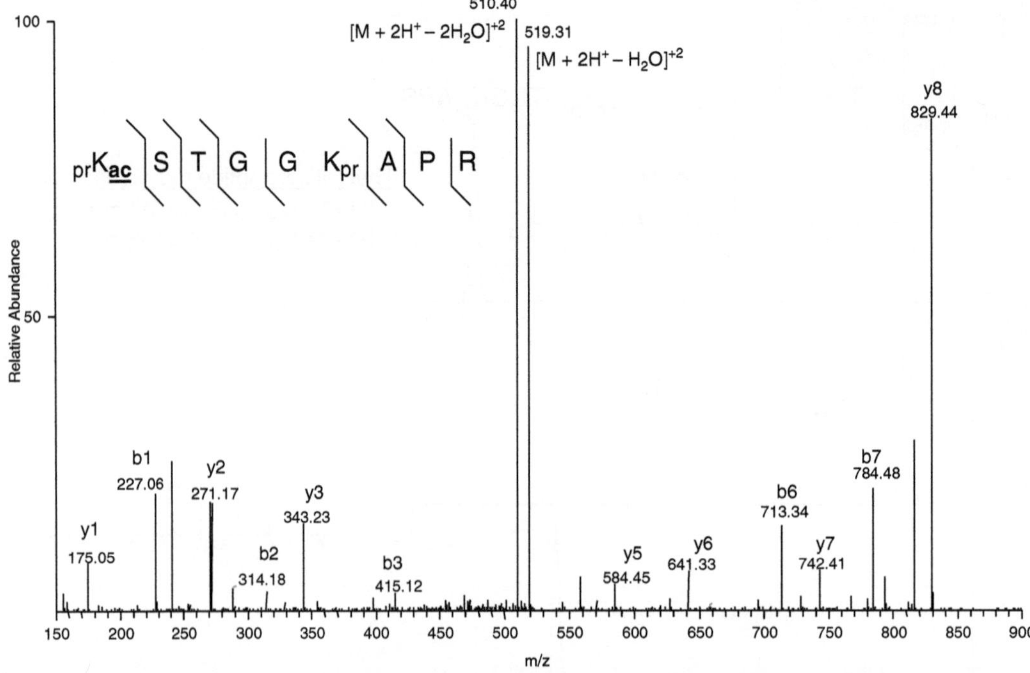

Fig. 2. Tandem mass spectrum of the monoacetylated H3 histone 9–17 peptide, K_{ac}STGGKAPR, with the expected and observed mass of the doubly charged molecular ion ($M + 2H^+$) provided. The MS/MS was acquired from CID fragmentation in the linear ion trap, while the precursor peptide mass ($M + 2H^+$) was determined from the orbitrap. *Hatched lines* above and below the peptide sequence correspond to b- and y-ions, respectively, that were positively annotated in the tandem mass spectrum. Abbreviation: $(M + 2H^+ - H_2O)^{+2}$ = doubly charged ion with loss of 1 water molecule, $(M + 2H^+ - 2H_2O)^{+2}$ = doubly charged ion with loss of 2 water molecules.

3. One does not have to use HPLC-grade acetonitrile for gel desiccation. Standard ACS-grade acetonitrile reagent is sufficient.
4. Desalting helps prevent the reduction in ionization efficiency during electrospray ionization and reduce the likelihood of salt adducts (such as Na^+ or K^+ or buffer salts) forming with the peptide ions. One should be mindful that the desalting resin used should be the same as or similar to the resin used for online LC separation of the peptides during MS analysis. Desalting should only be omitted if one knows a priori that the peptide(s) of interest is unusually hydrophilic (and thus will not be retained on the resin) or that the peptide(s) is of very low abundance (and thus more likely to suffer from sample loss during the desalting procedure).
5. To help maintain anhydrous conditions for isopropanol and propionic anhydride, one can purge the reagent containers with nitrogen (N_2) or argon every time the container is opened, and securely seal the lid. Excess moisture in the reagents will reduce the efficiency of derivatization.

6. Be mindful that the actual and apparent molecular weights of proteins are not necessarily equal. The migration distance of proteins in SDS-PAGE, although determined mostly on the basis of size, can be influenced by other parameters such as phosphorylation and overall charge state.

7. If one does not have access to a vacuum centrifuge, one could leave the samples to air-dry at room temperature for an extended period of time.

8. The optimal digestion temperature of sequencing-grade trypsin is approximately 37°C. However, trypsin digestion can proceed in a relatively broad range of temperatures. If one knows that the ambient temperature in the laboratory has a relatively wide fluctuation, one should digest the samples in a 25–37°C heat block or water bath.

9. If one digests the protein or histone samples with an alternative protease, such as Arg-C, a different fragmentation method may be more suitable for peptide sequencing. Proteases such as Arg-C have an increased likelihood to produce peptides of higher charge states than trypsin due to the presence of internal lysines on the peptides. Higher charge states ($z \geq 3$) lead to increased likelihood for internal CID fragmentation of the peptide, as opposed to fragmentation from the N- or C-terminus, and vastly complicate the sequencing process. Furthermore, because CID is known as a non-ergodic process, the efficiency of CID fragmentation decreases as peptide length increases. An alternative fragmentation scheme potentially more conducive to ArgC digests (or other similarly non-tryptic proteases) is electron transfer dissociation (ETD), which is less sensitive to longer peptide length and more conducive towards higher charged peptides.

10. For bioinformatic analysis, one needs to specify the addition of acetamide (+57.043 Da) on cysteines in the program used; the possible oxidation (+15.995 Da) of methionine, which can arise during sample preparation or electrospray ionization under high voltage; and the possible acetylation (+42.010 Da) on lysine. For propionylated histone samples, one should specify the addition of a propionyl group (+56.026 Da) on the N-terminus, on unmodified (+56.026 Da) and monomethylated lysines (+70.042 Da = 56.026 Da + 14.016 Da) rather than the acetamide derivatization of cysteines. It is important to allow for several (e.g., 1–4) miscleavages by trypsin, since the protease will not cleave at an acetylated lysine.

11. If one intends to use a hybrid linear ion trap-orbitrap mass spectrometer for peptide sequencing, one should perform external mass calibration prior to the experiments in order to take advantage of the mass accuracy of the precursor peptide ion.

In addition, one can use internal ion standard(s) with known masses that allow for mass calibration while the data is being collected (10).

12. It is possible that the isolation width for acquiring the MS/MS is wide enough to pick up multiple different peptides, resulting in a mixed MS/MS of several ion species. One may solve this with a narrower isolation width or a different LC gradient for improved peptide ion resolution.
13. Generally, the presence of an acetylated lysine is accompanied by an unmodified lysine. As described for histones, one can rationalize whether the putative acetylated histone peptide has the expected chromatographic behavior relative to its respective unmodified peptide. For acetylated peptides in general, the unmodified peptide will be longer due to the miscleavage of the acetylated lysine and the peptide chromatographic behavior will be harder to predict. For instance, if one detects the acetylated peptide $PEAK_{ac}ATR$, the unmodified peptides would be $PEAK_{un}$ and ATR. Thus, one should use an alternative protease such as Arg-C that will digest an unmodified and acetylated protein into the same peptide sequence. In this case unlike the histones, the acetylated peptide should elute *later* than the unmodified peptide.
14. Lysine acetylation is a reversible process mediated by acetyltransferases and deacetylases. If no acetylation is found, confirm that deacetylase inhibitors such as sodium butyrate were included in the protein isolation and purification steps. Conversely, the presence of protease inhibitor cocktails commonly used in protein isolation and purification will negatively affect the trypsin digestion step. Thus one should either omit protease inhibitors in the final step of purification prior to digestion or dialyze the proteins away from the inhibitors.

Acknowledgment

BMZ is supported by the NSF Graduate Research Fellowship Program and BAG gratefully acknowledges support from a National Science Foundation (NSF) Early Faculty CAREER award, an NIH Innovator award (DP2OD007447) from the Office of the Director, NIH, and an American Society for Mass Spectrometry research award sponsored by the Waters Corporation. We also thank all members of the Garcia lab for helpful discussion during the composition of this chapter.

References

1. Tolkovsky AM, Wyttenbach A (2009) Differential phosphoprotein labeling (DIPPL) using ^{32}P and ^{33}P. Methods Mol Biol 527:21–29
2. Archuleta AJ, Stutzke CA, Nixon KM, Browning MD (2011) Optimized protocol to make phosphor-specific antibodies that work. Methods Mol Biol 717:69–88
3. Steinberg TH, Agnew BJ, Gee KR, Leung WY, Goodman T, Schulenberg B, Hendrickson J, Beechem JM, Haugland RP, Patton WF (2003) Global quantitative phosphoprotein analysis using Multiplexed Proteomics technology. Proteomics 3:1128–1144
4. Villén J, Gygi SP (2008) The SCX/IMAC enrichment approach for global phosphorylation analysis by mass spectrometry. Nat Protoc 3:1630–1638
5. Norris KL, Lee J, Yao T (2009) Acetylation goes global: the emergence of acetylation biology. Sci Signal 2:76
6. Choudhary C, Kumar C, Gnad F, Nielsen ML, Rehman M, Walther TC, Olsen JV, Mann M (2009) Lysine acetylation targets protein complexes and co-regulates major cellular functions. Science 325:834–840
7. Huttlin EL, Jedrychowski MP, Elias JE, Goswami T, Rad R, Beausoleil SA, Villén J, Haas W, Sowa ME, Gygi SP (2010) A tissue-specific atlas of mouse protein phosphorylation and expression. Cell 143:1174–1189
8. Mann M, Jensen ON (2003) Proteomic analysis of post-translational modifications. Nat Biotechnol 21:255–261
9. Zee BM, Levin RS, DiMaggio PA, Garcia BA (2010) Global turnover of histone post-translational modifications and variants in human cells. Epigenet Chromatin 3:22
10. Olsen JV, de Godoy LM, Li G, Macek B, Mortensen P, Pesch R, Makarov A, Lange O, Horning S, Mann M (2005) Parts per million mass accuracy on an orbitrap mass spectrometer via lock mass injection into a C-trap. Mol Cell Proteomics 4:2010–2021

Chapter 2

Application of the CIRAD Mass Spectrometry Approach for Lysine Acetylation Site Discovery

Caroline A. Evans, Saw Yen Ow, Duncan L. Smith, Bernard M. Corfe, and Phillip C. Wright

Abstract

Mass spectrometry (MS)-based methods typically assess acetylation by detection of a diagnostic ion at 126.1 *m/z*, corresponding to the immonium ion of acetyl-lysine $-NH_3$, which is generated by collisionally induced dissociation. A novel implementation of this approach, based on the accurate mass and retention time technique, couples high mass resolution measurement with rapid cycling between low and elevated collision energies to generate intact and fragment high-resolution mass spectra. This allows acetyl lysine diagnostic ions at 126.1 *m/z* to be monitored and aligned to the precursor *m/z* based on retention time profile. The technique is termed *C*ollisionally *I*nduced *R*elease of *A*cetyl *D*iagnostic. Sequence information is also obtained for acetylation site assignment. This technique to identify acetylation species is information independent as it does not require the sequence of the protein/peptides to identify acetylation, and thus complementary to data-dependent methods. It is suitable for analysis of acetylated peptides, or proteins enriched by immunoprecipitation with acetyl lysine-specific antibodies.

Key words: Lysine acetylation, Site-specific analysis, Mass spectrometry, CIRAD

1. Introduction

Protein acetylation of the ε side chain of lysine residues occurs on a range of proteins, typically macromolecular complexes—and has emerged a key regulatory modification for many proteins—including histone and nonhistone proteins (1). There are several protein acetylation-specific workflows, which are based on mass spectrometry (2). Lysine acetylation can be analyzed by detection of the acetyl lysine diagnostic ion of *m/z* 126.1 (3–5). This chapter focuses on a novel MS method, *C*ollisionally *I*nduced *R*elease of

Sandra B. Hake and Christian J. Janzen (eds.), *Protein Acetylation: Methods and Protocols*, Methods in Molecular Biology, vol. 981, DOI 10.1007/978-1-62703-305-3_2,

*A*cetyl *D*iagnostic (CIRAD), for the analysis of lysine acetylation of peptides and proteins using an information-independent approach, which is distinct from, but complementary to, other MS-based methods describing acetylation site analysis in this volume in Chapters 1 and 3.

The key features of CIRAD are the following:

- A method for targeted analysis of lysine acetylation.
- Based on analysis of peptides derived by proteolytic cleavage of the protein of interest.
- A data-independent MS method—It is not essential to know the a priori identity of the target protein or proteolytic fragment *m/z* values.
- Specificity is achieved by high-resolution and high-accuracy mass measurement of the 126.09 acetyl lysine diagnostic ion and the precursor *m/z* (better than 2 ppm), high confidence identification, as well as high specificity ion extraction (hrEIC).
- Correlation of the acetyl diagnostic ion to the precursor *m/z* based on the time-overlap and elution pattern of chromatographic profile; no assumption is made as to the presence (or not) of other PTMs or the sequence in the peptide to enable the detection of acetylation.
- Application to a protein gel "band" obtained by SDS-PAGE, or mixtures of acetylated peptides enriched by acetyl lysine immunoprecipitation and analyzed by LC–MS/MS.

1.1. Principle of CIRAD

Analysis of peptides by MS generates a survey (or MS1) scan with *m/z* ratios plotted against ion current. Amino acid sequence can be determined by fragmentation along the peptide backbone, promoted by collision with an inert gas at low pressure, and termed collision-induced dissociation (CID). The resulting MS/MS (or MS2) product ion spectrum shows the *m/z* ratios for different fragments of the selected peptide precursor. Conventionally, MS/MS data for a given precursor ion population present in the survey scan is acquired by sequential selection and fragmentation of (typically) the most intense precursor ions (6). In this data-dependent acquisition (DDA) mode, fragment ions are generated from a single precursor ion population. Peptide sequences and sites of modification can be assigned by matching experimental MS/MS spectra to theoretical fragment masses (generated in silico for any given protein database) via protein database searching (7).

Acetylated lysine residues can generate product ions of 143.1 *m/z* and 126.1 *m/z* following peptide fragmentation (CID) (3–5). The 143.1 *m/z* and 126.1 *m/z* ions correspond to the acetyl lysine immonium ion and the immonium ion $-NH_3$, respectively. While the 143.1 *m/z* is only semi-diagnostic, the release of 126.1 *m/z* ion is diagnostic since it is acetylation specific (see Notes 1 and 2).

1.2. CIRAD: Application to Analysis of Acetylated Peptides

The CIRAD technique utilizes the Broad Band CID (bbCID) functionality of the MS instrumentation to simultaneously fragment all precursors observed in the survey scan. Alternating between MS survey (low energy) scans and bbCID (high collision energy) at rapid intervals enables generation of MS scans for all precursor ions and their fragments. Data then can be correlated to link individual co-eluting precursors and their corresponding product ions without the need for specific *m/z*-based precursor selection (see Note 3). The method essentially performs data-independent acquisition and as such is distinct from conventional DDA analysis; the first quadrupole of the MS simply functions in RF mode only to transfer ions of all *m/z* values ("broadband" ion transmission) to the collision cell where low and high alternating collision energies are applied. The presence of acetylated peptides is assessed based on detection of the acetyl lysine diagnostic immonium ion (minus NH_3) at *m/z* 126.092, analyzed by high-resolution extracted ion chromatogram (hrEIC) to "search" for the potential acetylated precursors. The use of an ultra high-resolution time-of-flight instrument generates high-resolution, high-mass-accuracy data, which allows both the use of providing unambiguous precursor and product *m/z* values for peptide identification.

Coupling online to high-performance liquid chromatography (HPLC) peptide fractionation simplifies the mixture of peptides for analysis. Peptides are resolved by HPLC, using reversed-phase chromatography at nanolitre per minute flow rates (200–300 nL/min "nanoflow"). The column eluent is directly coupled to the MS and gas-phase ions are generated and introduced to the MS via nano electrospray ionization (nano ESI). Alternating low and high energy data is acquired at a rate of 1 Hz (1 scan recorded per second) over the course of the RP-HPLC chromatography run, which typically lasts for 90–120 min. Processing of the LC/MS raw data with a specific extraction algorithm (the Dissect algorithm) links precursors and product ions of similar chromatographic behavior, traced across the LC elution peak using both the high and low energy data. A mascot generic format file (with ".mgf" extension) is generated for database searching and sequence assignment. The workflow schematic for CIRAD is shown in Fig. 1.

2. Materials

1. Standard proteins for method setup: Acetylated protein standard, for example acetylated bovine serum albumin (BSA) (Invitrogen, Paisley, UK, or Sigma, Poole, UK).
2. Trypsin: Proteomics grade. Perform solution digest with a ratio of trypsin:acetylated BSA of 1:50 in 25 mM ammonium bicarbonate buffer.

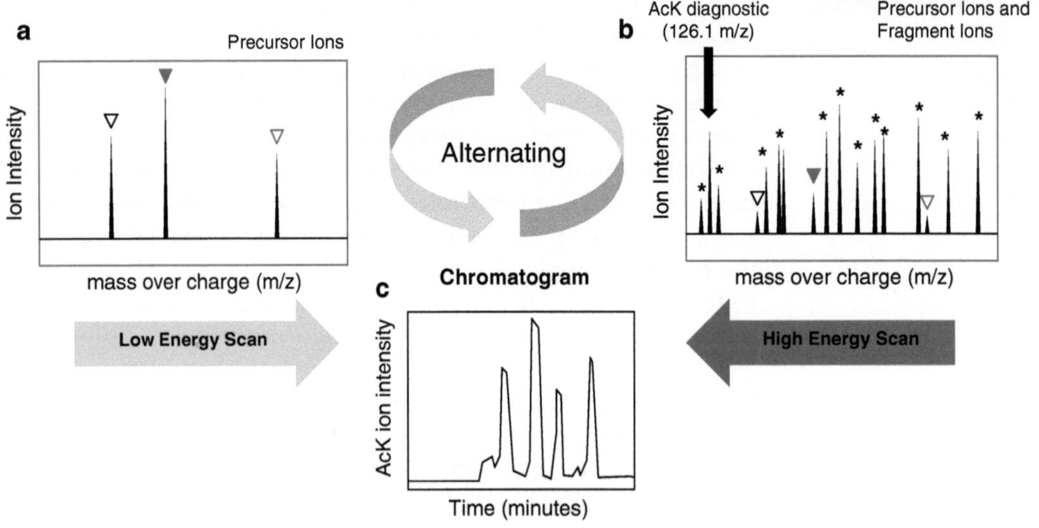

Fig. 1. CIRAD workflow for acetylation site analysis. The CIRAD technique utilizes the Broad Band CID (bbCID) functionality of the MS instrumentation to simultaneously fragment all precursors observed in the survey scan. Alternating between MS survey (low energy) scans and bbCID (high collision energy) at rapid intervals enables generation of MS scans for all precursor ions and their fragments. (**a**) Three different precursors are shown (*triangles*) together with their precursors, including an acetylated precursor (*black triangle*) in the lower energy scan. (**b**) Alternating to high energy promotes collisionally induced dissociation to fragment ions (*asterisk*), which for an acetylated precursor generates the acetyl lysine diagnostic ion at 126.1 *m*/*z*. Sequence information is provided by other *m*/*z* fragments. (**c**) Precursor and product ion connectivity are determined based on common chromatographic elution (retention time) profiles. The presence of acetylated peptides is indicated by use of high-resolution extracted ion chromatograms (hrEIC) for 126.09.

3. Reversed phase-HPLC column for nanoLC and micro precolumn C18 for online sample loading and desalting at flow rates of 300 nL/min and 30 μL/min, respectively.
4. HPLC solvents: Acetonitrile and water (HPLC grade), formic acid.
5. Glass sample vials with lids, 300 μL glass insert, fused into a 2 mL crimp top vial (Chromacol (ThermoFisher), Welwyn Garden City, UK). For sample pick up prior to RP-HPLC analysis.
6. Mass spectrometer, maXis™ UHR-Qq-ToF ToF (Bruker), fitted with electrospray source and EasyNano electrospray needle. Data Analysis software 4.1, with the Dissect Algorithm (see Note 3).

7. Electrospray tunemix ESI-L (Agilent, West Lothian, UK), directly infused at 300 nL/min using a Hamilton syringe matching pump (KD Scientific, MA, USA).
8. NanoLC system: Ultimate U3000 (Thermo Fisher, Hemel Hempstead, UK) modular system (solvent degasser, micro and nanoflow pumps, flow control module and autosampler) or equivalent.

3. Methods

3.1. CIRAD Workflow

The method requires both detection of the 126.1 *m/z* acetyl lysine diagnostic ion and higher mass fragment ions (up to 1,200 *m/z*) to obtain sequence information and this assigns the site of acetylation. Representative CIRAD data for an acetylated BSA peptide (MS survey and associated bbCID CID spectra from sequential scans) is shown in Fig. 2.

3.1.1. MS Method Setup for maXis™ UHR ToF

In micrOToF control software, open the default.m method, under the "Method" tab, and then save the method to another filename so that this method can be edited, leaving the default.m method for day-to-day instrument calibration and tuning.

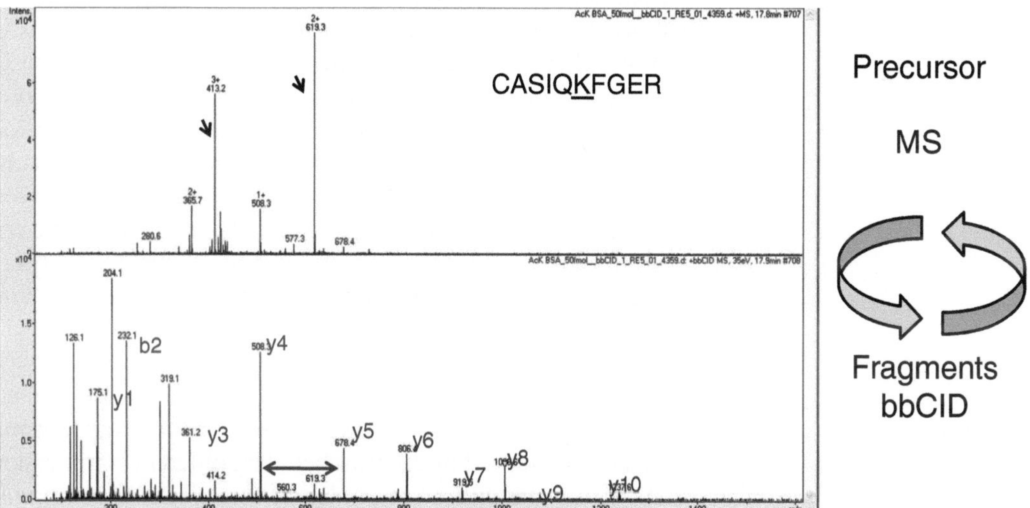

Fig. 2. A representative MS/MS spectrum for an acetylated bovine serum albumin peptide. The MS survey data and product ion spectrum obtained by switching between low and high collision energies are shown for an acetylated peptide. The precursor present as the doubly and triply charged forms. The high energy spectrum is annotated for the y ion series of peptide fragments. The mass difference of 170 Da between the y4 and y5 observed fragments is confirmatory of acetylation and corresponds to the mass of lysine plus acetylation: 128 + 42 Da, respectively. The 126.1 acetyl lysine immonium is present in this spectrum.

3.1.2. General Instrument Settings

1. The maXis is run in the positive mode, with the following settings (as viewed and set in the "Source" tab of micrOTOF control software v3.0): As a guide, our settings are shown, as listed from top to bottom on the left-hand side in the "Source" area of the window, EndPlate Offset –500 V, capillary voltage 5,000 V, nebulizer gas 0.4 bar, dry gas 6.0 L/min, and dry temperature 150°C.

3.1.3. Optimization of Ion Transmission for CIRAD

Optimization can be simply performed using a low concentration (100-fold dilution) of an electrospray tunemix containing ten defined compounds that ionize to cover the 118–2,722 *m/z* range. This is convenient, as this is typically used for mass calibration of the maXis™ on a regular basis. This approach is described in detail by Ow et al. (8), in this case for the use of iTRAQ reagents, where similar considerations apply to optimization for acetylation analysis, i.e., improved transmission and sensitivity towards low *m/z* ions (in this case, 126.09 *m/z* acetylation diagnostic ion), but retaining a sufficient population of higher mass sequence ions.

In order to improve detection of the acetylation derived ions, guide voltages and storage times from the collision cell down to the ion cooler are adjusted to give the most appropriate balance between improved sensitivity of the 126.09 *m/z* diagnostic ion versus the stability of other larger fragment ions. Four key parameters as noted by Ow et al. (8) can be adjusted here for the transmission of the collision cell and the ion cooler cell (RF guide voltages CC_{RF} and IC_{RF}, transfer time IC_{TT}, and pre-pulse time IC_{PP}).

In short, varying the guide voltage of the collision cell, CC_{RF} (affecting fragment ion stability), and ion cooler, IC_{RF} (cooling of ion beam), will adjust the net balance of ion population transferred. More generally, decreasing the guide voltage value for the collision cell and ion cooler allows lower *m/z* ions to be effectively transmitted, thus improving CIRAD sensitivity.

Likewise, the ion cooler voltage (IC_{RF}), together with the co-balancing of ion cooler transfer (IC_{TT}) and of pre-pulse times (IC_{PP}), can be co-adjusted to enable improved transmission and sensitivity towards low *m/z* ions.

3.1.4. Implementation of Collision Sweeping

This allows linked operation of different IC_{RF} transmission parameters (either those which focus squarely on low mass or those that transmit high mass) and is achieved by evenly scheduling different transmission parameters across the entire accumulation time. This is commonly termed as "collision sweeping." This implementation allows up to five steps of IC_{RF} to be made during the course of the scan allowing different IC_{RF} adjustments that are optimized for different mass ranges.

3.2. CIRAD Method Execution

1. For method setup using acetylated BSA digest, analyze at least 50 fmol on column, for initial evaluation. Recommended RP-HPLC buffers are 2% acetonitrile (v/v):0.1% formic acid (buffer A) and 80% acetonitrile (v/v):0.1% formic acid (buffer B), with peptide separation achieved using a gradient of 5% buffer B

to 40% buffer B over 40 min at a flow rate of 300 nL/min. Perform sample desalting by in-line micro-precolumn with loading solvent (0.1% formic acid) delivered at 30 μL/min for 4 min. The LC program (Chromeleon, LC Packings, The Netherlands) and MS operating parameters are integrated via HyStar software.

2. The maXis™ MS parameters are specified in the micrOTOF control software (v 3.0) (Bruker), run in the positive mode, and use settings optimized for ion transmission as detailed above in Subheading 3.1.2 and accessible under the "Source" tab. Set the mass range (right-hand panel) to 50–1500 *m/z*.

   The "Transfer" settings are as follows:

   Funnel RF, 400.0 Vpp; ISCD energy, 0.0 eV; multiple RF 400 Vpp.

   Quadrupole ion energy, 5.0 eV; low mass, 100.00 *m/z*.

   Collision cell settings: collision energy, 10.0 eV; collision RF, 600 Vpp.

   The "Ion Cooler" settings are as follows:

   Ion cooler RF, 400.0 Vpp, transfer time, 65 μs; pre-pulse storage, 5.0 μs (see Note 4). Data acquisition rate is 1 Hz (see Note 5).

3. Click on "MS/MS" tab, selecting MRM option in the left-hand section. Check the boxes for both MS/MS (MRM) and Alternate Collision Energies in the Scan Mode window. This enables the run to be performed with alternating low and high collision energies in bbCID mode. Set the collision energy parameters to 10 and 35 eV in the first and second lines of the table, to specify low and high collision, respectively. Set the Acquisition Factor to 2 for high collision energy; this is set to 1 for the low collision energy as the default (non-editable) value.

4. In the Hystar Acquisition table, specify the CIRAD method as the MS method option and select an LC gradient when using online RP-HPLC. A data file name and storage location will also need to be specified.

3.3. Data Analysis

1. Within Compass Data Analysis v. 4.1 software (Bruker Daltonics), open the data file via the File menu. The file will be visible in the Analysis list, and click on the + symbol to the left of the file to view the chromatograms of total ion current (TIC); these are TIC +All MS and TIC +bbCID. These can be visualized in the Chromatogram window by checking the selection box for each chromatogram. The data in the TIC can be viewed for the duration of the MS, on a spectra-by-spectra basis by highlighting the "arrow" in the tool bar and then clicking on a particular point of interest on the TIC (Fig. 2). The TIC+ All MS shows sequential low and high energy scans and TIC +bbCID shows the high energy scans.

2. If the MS acquisition were performed using a calibrant mass, recalibration can be performed, for fine adjustment of mass calibration. For example, the calibrant mass of 1221.9906 (Molecular composition $C_{24}H_{19}F_{36}N_3O_6P_3$) can be used for recalibration. Calibration in Data Analysis v 4.1 can be performed by clicking on the "Calibrate" tab and selecting "Apply lock mass" or "Automatic internal" to recalibrate either with lock mass calibration (which performs a scan-to-scan correction) or automatic internal algorithms (by also specifying a window of which calibrant signals are averaged and used). If using, the 1221 calibrant mass, this can be selected for the reference list as a component of the "Tuning Mix ESI-ToF (pos)" which is accessed via the software query "Use of specify reference list."
3. To visualize data containing the acetyl lysine diagnostic ion, perform an hrEIC: Check the TIC+bbCID data, right click in this window, and select "edit chromatograms." From the drop-down menu, select the following options for Type and Filter: Extracted Ion Chromatogram and All MS, respectively. Type in the mass 126.09, with a mass selection window of 0.05 Da. Select polarity as positive and background to none and edit color of chromatogram if required. Click Add, and then Apply.
4. To link precursor m/z to associated product ions based on LC elution profile: Go to "Find" and select Compounds-Dissect option. Precursors and products are linked by elution time. Creation of the Dissect.mgf allows data to be processed for sequence information by submitting the mgf peak lists generated via "Dissect" to a search tool such as MASCOT (9). The mgf is created by exporting the compounds to the mgf file, via File and then Export Compound.

3.4. Sample Analysis: Beyond Protein Standards

A key technical consideration is the low stoichiometry of PTMs in general, including acetylation, which results in acetylated peptides being difficult to detect without targeted MS or specific enrichment strategies. For lysine acetylation, enrichment is typically achieved by affinity capture of protein or peptides by use of acetyl lysine-specific antibodies (1, 10). In terms of acetylation analysis using CIRAD, a preparation of either acetylated peptides enriched by immunoaffinity capture or immunoaffinity-enriched proteins is recommended. Alternatively, enrichment of specific proteins based on subcellular localization can be performed (11). In a typical experiment, to analyze specific proteins of interest, the protein complex is purified and subsequently resolved by SDS-PAGE followed by Coomassie blue staining to visualize the protein. In order to generate peptides, the gel band needs to be excised from the gel, destained to remove Coomassie blue, and then be subjected to tryptic digestion. Typically this would provide sufficient material for several CIRAD runs.

3.5. Acetylation Site Assignment

When assigning acetylation to a particular peptide sequence, the presence of the 126.1 *m/z* acetyl lysine ion is diagnostic, and where present, the 143.1 *m/z* provides additional confirmation. The site of acetylation can be assigned on y ions and b ions a little distant from the modification site; a full sequence is not essential to infer sequence or site of assignment in a database search. A mass difference between two product ions of 170 Da corresponding to an acetylated lysine residue confirms acetyl lysine site assignment (Fig. 2) and in an ideal scenario these ions would be detected.

CIRAD is of particular value for proteins that are co-modified by acetylation and, e.g., phosphorylated or unknown modification, where targeted analysis (assuming the presence of acetylation only) would fail to identify since it is based on a precursor *m/z* that is not theoretically present in this scenario. As research into PTM continues apace, it has become evident that different PTMs enable cross talk (12) and that different PTMs can occur within the same amino acid sequence, making this feature valuable to acetylation (see Note 6). This approach also provides precursor–fragment pair information for targeted analysis via selected reaction monitoring (SRM) for further PTM analysis, e.g., by MIDAS (13–15) or quantification (16), and thus is complementary to other MS-based techniques. The CIRAD technique is potentially applicable to other modifications, e.g., acylation by propionylation which generates a diagnostic *m/z* at *m/z* 140.1063 on fragmentation (17).

4. Notes

1. However, the 143 Da is not of itself truly diagnostic, since it can also be associated with mono-methylated arginine, and thus lacks specificity for acetyl lysine. However, the presence of 143.1 Da can be confirmatory of lysine acetylation when 126.1 Da is also present in the product ion spectrum.
2. Lysine acetylation can be difficult to distinguish from isobaric lysine trimethylation at the MS level, since they both result in an addition of 42 Da to the precursor *m/z* relative to the unmodified counterpart (monoisotopic masses of 42.0105 and 42.0469 Da, respectively). The specificity of the 126.1 ion for lysine acetylation enables discrimination between acetylation and trimethylation of lysine (5).
3. This protocol can also be performed using MS instrumentation other than the maXis™ that has bbCID functionality and is essentially an MS^E time approach, as used for "label-free" protein quantification (18).

4. These parameters are maXis™ specific. MS instruments can vary and we have found this to be the case for the maXis as well (8). The guidelines thus provide a starting point for setting up CIRAD, for maXis™ instrumentation, with the rationale on how to optimize parameters in Subheading 3.1.
5. This rate can be faster. Essentially the faster the rate the more data will be sampled across the elution peak; the signal intensity will be lower in this case, and the settings given are assuming an elution peak width of 30 s.
6. PhosphoSitePlus® is an online database (www.phosphosite.org), listing manually curated PTMs, providing information on sites of phosphorylation and other commonly studied PTMs including acetylation, glycosylation which can be useful in assessing the current information available for a specific protein, in terms of known PTMs.

Acknowledgements

Financial support from the Engineering and Physical Sciences Research Council, ChELSI initiative EP/E036252/1 (CE, OSY, PCW), Cancer Research UK (DS), and Biochemical Biophysical Research Council (BMC). We thank the Bruker Daltonik GmbH research team, in particular Peter Sander, for provision and discussion on the Dissect algorithm.

References

1. Choudhary C, Kumar C, Gnad F, Nielsen ML, Rehman M, Walther TC, Olsen JV, Mann M (2009) Lysine acetylation targets protein complexes and co-regulates major cellular functions. Science 325:834–840
2. Mischerikow N, Heck AJ (2011) Targeted large-scale analysis of protein acetylation. Proteomics 11:571–589
3. Borchers C, Parker CE, Deterding LJ, Tomer KB (1999) Preliminary comparison of precursor scans and liquid chromatography–tandem mass spectrometry on a hybrid quadrupole time-of-flight mass spectrometer. J Chromatogr A 854:119–130
4. Kim JY, Kim KW, Kwon HJ, Lee DW, Yoo JS (2002) Probing lysine acetylation with a modification specific marker ion using high-performance liquid chromatography/electrospray-mass spectrometry with collision-induced dissociation. Anal Chem 74:5443–5449
5. Dormeyer W, Ott M, Schnolzer M (2005) Probing lysine acetylation in proteins: strategies, limitations, and pitfalls of in vitro acetyltransferase assays. Mol Cell Proteomics 4: 1226–1239
6. Domon B, Aebersold R (2006) Mass spectrometry and protein analysis. Science 312:212–217
7. Walther TC, Mann M (2010) Mass spectrometry-based proteomics in cell biology. J Cell Biol 190:491–500
8. Ow SY, Noirel J, Salim M, Evans C, Watson R, Wright PC (2010) Balancing robust quantification and identification for iTRAQ: application of UHR-ToF MS. Proteomics 10:2205–2213
9. Perkins DN, Pappin DJ, Creasy DM, Cottrell JS (1999) Probability-based protein identification by searching sequence databases using mass spectrometry data. Electrophoresis 20:551–567
10. Kim SC, Sprung R, Chen Y, Xu Y, Ball H, Pei J, Cheng T, Kho Y, Xiao H, Xiao L, Grishin NV, White M, Yang XJ, Zhao Y (2006) Substrate and functional diversity of lysine acetylation revealed by a proteomics survey. Mol Cell 23:607–618

11. Leech SH, Evans CA, Shaw L, Wong CH, Connolly J, Griffiths JR, Whetton AD, Corfe BM (2008) Proteomic analyses of intermediate filaments reveals cytokeratin8 is highly acetylated—implications for colorectal epithelial homeostasis. Proteomics 8:279–288
12. Yang XJ, Seto E (2008) Lysine acetylation: codified cross talk with other post translational modifications. Mol Cell 31:449–461
13. Unwin R, Griffiths J, Leverentz M, Grallert A, Hagan I, Whetton A (2005) Multiple reaction monitoring to identify sites of protein phosphorylation with high sensitivity. Mol Cell Proteomics 4:1134–1144
14. Mollah S, Wertz IE, Phung Q, Arnott D, Dixit VM, Lill JR (2007) Targeted mass spectrometric strategy for global mapping of ubiquitination on proteins. Rapid Commun Mass Spectrom 21:3357–3364
15. Griffiths J, Unwin R, Evans C, Leech S, Corfe B, Whetton A (2007) The application of hypothesis driven strategy to the sensitive detection and location of acetylated lysine residues. JASMS 18:1423–1428
16. Lange V, Picotti P, Domon B, Aebersold R (2008) Selected reaction monitoring for quantitative proteomics: a tutorial. Mol Syst Biol 4:222–235
17. Liu B, Lin Y, Darwanto A, Song X, Xu G, Zhang K (2009) Identification and characterization of propionylation at histone H3 lysine 23 in mammalian cells. J Biol Chem 284:32288–32295
18. Li GZ, Vissers JP, Silva JC, Golick D, Gorenstein MV, Geromanos SJ (2009) Database searching and accounting of multiplexed precursor and product ion spectra from the data independent analysis of simple and complex peptide mixtures. Proteomics 9(6):1696–1719

Chapter 3

Application of the MIDAS Approach for Analysis of Lysine Acetylation Sites

Caroline A. Evans, John R. Griffiths, Richard D. Unwin, Anthony D. Whetton, and Bernard M. Corfe

Abstract

*M*ultiple Reaction Monitoring *I*nitiated *D*etection *a*nd *S*equencing (MIDAS™) is a mass spectrometry-based technique for the detection and characterization of specific post-translational modifications (Unwin et al. 4:1134–1144, 2005), for example acetylated lysine residues (Griffiths et al. 18:1423–1428, 2007). The MIDAS™ technique has application for discovery and analysis of acetylation sites. It is a hypothesis-driven approach that requires a priori knowledge of the primary sequence of the target protein and a proteolytic digest of this protein. MIDAS essentially performs a targeted search for the presence of modified, for example acetylated, peptides. The detection is based on the combination of the predicted molecular weight (measured as mass–charge ratio) of the acetylated proteolytic peptide and a diagnostic fragment (product ion of m/z 126.1), which is generated by specific fragmentation of acetylated peptides during collision induced dissociation performed in tandem mass spectrometry (MS) analysis. Sequence information is subsequently obtained which enables acetylation site assignment. The technique of MIDAS was later trademarked by ABSciex for targeted protein analysis where an MRM scan is combined with full MS/MS product ion scan to enable sequence confirmation.

Key words: Lysine acetylation, Site-specific analysis, Mass spectrometry, MIDAS

1. Introduction

Protein acetylation can occur on the N-terminal α-amino group or on the ε side chain of lysine residues. This article is directed to mass spectrometry (MS) based analysis of lysine acetylation, a post translational modification (PTM) which occurs on a range of proteins to regulate protein function and which can act in concert with other PTMs such as phosphorylation (1). Mass spectrometry analysis has widespread application in analysis of the proteome and is

Sandra B. Hake and Christian J. Janzen (eds.), *Protein Acetylation: Methods and Protocols*, Methods in Molecular Biology, vol. 981, DOI 10.1007/978-1-62703-305-3_3, © Springer Science+Business Media, LLC 2013

based on accurate measurement of the mass of charged peptide ions, measured as a mass–charge ratio (m/z). There is currently no single method for global profiling of protein abundance, PTM status and subcellular localization, rather a series of workflows directed to specific challenges (2).

There are several protein acetylation-specific MS workflows, which have been reviewed in a recent article by Mischerikow and Heck (3). These include MIDAS for lysine acetylation site mapping. We developed a MIDAS workflow for targeted acetylation analysis and compared this method to a precursor ion scanning of the 126.1 product ion on the same MS instrumentation. MIDAS was shown to be in the order of ten times more sensitive in this comparison, based on the use of an acetylated protein standard, and thus was shown to offer significant advantages over existing methods (4). We subsequently employed the MIDAS workflow for analysis of keratins 8 and 18 intermediate filament proteins (5) and performed follow-up studies to investigate the functional significance of these findings using high content analysis (HCA) (6). The technique of HCA as applied to analysis of protein acetylation is described in Chapter 4.

Key features of MIDAS

- A method for targeted analysis of PTMs such as acetylation for a single known protein.
- A data dependent MS method requiring a priori knowledge of the identity of the target protein.
- Based on analysis of peptides derived by proteolytic cleavage of the protein of interest.
- Typically applied to a protein gel "band" obtained by SDS-PAGE following immunoprecipitation.
- Sensitivity is achieved due to the selectivity of the method, which leads to enhanced signal to noise.
- Low false-positive rate.

1.1. Principle of MIDAS

1.1.1. Background Information

In order to explain the MIDAS technique, it is first necessary to outline the typical approach taken for the analysis of peptides using MS. Generally, peptides are derived from proteins by proteolytic digestion, typically using trypsin. Trypsin is the enzyme of choice since tryptic peptides (containing C terminal lysine or arginine basic residues) are amenable to MS analysis due to their ability to ionize in the gas phase, which is a key requirement for entry into the mass analyser. MS analysis generates a MS-spectrum of peptides, with mass-to-charge ratios plotted against ion current (or signal). The amino acid sequence of a peptide can be determined by fragmentation along the peptide backbone, promoted by collision with an inert gas such as helium or nitrogen at low pressure, so-called collision induced dissociation (CID). The resulting MS/MS spectrum shows the m/z ratios for different fragments of selected peptide precursor, and the mass difference between fragments can correspond to the specific mass of one amino acid. The sequence of the peptide precursor can essentially be reconstructed by connecting the fragments with increasing size from the N terminus (b-ion series) or C terminus (y-ion series). Peptide sequencing can be performed manually or via database searching where the experimentally derived fragment and precursor m/z values are compared to those theoretically possible for a digest of the protein by a specific proteolytic enzyme with defined consensus site. Under appropriate conditions during MS analysis, acetylated lysine residues can generate strong product ions of 143.1 and 126.1 Da following peptide fragmentation (CID) (7–9). The 143.1 and 126.1 Da ions correspond to the acetyl lysine immonium ion, and the derivative ion resulting from further elimination of ammonia from the 143.1 ion respectively. The 126.1 ion is the ion of choice for diagnostic purposes since it is acetylation specific.

However, The 143.1 Da is not of itself truly diagnostic, since it can also be associated with monomethylated arginine, and thus lacks specificity for acetyl lysine. However, the presence of 143.1 Da can be confirmatory of lysine acetylation when 126.1 Da is also present in the product ion spectrum. Lysine acetylation can be difficult to distinguish from isobaric lysine trimethylation at the MS level, since they both result in an addition of 42 Da to the precursor m/z relative to the unmodified counterpart (monoisotopic masses of 42.0105, 42.0469 Da, respectively). For MIDAS this is not an issue due to specificity of 126.1 ion for lysine acetylation: trimethyl lysine does not generate ions at 143.1 and 126.1 (9) and as such MIDAS can provide a means to differentiate between these modifications.

1.2. MIDAS: Application to Analysis of Acetylated Peptides

MIDAS essentially performs a targeted search for the presence of putative acetylated peptides (4). The detection is based on the combination of the m/z value of the predicted acetylated peptide and the acetylation diagnostic fragment (product ion of m/z 126.1), which is generated by specific fragmentation of peptides by CID. The precursor-product pair represents the "126.1" transition. In order to achieve this, the MS operates in "Multiple Reaction Monitoring" (MRM) mode. In the instrument to be used for these experiments, there are three analytical quadrupoles through which ions transit in sequence (Q1–Q3). In MRM mode, Q1 is set to sequentially transmit a group of predefined precursor masses (in this case, putative tryptic acetylated peptides), the peptides undergo CID in Q2, the collision cell, and the fragments are transmitted to Q3 which is held static at a suitable voltage regime to only allow the transmission of the 126.1 Da product ion. Detection of the 126.1 Da ion, i.e., presence of a putative acetylated peptide ion, triggers the instrument to switch modes such that Q3 now functions as a linear ion trap in order to collect data on all product ions associated with the putative acetylated precursor. The resulting enhanced product ion spectrum enables amino acid sequence and acetylation site assignment to be inferred for the selected peptide. In order to perform MIDAS, the protein sequence of the target protein is required. A list of potential MRM transitions (for theoretical acetylated peptides, for example) is generated using specific software. In order to simplify the peptide mixture, the sample is pre-fractionated upstream of MS, by high performance liquid chromatography (HPLC), using reversed phase chromatography at nanoliter flow rates (200–300 nL/min; "nanoflow"). The column eluent is directly coupled to the MS and gas-phase ions are generated and introduced to the MS via electrospray ionization. The workflow schematic is shown in Fig. 1.

2. Materials

1. Standard proteins for method setup (see Note 1)—acetylated protein standard, for example acetylated bovine serum albumin (BSA) (Invitrogen, Paisley, UK or Sigma, Poole, UK).
2. Trypsin: Proteomics grade. Perform solution digest with a ratio of trypsin–acetylated BSA of 1:50 in 25 mM ammonium bicarbonate buffer.
3. Reversed phase-HPLC column for nanoLC and micro precolumn C18 for on line sample loading and desalting at flow rates of 300 nL/min and 30 μL/min respectively.
4. HPLC solvents: Acetonitrile and water (HPLC grade), formic acid.
5. Glass sample vials with lids, 300 μL glass insert, fused into a 2 mL crimp top vial (Chromacol (ThermoFisher), Welwyn Garden City, UK). For sample pick up prior to RP-HPLC analysis.
6. Electrospray emitters for sample introduction to the MS (New Objective, cat. no. FS360-20-10-CE-50).
7. Mass spectrometer, 4000 Q-TRAP (AB Sciex), fitted with the standard NanoSpray II source and MicroIonSpray II head (see Note 2). Analyst 1.4 software and MIDAS™ Workflow Designer v1.1 or Analyst 1.5 software used with either MIDAS™ Workflow Designer v1.1 or MRM Pilot software (ABSciex).
8. NanoLC system: Standard LC Packings Ultimate Pump and FAMOS autosampler and Switchos Units or equivalent.

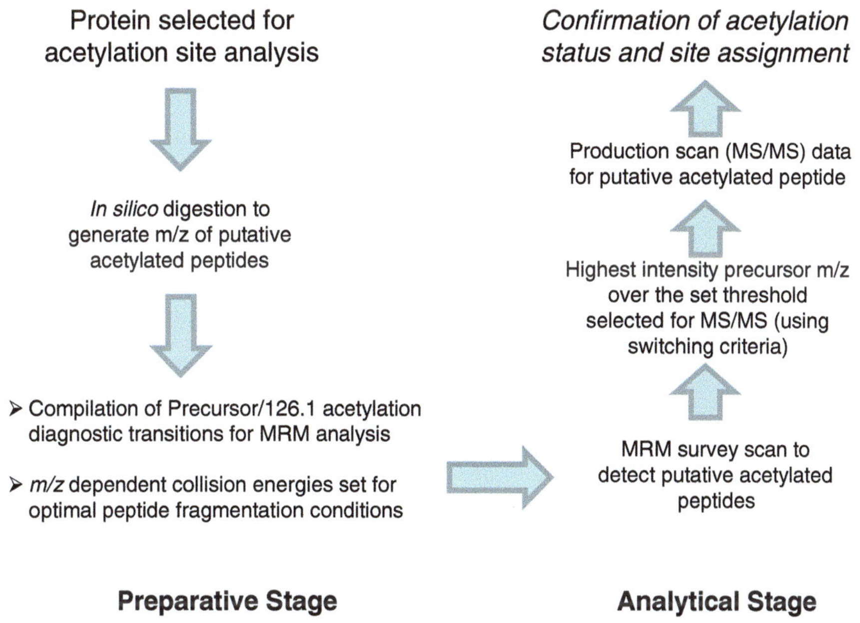

Fig. 1. MIDAS workflow for acetylation site analysis. MIDAS analyzes a target protein of known amino acid sequence. In the first preparative stage, a group of precursor masses (i.e., putative tryptic acetylated peptides) is predefined together with optimal collisions energies to ensure generation of the diagnostic 26.1 Da product ion. The MS is then run in MRM mode to scan for "126.1 transitions." Detection of the 126.1 Da ion triggers the instrument to switch modes in order to collect data on all product ions associated with the putative acetylated precursor. The resulting enhanced product ion spectrum enables amino acid sequence and acetylation site assignment to be inferred for the selected peptide. This workflow utilizes the MRM and linear ion trap modalities of the 4000 QTRAP MS instrumentation.

3. Methods

3.1. MIDAS Workflow

There are five key steps required to perform MIDAS analysis, these guidelines are adapted from the protocol recommended for MIDAS-directed phosphosite discovery (10) and based on instrument parameters and workflow outlined (4). A representative MS/MS spectrum for an acetylated BSA peptide is shown in Fig. 2.

1. Obtain the sequence of the target protein of interest. Several databases exist to obtain this information, such as Entrez protein (http://www.ncbi.nlm.nih.gov/protein) or UniProt (http://www.uniprot.org/help/uniprotkb). If performing the search in UniProt, click the "FASTA" link above the sequence in the protein entry to obtain the amino-acid sequence without formatting. Copy and paste the sequence into Notepad and save the document (see Notes 3 and 4).
2. Create or select a template MIDAS method. This can be done either by (A) creating a template in Analyst 1.4 software or (B)

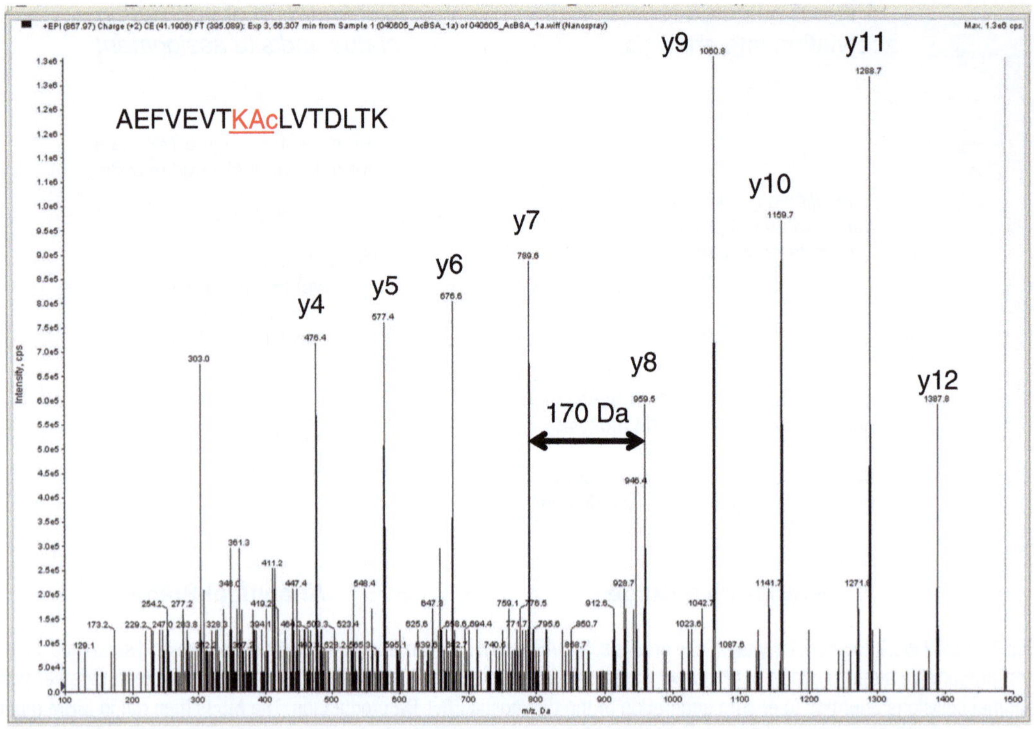

Fig. 2. A representative MS/MS spectrum for an acetylated bovine serum albumin peptide. The product ion spectrum is annotated for the y ion series of peptide fragments. The mass difference of 170 Da between the y7 and y8 observed fragments is confirmatory of acetylation and corresponds to the mass of lysine plus acetylation: 128 + 42 Da respectively. The 126.1 acetyl lysine immonium is present but at low intensity in this spectrum.

selecting an existing working standard identification. Option A is required if this is a new experiment.

3.1.1. How to Create a MIDAS Template (Required if the Experiment Is New)

1. Select "Build acquisition method" from the left-hand menu.
2. Change the experiment type to MRM, with positive polarity.
3. Select the "Advanced MS" tab, set the quadrupole resolutions to either LOW or UNIT and UNIT, respectively.
4. In the Experiment overview, right-click on "+MRM" and select "Add IDA Criteria." Set the method to select the single most intense peak. Set the threshold for switching (e.g., 100 counts). Select collision energy (CE) to be rolling and set up the values to those used for standard peptide fragmentation. These are likely to vary between instruments, but a slope of 0.044 and intercept 3 for 2+ precursors, and values of 0.04 and 2 for 3+ precursors is a reasonable starting point.
5. In Experiment overview, right-click on "IDA Criteria" and select "Add experiment." Select "Enhanced Product Ion" scan, in positive mode with a product of 30. Set the scan range 140–1,400 so that upon clicking on "optimize masses," only

two mass ranges are shown. Sum at least three scans. Select the Edit parameter button and enter the optimum values for ion spray and source voltages and gas pressures. Apply to all other experiments in the method. In the "Advanced MS" tab, set the quadrupole resolution to LOW and the scan speed to 4,000 amu/s and ensure that dynamic fill time is selected.

3.1.2. Editing an Existing Working Standard Identification Method

1. Select an existing working standard identification method.
2. Adjust the first experiment to specify "MRM."
3. Alter the switching threshold in the IDA criteria to 100 counts (selecting the single most abundant peak to switch) and adjust the scan range in the product ion experiment to 140–1,400, ensuring only two mass ranges to scan.

3.2. Building a MIDAS Method

This is done using the MIDAS™ Workflow Designer.

1. Select the MIDAS™ Workflow Designer from the left-hand menu in Analyst 1.4.1.
2. To specify the protein sequence, locate the amino acid sequence (as detailed in step 1 above), copy it and select "Add" (above "Protein Sequence" box), then paste in the sequence. Click "Add," then, if prompted, remove all non-alphabetical characters to ensure correct sequence information is provided.
3. Select sample details to specify the parameters: enzyme used for digestion, number of missed cleavages allowed (see Note 5).
4. In terms of specifying modifications: "Fixed" modifications should be selected dependent on sample preparation (alkylation of cysteine using iodoacetamide post reduction—"carbamylation") as fixed. For variable modifications, ensure that "acetylation" is selected. If protein sequence contains methionine, select oxidation of methionine.
5. Provision of information for "Acquisition Method Details" section: select "acetylation" to generate a list of amino acid residues affected and, in grey text, specify the diagnostic fragment as 126.1. Assign collision energy (CE) to ranging between 50 and 80 eV for each transition, based on precursor m/z value. Suggested values are those applied in our setup work to apply MIDAS for acetylation analysis.
6. m/z value 400–600, CE 50 eV; m/z 600–700, CE 60 eV.
7. m/z 700–800, CE 70 eV; m/z >800, CE 80 eV.
8. Select the method template generated in Subheading 3.1 as the "Starter Method," make an entry to name for the method. It is necessary to select the charge states to scan for (2+, 3+ as a starting point), and the total number of modifications (suggesting starting value of 2). If this generates too large a list of MRMs for a single experiment, it is simple to come back and

narrow these down later on, or split the list into two distinct methods. Click "Next >."

9. Select the transitions to comprise the MIDAS experiment. The eventual aim is to obtain maximum protein/peptide coverage with the minimum MRM transitions as possible. The number of transitions selected is shown in the bottom left-hand corner of the screen. Transitions can be filtered using the options in the top right-hand corner. We recommend initially using transitions with $450 < Q1 < 1{,}200$, and between 8 and 25 amino acids in length, as these represent the most likely peptides to generate MS/MS data of sufficient quality. Additional transitions can be selected/deselected using the checkboxes, or by altering these filters (see Note 6).
10. Select the maximum number of MRMs in a method and the total scan time. The MRM dwell time requires these values for calculation. Aim typically for 100 MRMs per method. This number can be higher to accommodate all transitions in a single run; however, the dwell time/transition will be less, leading to reduced sensitivity (see Note 7).
11. A collision energy (CE) for each transition is calculated using an optimized linear formula based on precursor mass and charge where for acetyl lysine, CE is m/z value ×0.1, for both 2+ and 3+ ions (4). After selection, click "Finish."
12. Method files will be generated and saved into the same project as the Starter method. If more than one method is generated, it will be given the same name followed by 2, 3, etc. Check the final method timings by opening the method in Analyst 1.4. Ensure that the cycle time is around 10 s: if longer, reduce the dwell times on each transition (no lower than 40 ms), reduce the number of scans to sum in the product ion method, or modify the method to include fewer transitions. If the cycle time is <10 s, increase the dwell time on each transition (up to 200 ms) or the number of scans to sum in the product ion scan to increase sensitivity. Global changes can be made by copying and pasting the MRM list into (and out of) Microsoft Excel using the Ctrl + C and Ctrl + V functions. Save any changes to the method (see Notes 8–10).

3.3. MIDAS Method Execution

1. For method set up using acetylated BSA digest, analyze at least 10 fmol on column, for initial evaluation. Recommended RP-HPLC buffers are 2% acetonitrile (v/v):0.1% formic acid (buffer A) and 80% acetonitrile (v/v):0.1% formic acid (buffer B), with peptide separation achieved using a gradient of 5% buffer B to 40% buffer B over 40 min at a flow rate of 300 nL/min.

2. Perform sample desalting by in-line micro precolumn with wash solvent (0.1% formic acid) delivered at 30 μL/min for 4 min.
3. 4000 QTRAP settings: Keep settings constant for each scan mode. Interface temperature is 155°C, ion spray voltage 2,300 V in positive mode, source gas setting 3 (see Note 11).

3.4. Data Analysis

The quickest way to analyze the data is to submit peak lists from the MS/MS to a search tool such as MASCOT (11). Users should note that precursor ion *m/z* value is accurate, since it is determined by the Q1 mass of the MRM transition. Database searches should not however be performed using narrow peptide mass tolerances as the resolution of Q1 determines what is being analyzed in the MRM transition and subsequent MS/MS analysis. Peptide sequences for each 126.1 transition that generate product ion spectra but no identification must be manually inspected to confirm acetylation status.

3.5. Applications of MIDAS

An important technical consideration here is that PTM occur in low stoichiometry and in general, analysis of PTM requires specific enrichment strategies. For example analysis of phosphorylation typically employs workflows with specific enrichment steps: use of metal ion/metal oxide chelating of phosphoproteins or phosphopeptides which can be combined with liquid chromatography where phosphopeptides have altered chromatographic retention times relative to non-phosphorylated peptides (12) or antibody based affinity capture based methods (12–14).

For lysine acetylated protein analysis there is currently no chromatographic enrichment technique. Instead, enrichment is typically achieved by affinity capture of protein or peptides by use of acetyl lysine specific antibodies (15, 16). In terms of acetylation analysis using MIDAS, a preparation of a single protein is the preferred option and this can be obtained by immunoaffinity-based enrichment or enrichment of specific proteins based on subcellular localization as described by Leech et al. (5). In a typical experiment, a protein of interest is purified and subsequently resolved by SDS-PAGE followed by Coomassie blue staining to visualize the protein. Typically this would provide sufficient material for several MIDAS runs. Prior to acetylation analysis, it is essential to check that the target protein is present in the sample and represents the major protein (see Note 12). For further information on use of MIDAS for discovery mode, the user is directed to our study on the application of MIDAS to acetylation site analysis of cellular proteins, specifically keratins 8 and 18 in a colon cancer cell line (5) (see Note 13). The MIDAS approach has been applied to other PTM including phosphorylation and ubiquitination (17).

4. Notes

1. In principle, any acetylated protein standard can be used but acetylated-BSA is suggested since method setup and optimization are described in detail in Griffiths et al. (4).
2. This protocol can also be performed using the QTRAP 5500 next generation QTRAP (ABSciex), which has improved sensitivity in both MRM (5–10 fold) and linear ion trap (10–100×) modes. MIDAS can, in principle, be performed on other triple quadrupole mass spectrometers; however, the unique combination of modes available on the QTRAP hybrid makes it highly suited to this application. Use of MS instrumentation other than the QTRAP 4000 would require optimization.
3. It is essential at this stage to ensure that the correct species has been selected for the target protein.
4. PhosphoSitePlus® is an online database (www.phosphosite.org), listing manually curated PTMs, including phosphorylation and other commonly studied PTMs including acetylation, which can be useful is assessing the current status of the protein in terms of known PTM.
5. Lysine acetylation results in a missed cleavage due to prevention of cleavage C terminally of lysine.
6. If there is preexisting information on known acetylation sites, this should be used to aid experimental design.
7. Although the maximum cycle time allowed in the software is 4 s, we routinely allow much longer for the MRM experiment to increase sensitivity. To overcome this, dwell times are altered manually in the final method.
8. Ensure that the number of transitions allows for a sufficient dwell time for each transition (ideally at least 40 ms) while still enabling the entire experiment (MRMs plus product ion scan) to be completed in around 10 s.
9. If the MRM list is large and has to be spread across two methods, set the "Maximum MRMs in Method" to approximately 50% of the number of transitions split equally over 2 methods.
10. If using QTRAP 5500, the CE settings would be similar but faster dwell times can be employed due to the higher sensitivity that can be achieved, resulting from improved ion optics relative to the 4000 QTRAP.
11. These are generic settings and may require tuning for individual instruments.

12. In order to generate peptides, the gel band needs to be excised from the gel, destained to remove Coomassie blue and then be subjected to tryptic digestion (5, 17).
13. Manual inspection of the product ion spectra is generally recommended to confirm acetylation status. Here the presence of the 126.1 *m/z* acetyl lysine ion is diagnostic (Fig. 2). The co-presence of 143.1 with 126.1 is additional confirmation as the presence of a mass difference between two product ions of 170 Da corresponding to an acetylated lysine residue.

Acknowledgments

Financial support from Leukaemia Lymphoma Research, UK (ADW, CE, JG), Biotechnology and Biological Sciences Research Council (BC), Cancer Research, UK (ADW, JRG), and NIHR Manchester Biomedical Research Centre (RDU). Financial support from Engineering and Physical Sciences Research Council, ChELSI initiative EP/E036252/1 (CE).

References

1. Yang XJ, Seto E (2008) Lysine acetylation: codified cross talk with other post translational modifications. Mol Cell 31:449–461
2. Walther TC, Mann M (2010) Mass spectrometry-based proteomics in cell biology. J Cell Biol 190:491–500
3. Mischerikow N, Heck AJ (2011) Targeted large-scale analysis of protein acetylation. Proteomics 11:571–589
4. Griffiths J, Unwin R, Evans C, Leech S, Corfe B, Whetton A (2007) The application of hypothesis driven strategy to the sensitive detection and location of acetylated lysine residues. JASMS 18:1423–1428
5. Leech SH, Evans CA, Shaw L, Wong CH, Connolly J, Griffiths JR, Whetton AD, Corfe BM (2008) Proteomic analyses of intermediate filaments reveals cytokeratin 8 is highly acetylated—implications for colorectal epithelial homeostasis. Proteomics 8:279–288
6. Drake PJ, Griffiths GJ, Shaw L, Benson RP, Corfe BM (2009) Application of high-content analysis to the study of post-translational modifications of the cytoskeleton. J Proteome Res 8:28–34
7. Borchers C, Parker CE, Deterding LJ, Tomer KB (1999) Preliminary comparison of precursor scans and liquid chromatography-tandem mass spectrometry on a hybrid quadrupole time-of-flight mass spectrometer. J Chromatogr A 854:119–130
8. Kim JY, Kim KW, Kwon HJ, Lee DW, Yoo JS (2002) Probing lysine acetylation with a modification specific marker ion using high-performance liquid chromatography/electrospray-mass spectrometry with collision-induced dissociation. Anal Chem 74:5443–5449
9. Dormeyer W, Ott M, Schnolzer M (2005) Probing lysine acetylation in proteins: strategies, limitations, and pitfalls of in vitro acetyltransferase assays. Mol Cell Proteomics 4:1226–1239
10. Unwin RD, Griffiths JR, Whetton AD (2009) A sensitive mass spectrometric method for hypothesis-driven detection of peptide post-translational modifications: multiple reaction monitoring-initiated detection and sequencing (MIDAS). Nat Protoc 4:870–877
11. Perkins DN, Pappin DJ, Creasy DM, Cottrell JS (1999) Probability-based protein identification by searching sequence databases using mass spectrometry data. Electrophoresis 20:551–567
12. Lee CF, Griffiths S, Rodríguez-Suárez E, Pierce A, Unwin RD, Jaworska E, Evans CA,

Gaskell SJ, Whetton AD (2010) Assessment of downstream effectors of BCR/ABL protein tyrosine kinase using combined proteomic approaches. Proteomics 10:3321–3342

13. Pandey A, Podtelejnikov AV, Blagoev B, Bustelo XR, Mann M, Lodish HF (2000) Analysis of receptor signaling pathways by mass spectrometry: identification of vav-2 as a substrate of the epidermal and platelet-derived growth factor receptors. Proc Natl Acad Sci U S A 97:179–184

14. Gronborg M, Kristiansen TZ, Stensballe A, Andersen JS, Ohara O, Mann M, Jensen ON, Pandey A (2002) A mass spectrometry-based proteomic approach for identification of serine/threonine-phosphorylated proteins by enrichment with phospho-specific antibodies: identification of a novel protein, Frigg, as a protein kinase A substrate. Mol Cell Proteomics 1:517–527

15. Kim SC, Sprung R, Chen Y, Xu Y, Ball H, Pei J, Cheng T, Kho Y, Xiao H, Xiao L, Grishin NV, White M, Yang XJ, Zhao Y (2006) Substrate and functional diversity of lysine acetylation revealed by a proteomics survey. Mol Cell 23:607–618

16. Choudhary C, Kumar C, Gnad F, Nielsen ML, Rehman M, Walther TC, Olsen JV, Mann M (2009) Lysine acetylation targets protein complexes and co-regulates major cellular functions. Science 325:834–840

17. Unwin R, Griffiths J, Leverentz M, Grallert A, Hagan I, Whetton A (2005) Multiple reaction monitoring to identify sites of protein phosphorylation with high sensitivity. Mol Cell Proteomics 4:1134–1144

Chapter 4

Application of High Content Biology to Yield Quantitative Spatial Proteomic Information on Protein Acetylations

Bernard M. Corfe, Josephine Kilner, Joanna Chowdry, Roderick S.P. Benson, Gareth J. Griffiths, and Caroline A. Evans

Abstract

High content analysis (HCA; also referred to as high content biology) is a quantitative, automated, medium-throughput microscopy approach whereby cell images are segmented into relevant compartments (nuclei, cytoplasm) and the staining in each compartment quantified by computer algorithms. The extraction of quantitative information from the cell image generates a wealth of data which contributes significantly to the acceleration of drug discovery and biological research. Here we have adapted HCA to analyze protein acetylations in the cytoskeleton. This approach yields associative information on the link between acetylation and cytoskeletal organization. The protocol also describes optimization steps for cytoskeletal analysis and its application across different cell types, and HCA platforms. The methods described herein are readily adaptable to non-cytoskeletal acetylations and have been applied to the analysis of transcription factors.

Key words: Acetylation, High content analysis, Cytoskeleton, Spatial proteomics

1. Introduction

High content analysis (HCA) is a rapidly emerging approach widely employed in the pharmaceutical sector and increasingly in academic environments (1). The platform combines the throughput of ELISA, the quantitation of flow cytometry and the organizational data of microscopy to yield a very detailed information profile on the effects of a treatment on multiple cell parameters. The approach is particularly effective when combined with quantitative proteomics to map spatial information onto changes observed at protein level or derived from pathways analysis. Our group has

Sandra B. Hake and Christian J. Janzen (eds.), *Protein Acetylation: Methods and Protocols*, Methods in Molecular Biology, vol. 981, DOI 10.1007/978-1-62703-305-3_4, © Springer Science+Business Media, LLC 2013

used HCA to analyze the effects of divergent histone deacetylase inhibitors (HDACi) on the acetylation of transcription factor Sp1 (2) to analyze the effects of different short-chain fatty acids (SCFA) on acetylations of keratin 8 (3) and to analyze the effect of the SCFA butyrate on cytoskeletal structure and organization (4).

A critical consideration for the HCA approach is the availability of antibodies. Specific antibodies for protein acetylations are not yet widely available with only a few target modifications currently represented in catalogues. Our work has been dependent upon custom polyclonal antibodies. These are raised to acetyl-epitopes linked to KLH. Sera from immunized animals are counter-purified against non-acetyl epitopes and then affinity purified against acetyl-epitopes to yield highly specific antibodies. Whilst this step is time-consuming, it is critical in the development of tools for analysis of protein acetylation and a number of commercial antibody manufacturers now offer this service. Just as the study of protein phosphorylation has been transformed with the onset of phospho-specific antibodies, so monoclonal programs are now underway which will in time yield equivalent resources for the acetylation community.

A particular advantage of HCA is the ability to measure changes in acetylation and through time—or concentration response—link these to other cellular outcomes. This provides associative or causal evidence for a role of acetylation in a particular outcome. We have previously reported the use of HCA to determine the EC50 for acetylation of transcription factor Sp1 in response to a number of different HDACi. We were then able to demonstrate sequential acetylation and upregulation of Sp1 target genes (p21) by gating subpopulations of cells (2). Likewise, we used antibodies to three different acetylation sites on keratin 8 in combination with butyrate treatment to show sequential rather than concomitant acetylation of each residue and the association of one acetylation with collapse of intermediate filaments (4). Large-scale proteomic efforts, coupled with pathways analysis often predict cellular outcomes of perturbations. We have undertaken such an analysis of proteomic datasets (5, 6) from cells treated with an SCFA. The data suggested alterations in composition of microtubules with potential consequences for cytoskeletal organization. HCA provided a rapid route to the orthogonal validation of proteomic data linked to cellular outcome. A summary of the complementarity between HCA and proteomics is shown in Fig. 1. The methods presented here summarize approaches to quantify the change in acetylation of cytoskeletal proteins and the impact on cytoskeletal organization, but detail is included to allow the application of such approaches to analysis of HDACi upon the cytoskeleton or quantification of a novel acetylation.

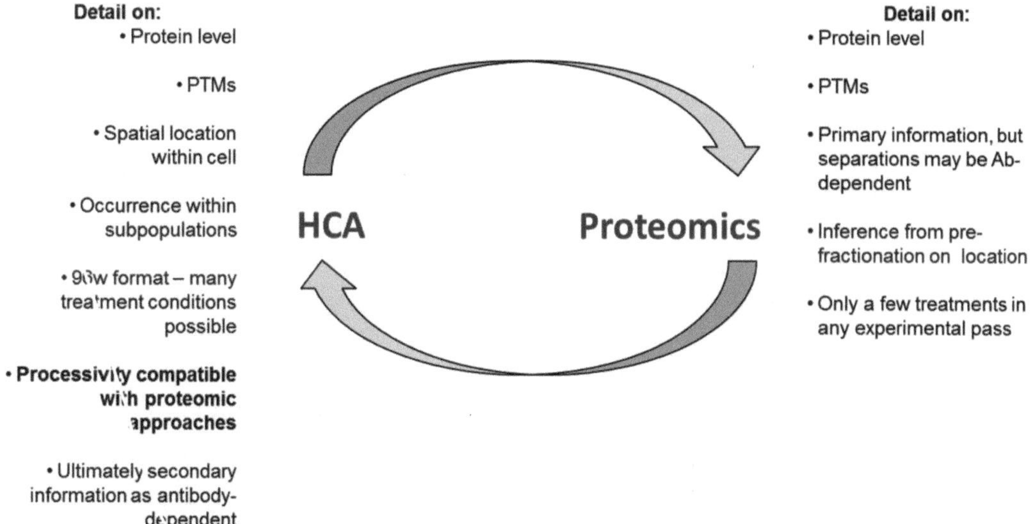

Fig. 1. Complementarity between proteomic and HCA discovery platforms. Proteomic approaches use pooled cells and generally quantify largest or most significant differences. The chemiproteomic sample preparation steps may allow inference on the subcellular location of a change of an alteration in, for example, charge, but loses information on individual cells and only allows for a few experimental passes. HCA allows analysis of specific changes, for which antibody or markers are available, in 96-well format, thereby allowing multiple repeats and a treatment range, providing cell-level information and allowing linking of multiple endpoints. In turn the HCA will refine and define further rounds of proteomic experimentation.

2. Materials

Solutions should be prepared using double-distilled, deionized water (to a sensitivity of 18 MΩ cm at 25°C). Use analytical grade reagents and store as specified by the supplier (e.g., room temperature; 4; −20; −80°C). All procedures are carried in out in laboratories dedicated for that purpose and maintained to required standards for cleanliness following laboratory waste disposal regulations. Follow appropriate local COSHH regulations.

2.1. General Cell Culture Reagents and Solutions

1. Cell lines used by our group include the following: HCT116, HT29, CaCo-2 colon carcinoma cell-lines; MCF-7 breast cancer cell lines; and HeLa cervical carcinoma cell lines.
2. Screen all cell lines bimonthly for mycoplasma contamination.
3. Store cells in incubators at 37°C, 5% CO_2 in humidified air.
4. Confluency, viability, and cell counts should be assessed using an optical microscope and hemocytometer (see Note 1).
5. Carry out cell culture in laminar flow cabinets decontaminated with multipurpose disinfectants, e.g., MicroSol3, Virkon, Teepol, or 70% ethanol. All reagent bottles and culture equipment is similarly cleaned.

6. Cell-culture media are specific to each cell line. Commonly used media include the following:
   (a) DMEM (Dulbecco/Vogt-modified Minimal Essential Medium) with 1 g/L D-glutamine, 4 mM L-glutamine, 110 mg/L sodium pyruvate, and 25 mM HEPES; supplemented with 5% (v/v) penicillin/streptomycin (10,000 units/mL) and streptomycin (10,000 μg/mL), and 10% (v/v) heat-deactivated fetal calf serum.
   (b) RPMI1640 supplemented with 500 units of penicillin/streptomycin, L-glutamine to 2 mM final concentration, and 10% (v/v) heat-deactivated fetal calf serum.
7. Use phosphate-buffered saline (PBS) solution to wash cells.
8. Media, PBS, and other cell-culture reagents should be pre-warmed to 37°C in water baths.
9. Cell culture protocols are specific for each cell type and described elsewhere (2, 6, 7).

2.2. Primary Antibodies Specific to Acetylation

1. Highly specific anti-acetyl antibodies are required to analyze protein acetylation. Only a limited number are commercially available (e.g., Acetylated Lysine Antibody KAP-TF120). For the generation of acetylation-specific antibodies see Chapter 11.
2. The majority are custom-generated polyclonal antibodies: double-affinity purified; counter-purified against non-acetyl epitopes; and positively purified against acetyl epitopes (2). A few manufacturers will accept commissions for custom anti-acetyl antibodies.
3. Both pan-specific anti-acetyl antibodies and those designed specifically for selected lysine sites can be raised.

2.3. HCA Cell Seeding and Treatment

1. Culture cells in Black-sided Costar 96-well culture plates treated for cell-culture.
2. Formalin/formaldehyde and methanol are the most commonly used fixing agents.
3. Prepare secondary antibodies, Alexa Fluor 555-red and Alexa Fluor 488-green (species depending on primary antibody), at a concentration of 1:400 in Hoechst (1:1,250). Hoechst stains nuclei blue allowing whole cells to be visualized.
4. Use Whole-Cell Red Stain to visualize cell shape and cytosolic area of immunostained cells.

2.4. HCB Platform, Software, and Algorithms for Organization

1. HCA image analysis and quantification is carried out using a Cellomics Arrayscan II platform (available at Imagen Biotech Ltd, Manchester) with Cellomics proprietary software (version 3.5.1.2). Other proprietary platforms are available.

2. The Morphology Explorer algorithm is used to create a mask so that only cytoplasmic staining is measured and any nuclear fluorescent signal is excluded. The Cellomics algorithm then provides a large number of measurements that can be derived from the regions of interest (ROIs). Use Mean Intensity Co-occurrence Contrast (MICOC—a type of texture measurement—see Note 4) to assess microtubule staining.
3. A cell cycle algorithm calculates the number of cells at each phase of the cell cycle in a population (sub-G1; G1; S; G2/M; post-G2/M). Include an Average Nuclear Intensity parameter to create gates by setting a minimum DNA content threshold. Confounding apoptotic cells can then be excluded from an analysis.
4. Use spot-fiber count and spot fiber total area analysis to identify pixels belonging to spots or fibers by evaluating the change in intensity over space within the object. A background correction is applied automatically to differentiate between genuine spots and fibers and intracellular noise.
5. "Prism" software (GraphPad) is used to provide quantification data and graphical output of cell-cycle and structural-parameter results.

3. Methods

3.1. HCB Cell Seeding and Treatment

1. Specify treatment conditions and primary antibody type for each experiment by drawing a plate map (see Note 2).
2. Seed cells in cell culture media at concentrations specific for each cell-line in each 100 μL well (as described in Subheading 2.3). Incubate plates at 37°C for 24 h to allow cells to adhere. Discard the old media and replace with 160 μL fresh media plus 40 μL of 5× stock treatment drug in PBS to give a final volume of 200 μL per well using a multichannel pipette.

3.2. HCB Cell Fixing and Antibody Staining

1. At the end of each treatment time, fix cells using reagents, times, and temperatures specific to each cell type and experiment. These include the following:
   (a) 4% Formalin for 15 min at room temperature.
   (b) Ice cold methanol for 5 min at –20°C.
2. Dilute 50 μL primary antibodies in appropriate permeabilization agent (see Subheading 3.2) (e.g., digitonin at 500 μg/mL in PBS), add to each well at concentrations specified by the plate map and incubate for 60 min at room temperature.
3. Wash cells three times with 100 μL PBS per well.

4. Stain cells by adding 50 μL per well of Alexa Fluor 555-red and/or Alexa Fluor 488-green secondary antibodies at a concentration of 1:400 in Hoechst (1:1,250) and incubate for 30 min at room temperature.
5. Stain immunostained cells with Cellomics Whole-Cell Red Stain in order to visualize cell shape and cytosolic area.
6. Wash cells three times with 100 μL PBS per well. Add 100 μL of PBS per well, seal the plates with a sealing film and store at 4°C prior to image analysis.

3.3. Optimization of Cell Fixation; Blocking; and Permeabilization Methods for Antibodies

Micrographs and quantification for an optimized protocol is shown in Fig. 2. The section below describes the parameters that the user should optimize in the development of each antibody for use in HCA protocols.

1. *Optimization of cell fixation*: Fixation should immobilize the protein of interest while retaining cellular and subcellular structure allowing access of antibody to all cells and subcellular compartments. Cross-linking reagents such as formaldehyde preserve cell structure but may reduce antigenicity components as cross-linking may obstruct antibody binding. Organic solvents such as methanol remove lipids but dehydrate the cells. The fixation method selected should produce a consistent cell count and sharper images.
2. *Permeabilization*: Intracellular epitopes require permeabilizing to facilitate antibody entry into the cell in order to detect the protein of interest. Permeabilization agents include the following: PBS control or 0.1% Triton X-100 made up in PBS. Milder membrane solubilizers such as digitonin at 500 μg/mL or saponin at 1 mg/mL produce pores large enough for the antibody to pass through without dissolving the plasma membrane. Ideally, a permeabilization agent should give uniform staining of the cellular architecture. Antibodies are diluted in a detergent or permeabilizing agent. Imagen Biotech's preferred permeabilization method is digitonin because it gives the antibody good access to all the relevant intracellular compartments while minimizing nonspecific staining. This means that often it is possible to use digitonin without a blocking agent (see below).
3. *Optimization of blocking*: Blocking prevents unspecific binding of the antibody. A blocking step can be applied prior to antibody staining to check for differences in to background staining intensities (see Note 3). Blocking agents include animal sera (1–10%) or bovine serum albumin (BSA) made up to different concentrations in PBS (e.g., 1, 2, or 5%). Blocking is carried out for 10–30 min.

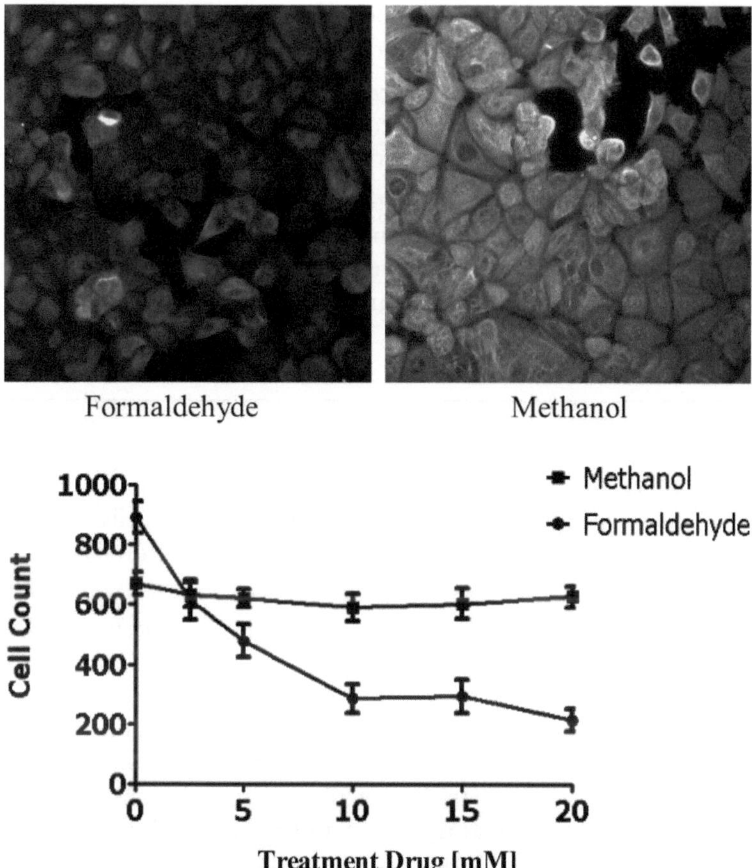

Fig. 2. Method optimization. Methanol fixation provides consistently sharper images for the intermediate filament cytoskeleton (micrographs, *top*) without a decrease in cell number (plot, *below*). Protocols are similarly optimized for permeabilization, antibody, and fluorescence.

3.4. HCB Quantification of Changes in Acetylation; Cytoskeletal Organization; and Links to Other Cell Parameters

1. *Quantification of changes in acetylation*: Use data generated using the Compartmental Analysis algorithm to quantify the total staining intensity of acetylation sites targeted by anti-acetyl antibodies. Acetylation sites can be visualized and changes in response to different treatments evaluated. Optimally antibodies to the non-acetyl form of the target protein should also be used in a separate channel to control for alterations in protein expression.
2. Anti-acetyl antibodies designed to target specific lysines, e.g., lysine 10, lysine 471, and the C-terminal lysine, allow the location of these lysines to be located (4).
3. *Quantification of cytoskeletal organization*: Use the morphology algorithm to measure texture by calculating a co-occurrence intensity of structural staining. High variations in intensity result in a high texture intensity measurement. Uniform staining gives a low value for this parameter, consistent with cellular depolymerization.

4. Perform spot-fiber counts and total area analysis using the analysis method described in Subheading 2.4.
5. *Links to other cell parameters*: The cell cycle algorithm is employed to calculate numbers of cells at each phase of the cell cycle in a population: sub-G1; G1; S; G2/M; post-G2/M (6).
6. The inclusion of the average nuclear intensity parameter enables the number of mitotic cells in a population of nuclei to be calculated to reveal dysmorphic mitotic nuclei, which may be associated with impaired cytoskeletal organization (4).

4. Notes

1. To distinguish viable from nonviable cells, cells are stained with Trypan Blue at a ratio of 1:1. A dilution factor of 2 needs to be included in the calculations.
2. Untreated samples are included as controls. Each treatment condition is applied in triplicate ($n=3$) for statistical analysis. Although multiple treatment conditions such as: different antibodies; sample or antibody concentrations; or staining reagents can be applied on a single plate, individual plates are required for each treatment time. This is because the plates are sealed after treating, fixing and staining prior to visualization.
3. A blocking step can be omitted if no significant differences in texture measurements are observed. This is to ensure mechanical manipulations are kept to a minimum and the quality of final images can remain uncompromised.
4. This measurement is derived from the probability of pixels with different intensities occurring next to each other: a matrix between the numbers of occurrences of all possible pairs of pixel intensities at either the x or y directions is computed and reported as a texture measurement. The contrast measures the strength of occurrences of pixels at disparate intensities that are adjacent to each other: the larger the disparity, the stronger the contrast. Thus, a highly textured image will have a higher MICOC than an image that is relatively uniform.

Acknowledgments

Our work is funded by BBSRC, FSA, University of Sheffield (BC) and by Engineering and Physical Sciences Research Council, ChELSI initiative EP/E036252/1 (CE).

References

1. Zock JM (2009) Applications of high content screening in life science research. Comb Chem High Throughput Screen 12:870–876
2. Waby JS, Chirakkal H, Yu CW, Griffiths GJ, Benson RSP, Bingle CD, Corfe BM (2010) Sp1 acetylation results in loss of DNA binding at promoters associated with cell cycle arrest and cell death in a colon cell line. Mol Cancer 9:275
3. Khan AQ, Shaw L, Drake PJM, Brown SR, Corfe BM (2012) Application of high content biology demonstrates differential responses of keratin acetylation to short chain fatty acids. J Integr Omics 1:253–259
4. Drake PJM, Griffiths GJ, Benson RP, Corfe BM (2009) Application of high-content analysis to the study of post-translational modifications of the cytoskeleton. J Proteome Res 8:28–34
5. Kilner J, Zhu L, Ow SY, Evans CA, Corfe BM (2011) Assessing information loss through application of the two-hit rule in iTRAQ datasets. J Integr Omics 1:124–134
6. Kilner J, Waby JS, Chowdry J, Wright PC, Corfe BM, Evans CA (2012) Use of proteomic approaches to identify differences in cellular response to butyrate, valerate and propionate in colon epithelial cells. Mol Biosyst. 8:1146–1156
7. Yu CW, Waby JS, Chirakkal H, Staton CA, Corfe BM (2010) Butyrate suppresses expression of neuropilin I in colorectal cell lines through inhibition of Sp1 transactivation. Mol Cancer 9:276

Chapter 5

Towards the N-Terminal Acetylome: An N-Terminal Acetylated Peptide Enrichment Method Using CNBr-Activated Sepharose Resin

Xumin Zhang and Peter Højrup

Abstract

Protein N-terminal acetylation (N^{α}-acetylation) is observed widely from prokaryotes to eukaryotes. It gains increased importance in biological field, due to its multiple roles in many aspects of the protein life, such as assembly, stability, activity, and location. Today, mass spectrometry (MS) has demonstrated its unprecedented ability in a variety of proteome-wide studies including the N^{α}-acetylome. The N^{α}-acetylated peptides are usually immerged into the massive amount of regular peptides and are further suppressed due to their reduced positive charge states during analysis; therefore, an efficient exploration of the N^{α}-acetylome necessitates a specific enrichment method. Here we describe a protocol for N^{α}-acetylated peptide enrichment using CNBr-activated sepharose resin, which has proved to be simple, sensitive, and highly reproducible.

Key words: Mass spectrometry, LC–MS/MS analysis, CNBr-activated sepharose, Chemical derivatization, Specific enrichment

1. Introduction

Posttranslational modifications (PTMs) of proteins play vital roles in all kinds of biological processes and pathways. Among the most common modifications in eukaryotes, N^{α}-acetylation occurs both co- and posttranslationally, and has been shown to be involved in protein assembly, stability, activity, and location (1–11). Therefore, it is of great importance to study N^{α} acetylation.

It has been estimated that the majority of proteins in eukaryotes possess N^{α}-acetylation, 56.9, 74.1, and 84.4% in yeast, fly, and

Sandra B. Hake and Christian J. Janzen (eds.), *Protein Acetylation: Methods and Protocols*, Methods in Molecular Biology, vol. 981, DOI 10.1007/978-1-62703-305-3_5, © Springer Science+Business Media, LLC 2013

human respectively, and particularly, 70.2 and 76.1% in fly and human are fully acetylated (12–14).

Mass spectrometry (MS) has been widely applied in different biological studies and plays an unparalleled role especially in high throughput proteome-wide associated studies. The most common approach for MS-based proteomics studies is the bottom-up method, in which peptides from enzymatic cleavage are analyzed instead of proteins. Despite the high occurrence of N^{α}-acetylation in protein level, the N^{α}-acetylated peptides contribute to a rather limited fraction in the peptide mixture after the protease digestion. As a result, the N^{α}-acetylome study would be greatly facilitated by a targeted enrichment prior to MS analysis.

To date a variety of methods have been developed for N^{α}-acetylated peptide enrichment. These methods can be classified into two categories on the basis of the involvement of chemical derivatization or not. The representative method in the chemical derivatization category is the combined fractional diagonal chromatography (COFRADIC), which employs two steps of chemical derivatization and two steps of reverse-phase high-performance liquid chromatography (RP-HPLC) separation (15). The other category of methods takes advantage of the physical or chemical difference between N^{α}-modified peptides and regular peptides. One method on the basis of the physical difference is the strong cation exchange chromatography (SCX) method, by which the N^{α}-modified tryptic peptides can be separated from regular tryptic peptides as they possess less positive charge and are thus eluted earlier (16, 17). The methods based on chemical differences mostly introduce an amine-reactive matrix in order to specifically couple regular peptides via their free α-amine (18–20). The selectivity for α-amine is critical for the enrichment, since the ε-amine on lysine may react with the amine-reactive matrix as well due to its chemical similarity.

Herein, we present a protocol for N^{α}-acetylated peptide enrichment using a commercially available amine-reactive matrix, cyanogen bromide CNBr (cyanogen bromide)-activated sepharose resin. The method is dramatically simpler than previous methods and mainly involves a double incubation at relatively low pH (20). In order to limit the reaction specificity to α-amine, we observed that by employing the incubation at pH 6.0, the regular tryptic peptides are quantitatively removed by coupling to the resin via a covalent bond, and the N^{α}-acetylated peptides survive the incubation regardless of the presence of lysine ε-amine. Another advantage is that the reaction efficiency is not affected by the ratio of digest to resin. Furthermore, to overcome the charge loss due to the blockage of N^{α}-amine, we proposed an MS data acquisition method, singly charged ion inclusion (SCII), to allow the singly charged precursor ions for MS/MS fragmentation analysis, by which N^{α}-acetylated peptides get a higher chance for identification, especially for low mass ones. Our study with HeLa cells has demonstrated its simplicity, efficiency, specificity and reproducibility by identifying more than

1,000 acetylated N-termini within a duplicate experiment using only 80 μg samples each. We also observed an inherent limitation of the method. The resin also couples to histidine via the imidazole group, and thus histidine containing N^{α}-acetylated peptides cannot be recovered by this approach. It is worth mentioning that using this approach we also discovered a novel mechanism for a specific α-amine blockage occurring during standard sample preparation, the cyclization of the N-terminal X-Asn-Gly motif (21).

2. Materials

Pure water was obtained from a PURELAB Ultra system (ELGA, UK) to attain a resistance of 18.2 M Ω cm at 25°C. GELoader tips were from Eppendorf (Hamburg, Germany). Poros R3 resin was from Invitrogen (Life Technologies/Invitrogen, USA). Empore C8 extraction disk was from 3M (St. Paul, MN, USA). All reagents and solvents were of analytical purity and purchased from Sigma-Aldrich (Steinheim, Germany) unless otherwise specified.

2.1. Protein Extraction and Digestion (See Note 1)

1. HEPES buffer: 25 mM HEPES sodium salt, pH 7.4.
2. Lysis buffer 1: 4 M guanidine hydrochloride (GndCl) and 10 mM dithiothreitol (DTT), prepared just before use.
3. pH adjusting buffer 1: 1 M Na_2HPO_4.
4. Alkylation solution: 1 M acrylamide.
5. Acetone (store at −20°C).
6. Lysis buffer 2: 4 M GndCl.
7. Trypsin solution: 100 ng/μL trypsin (Promega, Madison, WI, USA) in 1 mM HCl, stored on ice prior to use.
8. pH adjusting buffer 2: 1 M HCl.
9. pH adjusting buffer 3: 1 M NaOH.

2.2. Removal of N^{α}-Pyroglutamate (See Note 1)

1. pGAPase solution: 25 U/mL, TAGZyme™ (Qiagen, Hilden, Germany).
2. Qcyclase enzyme solution: 50 U/mL, TAGZyme™ (Qiagen, Hilden, Germany).
3. Activation buffer: 10 mM DTT + 20 mM Na_2HPO_4, pH 8.0 (prepared just before use).
4. pH adjusting buffer 1: 1 M Na_2HPO_4.
5. Reduction solution: 1 M DTT (prepared just before use).
6. Alkylation solution: 1 M acrylamide.
7. 1 M HCl.

2.3. N^{α}-Acetylated Peptide Enrichment (See Note 1)

1. CNBr-activated Sepharose (Sigma-Aldrich, St. Louis, MO, USA).
2. 1 mM HCl (ice-cold).
3. Incubation buffer: 200 mM Na_2HPO_4, pH 6.0.
4. Acetonitrile (ACN).
5. pH adjusting buffer 4: 10% FA.

2.4. Desalting

1. ACN.
2. 1% FA.
3. Elution solvent 1: 30% ACN + 0.1% FA.
4. Elution solvent 2: 60% ACN + 0.1% FA.

3. Methods

The main experimental scheme of the method is presented in Fig. 1. For the procedures after trypsin digestion, the volumes or amounts of the reagents in this protocol are specified for 100 μg peptide sample and should be scaled accordingly.

3.1. Protein Extraction and Digestion

1. Resuspend the HeLa cell pellet in 50× volume of HEPES buffer and then discard the supernatant after centrifugation at 4,000 × *g* for 5 min.
2. Repeat step 1.
3. Resuspend the washed cell pellet in 10× volume of lysis buffer 1 and adjust the pH to ~8.0 by the addition of 1 M Na_2HPO_4 (see Note 2).
4. Submit the solution to sonication. The sonication is carried out by 2 s sonication time with 5 s intervals and the total sonication time is 3 min.
5. Collect the supernatant after centrifugation at 20,000 × *g* for 20 min at 20°C and keep the solution at 37°C for 45 min to allow for reduction.
6. Add the alkylation solution to obtain the acrylamide concentration of 30 mM and keep the sample solution at room temperature for 1 h to allow for alkylation (see Note 3).
7. Mix 4× volume of acetone (–20°C) with the sample solution and keep the solution at –20°C for 2 h.
8. Remove the supernatant after centrifugation at 20,000 × *g* for 20 min at 4°C.
9. Wash the pellet twice with acetone (–20°C) and remove the acetone after centrifugation at 20,000 × *g* for 20 min at 4°C.

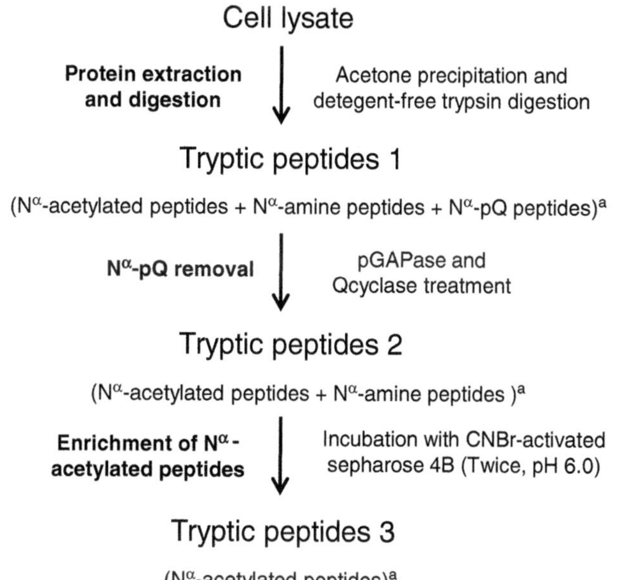

Fig. 1. The experimental scheme for the enrichment of N^{α}-acetylated peptides. The parentheses shows the major components expected in the sample after the corresponding treatments.

10. Dry the pellet completely in a vacuum centrifuge.
11. Transfer a small portion of the pellet to a new Eppendorf tube. Weigh and dissolve it in lysis buffer 2 for protein determination by Bradford assay (see Note 4).
12. Weigh the remaining major pellet and resuspend it in the trypsin solution to obtain a trypsin-to-protein ratio of 1:60 (w/w).
13. Adjust the pH to 8.0 and submit the slurry solution to sonication as in step 4.
14. Incubate the solution in an Eppendorf Mixer (Eppendorf, Hamburg, Germany) at 37°C with a 600 rpm vortex for 4 h (see Note 5).
15. Add the same volume of trypsin as in step 12 to the digestion solution and repeat step 13 (see Note 6).
16. Incubate the solution in an Eppendorf mixer with a 600 rpm vortex for overnight digestion at 37°C (see Note 5).
17. Adjust the pH of the solution to ~1.5 with 1 M HCl, and collect the supernatant after centrifugation at 20,000×g for 20 min at 20°C (see Note 7).
18. Add an equal volume of 1 M NaOH to the supernatant to adjust the pH back to ~8.0.

3.2. Enzymatic Removal of N^{α}-Pyroglutamate (100 μg Peptide as the Study Material) (See Note 8)

1. Mix 4 μL pGAPase solution (100 mU) with an identical volume of activation buffer and incubate at 37°C for 10 min (see Note 9).
2. Add 100 μg peptide containing solution to the activated pGA-Pase solution, followed by addition of 6 μL Qcyclase solution (300 mU).
3. Adjust the pH to 8.0 with 1 M Na_2HPO_4 and add the reduction solution to obtain the DTT concentration of 5 mM (see Note 9).
4. Incubate the solution at 37°C for 1 h.
5. Add the alkylation solution to obtain the acrylamide concentration of 10 mM and keep at room temperature for 1 h to quench the DTT (see Note 10).
6. Adjust the pH to 6.0 with 1 M HCl.

3.3. N^{α}-Acetylated Peptide Enrichment (100 μg Peptide as the Study Material)

1. Incubate the CNBr-activated sepharose with 10× volume of 1 mM HCl (ice-cold) on ice for 10 min, with a gentle vortex every 2 min (see Note 11).
2. Remove the supernatant using a GELoader tip.
3. Repeat steps 1 and 2 twice.
4. Aliquot the rehydrated resin in 500 μL Eppendorf tubes with 100 mg in each.
5. Dilute the sample solution resulting from procedure 3.2 with incubation buffer to achieve a final peptide concentration of 1 μg/μL (see Note 12).
6. Add 100 μL of the diluted solution directly to the rehydrated resin (100 mg).
7. Spin the tube to remove any bubbles in the slurry solution and make sure that the resin is completely submerged by the solution.
8. Put the tube in an Elmi Intelli-Mixer RM-2 and incubate for 2 h at room temperature with a constant and slow end-over-end rotation (3 rpm).
9. Load the ~20 μL slurry solution onto a GELoader tip and press the solution out back to the same Eppendorf tube with a syringe (see Note 13).
10. Load all the slurry solution onto the GELoader tip and collect the filtrate with a new 500 μL Eppendorf tube.
11. Load 100 μL ACN onto the GELoader tip and collect the filtrate with the same Eppendorf tube in step 10.
12. Concentrate the solution to around 100 μL in a vacuum centrifuge to make sure that the ACN is completely removed.

13. Add the 100 μL solution to a new 100 mg rehydrated resin-containing Eppendorf tube, repeat steps 7–11.
14. Add 20 μL 10% FA to adjust the pH < 2.0.

3.4. Desalting

1. Punch a small plug of C8 material out of an Empore C8 extraction disk with an HPLC syringe needle and place it in the narrow end of the GELoader tip (see Note 14).
2. Mix the Poros R3 resin with ACN by a ratio of 1:20 (w/v). Aspirate and expel the slurry solution with a pipette to make the solution uniform, and load 40 μL of the slurry solution onto the GELoader tip.
3. Press the solvent out with a syringe.
4. Wash the micro-column with 20 μL ACN.
5. Wash the micro-column with 20 μL 1% FA.
6. Load the acidified sample solution onto the micro-column and slowly press the solution out.
7. Wash the micro-column twice with 10 μL 1% FA.
8. Elute the micro-column with 10 μL 30% ACN + 0.1% FA and collect the eluate with a new Eppendorf tube.
9. Elute the micro-column with 10 μL 60% ACN + 0.1% FA and collect the eluate with the same tube in step 8.
10. Dry the solution completely in a vacuum centrifuge.
11. Store the dried sample at −20°C and redissolve the dried peptides with 1% FA just before LC–MS/MS analysis.

3.5. LC–MS/MS Analysis

The standard LC–MS/MS analysis (see also Chapter 1) is used with two exceptions: the inclusion of singly charged precursor ions and the exclusion of known background ions for MS/MS fragmentation use (22).

4. Notes

1. To avoid possible amines that could compete with Nα-amine containing peptides in the coupling to CNBr-activated resin, all amine or ammonium containing reagents, e.g., NH_4HCO_3 and NH_3H_2O, should not be used in the procedures before and during the incubation. To avoid the artificial modification on the Nα-amine, no Nα-amine reactive reagents, e.g., urea, should be used.
2. The pH ~8.0 would guaranty the success of the reduction. The reason not to use pre-adjusted buffer is to avoid the use of

high concentration of phosphate salt, which may precipitate in the subsequent acetone precipitation.

3. The use of acrylamide instead of iodoacetamide as the alkylation reagent has the advantages that no cyclization occurs on alkylated Nα-cysteines and the alkylation process does not affect the solution pH (19).
4. The small portion of sample is used to determine the protein content in the pellet, since the digestion solution used in this approach cannot dissolve the pellet completely and is thus not suitable for protein determination.
5. The vortexing during digestion can help the solubility of the pellet and therefore improve the digestion efficiency.
6. The additional trypsin after 4 h digestion increases the digestion efficiency.
7. Removal of the acid insoluble fraction reduces the sample complexity and avoids the blockage of the micro-column during desalting under acidic condition.
8. The removal of Nα-pyroglutamate helps to a great extent the identification of Nα-acetylated peptide. If ignored, the Nα-pyroglutamate contained peptide may contribute around half of the identified Nα-amine blocked peptides (19).
9. The step is important since the pGAPase functions only in the reduced form.
10. The remaining DTT would react with CNBr-activated resin and reduce the coupling efficiency towards Nα-amine.
11. Incubation on ice maintains the maximum amine reactive capacity of the CNBr-activated beads.
12. Large peptides may be difficult to eliminate, possibly due to their affinity to the plastic material of the tube. Addition of 10–20% ACN to the final solution can help reduce the affinity.
13. It is not necessary to constrict the GELoader tip, since the rehydrated CNBr-activated sepharose beads, generally, are too large to pass through the very narrow end of the tip. However, due to the heterogeneity of the beads, a few small-size beads may leak at the beginning. In this case, it is suggested to initially load ~20 μL slurry solution to block the tip end and collect the filtrate with the same sample tube for reload.
14. This step is necessary to prevent the leakage of Poros R3 resin (23, 24).

Acknowledgment

This work as supported by the Danish Free Research Council.

References

1. Berger EM, Cox G, Weber L, Kenney JS (1981) Actin acetylation in Drosophila tissue culture cells. Biochem Genet 19:321–331
2. Tercero JC, Wickner RB (1992) MAK3 encodes an N-acetyltransferase whose modification of the L-A gag NH_2 terminus is necessary for virus particle assembly. J Biol Chem 267:20277–20281
3. Luka Z, Loukachevitch LV, Wagner C (2008) Acetylation of N-terminal valine of glycine N-methyltransferase affects enzyme inhibition by folate. Biochim Biophys Acta 1784:1342–1346
4. Arendt CS, Hochstrasser M (1999) Eukaryotic 20S proteasome catalytic subunit propeptides prevent active site inactivation by N-terminal acetylation and promote particle assembly. EMBO J 18:3575–3585
5. Siddig MA, Kinsey JA, Fincham JR, Keighren M (1980) Frameshift mutations affecting the N-terminal sequence of Neurospora NADP-specific glutamate dehydrogenase. J Mol Biol 137:125–135
6. de Haan EC, Wauben MH, Wagenaar-Hilbers JP, Grosfeld-Stulemeyer MC, Rijkers DT, Moret EE, Liskamp RM (2004) Stabilization of peptide guinea pig myelin basic protein 72-85 by N-terminal acetylation-implications for immunological studies. Mol Immunol 40:943–948
7. Pesaresi P, Gardner NA, Masiero S, Dietzmann A, Eichacker L, Wickner R, Salamini F, Leister D (2003) Cytoplasmic N-terminal protein acetylation is required for efficient photosynthesis in Arabidopsis. Plant Cell 15:1817–1832
8. Maytum R, Geeves MA, Konrad M (2000) Actomyosin regulatory properties of yeast tropomyosin are dependent upon N-terminal modification. Biochemistry 39:11913–11920
9. Setty SR, Strochlic TI, Tong AH, Boone C, Burd CG (2004) Golgi targeting of ARF-like GTPase Arl3p requires its Nalpha-acetylation and the integral membrane protein Sys1p. Nat Cell Biol 6:414–419
10. Behnia R, Panic B, Whyte JR, Munro S (2004) Targeting of the Arf-like GTPase Arl3p to the Golgi requires N-terminal acetylation and the membrane protein Sys1p. Nat Cell Biol 6:405–413
11. Hwang CS, Shemorry A, Varshavsky A (2010) N-terminal acetylation of cellular proteins creates specific degradation signals. Science 327:973–977
12. Brown JL, Roberts WK (1976) Evidence that approximately eighty per cent of the soluble proteins from Ehrlich ascites cells are Nalpha-acetylated. J Biol Chem 251:1009–1014
13. Arnesen T, Van Damme P, Polevoda B, Helsens K, Evjenth R, Colaert N, Varhaug JE, Vandekerckhove J, Lillehaug JR, Sherman F, Gevaert K (2009) Proteomics analyses reveal the evolutionary conservation and divergence of N-terminal acetyltransferases from yeast and humans. Proc Natl Acad Sci U S A 106:8157–8162
14. Goetze S, Qeli E, Mosimann C, Staes A, Gerrits B, Roschitzki B, Mohanty S, Niederer EM, Laczko E, Timmerman E, Lange V, Hafen E, Aebersold R, Vandekerckhove J, Basler K, Ahrens CH, Gevaert K, Brunner E (2009) Identification and functional characterization of N-terminally acetylated proteins in Drosophila melanogaster. PLoS Biol 7:e1000236
15. Gevaert K, Goethals M, Martens L, Van Damme J, Staes A, Thomas GR, Vandekerckhove J (2003) Exploring proteomes and analyzing protein processing by mass spectrometric identification of sorted N-terminal peptides. Nat Biotechnol 21:566–569
16. Gorman JJ, Shiell BJ (1993) Isolation of carboxyl-termini and blocked amino-termini of viral proteins by high-performance cation-exchange chromatography. J Chromatogr 646:193–205
17. Dormeyer W, Mohammed S, Breukelen B, Krijgsveld J, Heck AJ (2007) Targeted analysis of protein termini. J Proteome Res 6:4634–4645
18. Mikami T, Takao T (2007) Selective isolation of N-blocked peptides by isocyanate-coupled resin. Anal Chem 79:7910–7915
19. Zhang X, Ye J, Hojrup P (2009) A proteomics approach to study in vivo protein N(alpha)-modifications. J Proteomics 73:240–251
20. Zhang X, Ye J, Engholm-Keller K, Hojrup P (2011) A proteome-scale study on in vivo

protein Nalpha-acetylation using an optimized method. Proteomics 11:81–93

21. Zhang X, Hojrup P (2010) Cyclization of the N-terminal X-Asn-Gly motif during sample preparation for bottom-up proteomics. Anal Chem 82:8680–8685

22. Olsen JV, de Godoy LM, Li G, Macek B, Mortensen P, Pesch R, Makarov A, Lange O, Horning S, Mann M (2005) Parts per million mass accuracy on an Orbitrap mass spectrometer via lock mass injection into a C-trap. Mol Cell Proteomics 4:2010–2021

23. Gobom J, Nordhoff E, Mirgorodskaya E, Ekman R, Roepstorff P (1999) Sample purification and preparation technique based on nano-scale reversed-phase columns for the sensitive analysis of complex peptide mixtures by matrix-assisted laser desorption/ionization mass spectrometry. J Mass Spectrom 34:105–116

24. Rappsilber J, Ishihama Y, Mann M (2003) Stop and go extraction tips for matrix-assisted laser desorption/ionization, nanoelectrospray, and LC/MS sample pretreatment in proteomics. Anal Chem 75:663–670

Chapter 6

Identification and Analysis of *O*-Acetylated Sialoglycoproteins

Chandan Mandal and Chitra Mandal

Abstract

5-*N*-acetylneuraminic acid, commonly known as sialic acid (Sia), constitutes a family of *N*- and *O*-substituted 9-carbon monosaccharides. Frequent modification of *O*-acetylations at positions C-7, C-8, or C-9 of Sias generates a family of *O*-acetylated sialic acid (*O*-AcSia) and plays crucial roles in many cellular events like cell–cell adhesion, proliferation, migration, etc. Therefore, identification and analysis of *O*-acetylated sialoglycoproteins (*O*-AcSGPs) are important. In this chapter, we describe several approaches for successful identification of *O*-AcSGPs. We broadly divide them into two categories, i.e., invasive and noninvasive methods. Several *O*-AcSias-binding probes are used for this purpose. Detailed methodologies for step-by-step identification using these probes have been discussed. We have also included a few invasive analytical methods for identification and quantitation of *O*-AcSias. Several indirect methods are also elaborated for such purpose, in which *O*-acetyl group from sialic acids is initially removed followed by detection of Sias by several approaches. For molecular identification, we have described methods for affinity purification of *O*-AcSGPs using an *O*-AcSias-binding lectin as an affinity matrix followed by sequencing using matrix-assisted laser desorption/ionization-time of flight (MALDI-TOF-TOF) mass spectroscopy (MS). In spite of special attention, loss of *O*-acetyl groups due to its sensitivity towards alkaline pH and high temperature along with migration of labile *O*-acetyl groups from C7–C8–C9 during sample preparation is difficult to avoid. Therefore there is always a risk for underestimation of *O*-AcSias.

Key words: Achatinin-H, Acetylacetone method, Crab lectin, Fluorimetric high-performance liquid chromatography (HPLC), Influenza C hemagglutinin-esterase, Mild periodic acid Schiff (mPAS) staining, *O*-acetylated sialic acid, *O*-acetylated sialic acid-binding lectin, *O*-acetylated sialoglycoprotein (*O*-AcSGP), 9-*O*-acetylated sialoglycoprotein (Neu5,9Ac$_2$GP), Sialic acid-binding lectin, sequencing, Siglecs

Sandra B. Hake and Christian J. Janzen (eds.), *Protein Acetylation: Methods and Protocols*, Methods in Molecular Biology, vol. 981, DOI 10.1007/978-1-62703-305-3_6, © Springer Science+Business Media, LLC 2013

1. Introduction

5-*N*-acetylneuraminic acid (Neu5Ac), commonly known as sialic acid (Sia), constitutes a family of *N*- and *O*-substituted 9-carbon monosaccharides (1–4). Frequent modification of *O*-acetylations at positions C-7, C-8, or C-9 of Sias generates a family of *O*-acetylated sialic acid (*O*-AcSia) and plays crucial roles in many cellular events like cell–cell adhesion, proliferation, migration, etc. (1–8). Therefore, identification and analysis of *O*-acetylated sialoglycoproteins (*O*-AcSGPs) are important.

Identification and analysis of *O*-AcSGPs require several protections to avoid the loss of *O*-acetyl group. This group is very sensitive towards high temperature and alkaline pH. Introduction of *O*-AcSias to moderate alkaline conditions results in migration of acetyl groups from carbon-7 to carbon-8 or -9 (1–6). For this purpose temperature and pH of the buffer should be critically maintained during sample preparation, identification, quantitation, and purification. The overall flowchart for the identification of *O*-AcSias has been depicted in Fig. 1. The description of the probes has been shown in Table 1.

Following lectins are used to identify *O*-AcSGPs. The crab lectin purified from hemolymph of *Liocarcinus depurator* is specific towards *O*-AcSias (10). The lectins purified from hemolymph of *Cancer antennarius* (11, 12) and serum of freshwater crab, *Paratelphusa jacquemontii* (13), bind to both 4-*O*- and 9-*O*-AcSias. These lectins show specificity towards *O*-acetylated sialoglycocongugates irrespective of their subterminal linkages. Purified sialoglycoprotein from human placenta shows binding towards 4-*O*- and 9(7,8)-*O*-AcSias (14). Another lectin, Achatinin-H, purified from the hemolymph of African giant land snail, *Achatina fulica*, shows preferential specificity towards 5-*N*-acetyl-9-*O*-acetyl neuraminic acid (Neu5,9Ac_2) linked to α2-6 GalNAc (Fig. 2 and ref. 9). However, as the binding of this lectin towards 7-*O*- and/or 8-*O*-AcSias cannot be excluded, such linkages in *O*-AcSGPs may also be detected by this lectin (9, 17, 18).

Simple hemagglutination (HA) is performed to check the presence of *O*-AcSias on red blood cells using these lectins. This *O*-acetylated sialoglycotopes exclusively appear on erythrocytes of patients suffering from *visceral leishmaniasis* (VL, refs. 19, 20) and totally absent on normal human erythrocytes, hence can be used as diagnostic marker (19). This lectin is also used as coating antigen in enzyme-linked immunosorbent assay (ELISA) for quantitation of *O*-AcSias induced on VL-erythrocytes (19, 20).

Fluorescein isothiocyanate (FITC)-coupled lectin (FITC-Achatinin-H) is a powerful tool for identification of 5-*N*-acetyl-7(8),9-*O*-acetylated sialoglycoproteins (Neu5,9Ac_2GP) on intact

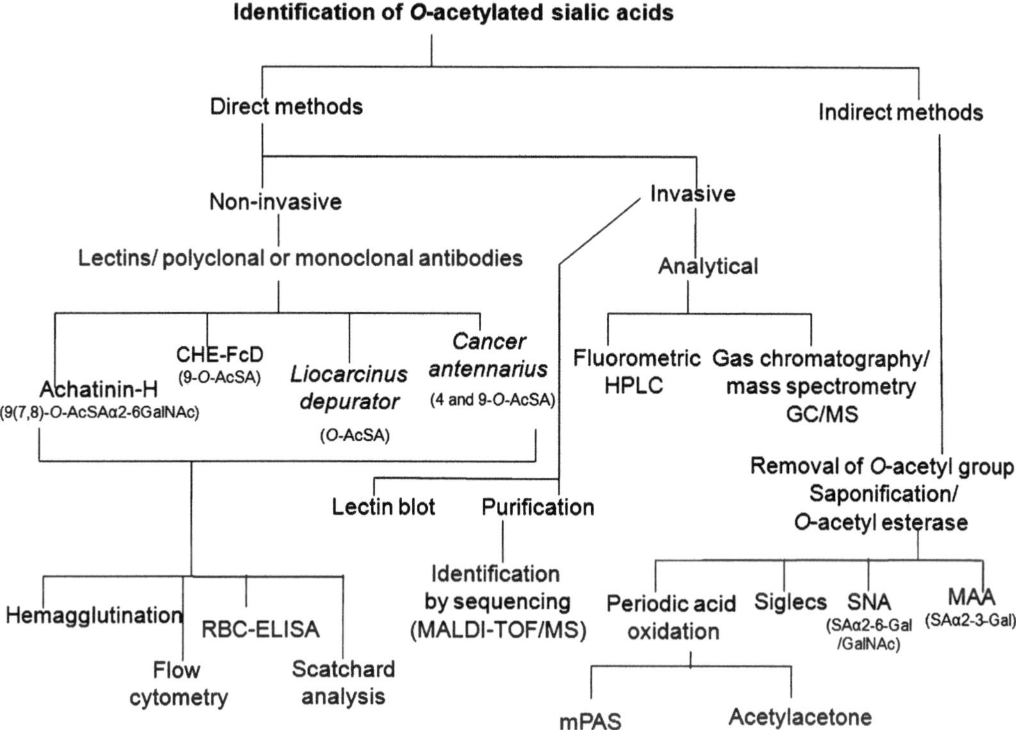

Fig. 1. Schematic diagram of the methods for identification of *O*-acetylated sialic acids. It includes the direct and indirect methods to identify the *O*-AcSGPs. Direct method is subdivided to noninvasive and invasive categories. Noninvasive methods include Hemagglutination, RBC-ELISA, and flow cytometry using several *O*-acetylated sialic acid (*O*-AcSA)-binding lectins and polyclonal or monoclonal antibodies. Identification of *O*-AcSGPs by lectin blot and purification of Neu5,9Ac$_2$GPs by lectin column are included in invasive method. Invasive method also comprises fluorimetric-HPLC and GC/MS analysis of purified *O*-AcSA. In indirect method, *O*-acetyl group is removed first either by saponification or *O*-acetyl esterase treatment. Subsequently sialic acid content of these samples is determined by periodic acid oxidation followed by mild periodic acid Schiff staining and acetylacetone methods. *O*-acetyl esterase-treated samples are identified by sialic acid-binding lectins like *Sambucus nigra agglutinin* (SNA) and *Maackia amurensis agglutinin* (MAA) and siglecs. Here CHE-FcD represents chimera of influenza C hemagglutinin-esterase (CHE) fused with Fc human IgG (CHE-Fc) inactivated esterase form.

cells by flow cytometry. FITC-Achatinin-H is successfully used for identification of Neu5,9Ac$_2$GPs specifically induced on lymphoblasts of childhood acute lymphoblastic leukemia (ALL, refs. 21–27) and therefore may be assigned as ALL-associated signature molecules (21). Interestingly, based on the minimal binding of Achatinin-H with peripheral blood mononuclear cells (PBMC) from normal healthy individuals or patients with other cross-reactive hematological disorders such as acute myelogenous leukemia, chronic myeloid leukemia, chronic lymphocytic leukemia, non-Hodgkin's lymphoma, thalassemia, and aplastic anemia, it may be stated that Neu5,9Ac$_2$GPs are minimally expressed.

FITC-Achatinin-H along with commercially available polyclonal/monoclonal antibodies against cluster differentiation (CD) antigens present on different immune cells is successfully used to monitor the differential expression of Neu5,9Ac$_2$GPs on lymphoblasts

Table 1
Lectins from different species and sugar specificity

Species	Scientific name	Source	Sugar specificity	References
	Snail, *Achatina fulica*	Hemolymph	*O*-AcSiaα2-6GalNAc	(9)
	Crustacean, *Liocarcinus depurator*	Hemolymph	*O*-AcSia	(10)
	Marine crab, *Cancer antennarius*	Hemolymph	9-*O*-AcSia 4-*O*-AcSia	(11, 12)
	Paratelphusa jacquemontii	Hemolymph	9-*O*-AcSia 4-*O*-AcSia	(13)
	Placenta from human	*Placenta*	*O*-AcSias	(14)
	Sambucus nigra	Bark	Sia α2-6 Gal/GalNAc	(15)
	Maackia amurensis	Seed	Sia α2-3 Gal	(16)

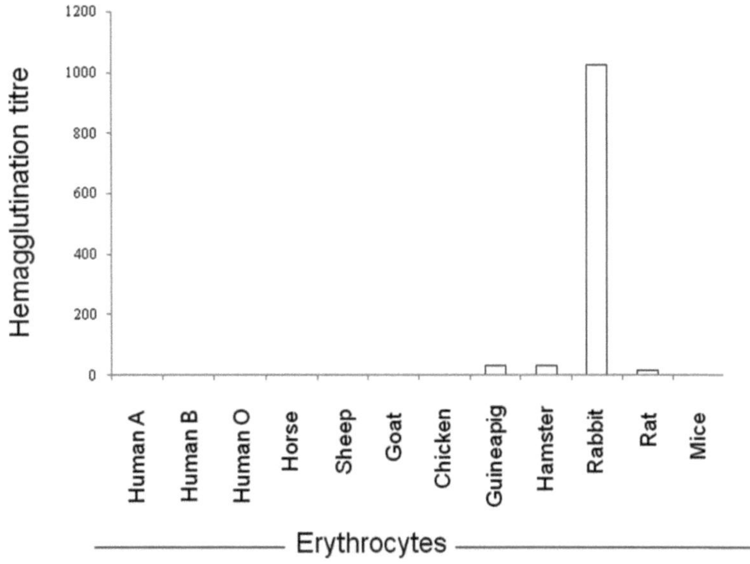

Fig. 2. HA titer of Achatinin-H with erythrocytes from different sources. Achatinin-H (25 μL) is serially diluted in wells in a round-bottomed microtiter plates and mixed with 25 μL Ca^{2+} with final concentration of 0.03 M. This lectin solution is incubated with 2% suspension of freshly isolated erythrocytes (25 μL) obtained from various species and HA titer is measured as described in Subheading 3.1.2. This data was adopted from Mandal et al. (17).

present both in peripheral blood (PB) and bone marrow (BM) of patients with B- or T-ALL and also for detection of minimal residual disease (MRD) (25). Such studies of cell surface expression of Neu5,9Ac$_2$GPs show a significant correlation with the disease progression. This powerful method can detect one cancer cells in 1,000 normal cells (25).

Labelling of Achatinin-H by radioactive iodine (^{125}I)Na is another method to quantitate the Neu5,9Ac$_2$GPs present on cells (20). Cells are incubated with increasing amount of (^{125}I)-Achatinin-H along with unlabelled Achatinin-H. The dissociation constant (K_d) and number of binding sites of Achatinin-H can be calculated by Scatchard plot by measuring the cell-associated radioactivity (28). Purified anti-Neu5,9Ac$_2$GP antibodies towards Neu5,9Ac$_2$GPs may also be used for this purpose in place of lectin. Molecular characterization of Neu5,9Ac$_2$GPs is possible by lectin blot and its visualization by using anti-lectin antibodies.

For further characterization, Neu5,9Ac$_2$GPs is purified by using lectin as an affinity matrix (see Fig. 3) The inhibition of binding between purified Neu5,9Ac$_2$GPs and Achatinin-H by several synthetic analogues of *O*-AcSias demonstrates that Neu5,9Ac$_2$ has the highest inhibitory capability whereas Neu5Ac and 5-*N*-acetyl-4-*O*-acetyl neuraminic acid (Neu5,4Ac$_2$) show no inhibition

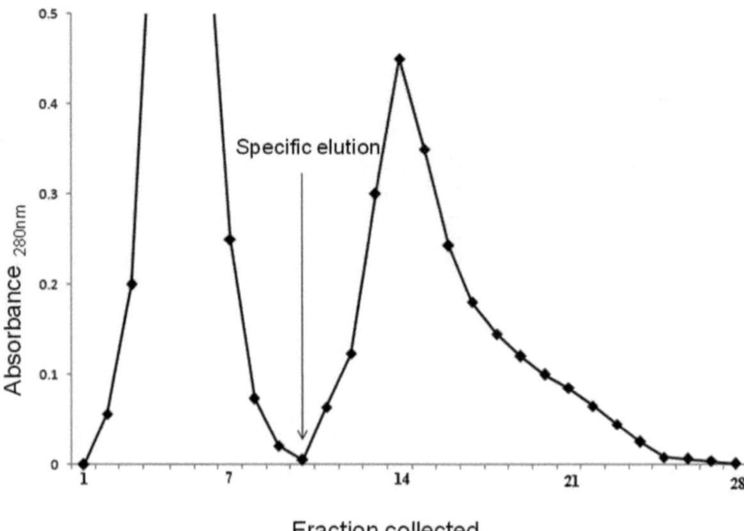

Fig. 3. A representative elution profile of Neu5,9Ac$_2$GPs from affinity matrix. Lymphoblasts present in bone marrow or peripheral blood from childhood ALL, at the presentation of disease, are separated. Membrane proteins are prepared from these lymphoblasts. Neu5,9Ac$_2$GPs present in membrane fraction are separated by using Achatinin-H conjugated to Sepharose 4B column as described in Subheading 3.3. The data was adopted from Pal et al. (22).

(see Fig. 4). Purified Neu5,9Ac$_2$GPs can be sequenced using MALDI-TOF-TOF/MS for subsequent identification (29).

Influenza C virus expresses a membrane-bound glycoprotein, namely, hemagglutinin-esterase (CHE) which can both detect and cleave 9-*O*-acetyl groups from Sias (30–32). However the intact influenza C virions are unstable, and create steric hindrance in binding therefore impractical to use. To overcome these problems, a recombinant soluble chimera of CHE is prepared. To apply in human system, CHE is fused to hinge and Fc human immunoglobulin domain to produce CHE-Fc. The purified CHE-Fc retains the specific 9-*O*-acetyl esterase activity. An irreversible inactivation of specific 9-*O*-acetyl esterase activity is done by the treatment of diisopropyl fluoriphosphates, a general serine hydrolase inhibitor, to produce CHE-FcD. This chimeric protein (CHE-FcD) is specific to detect 9-*O*-AcSias on intact cells (32–36).

Several polyclonal as well as monoclonal antibodies can also be used for identification of *O*-AcSGPs. A few natural occurring antibodies bind to *O*-AcSias. A polyclonal IgG2 fraction, purified from normal human serum, shows specific binding to 9-*O*-AcSias (37). Cytosolic *O*-AcSias-binding antibodies in both IgG1 and IgG2 subtypes are induced in the serum of childhood ALL (38) and VL (39). These antibodies can be utilized to detect the presence of Neu5,9Ac$_2$GP by flow cytometry and ELISA (38, 40, 41).

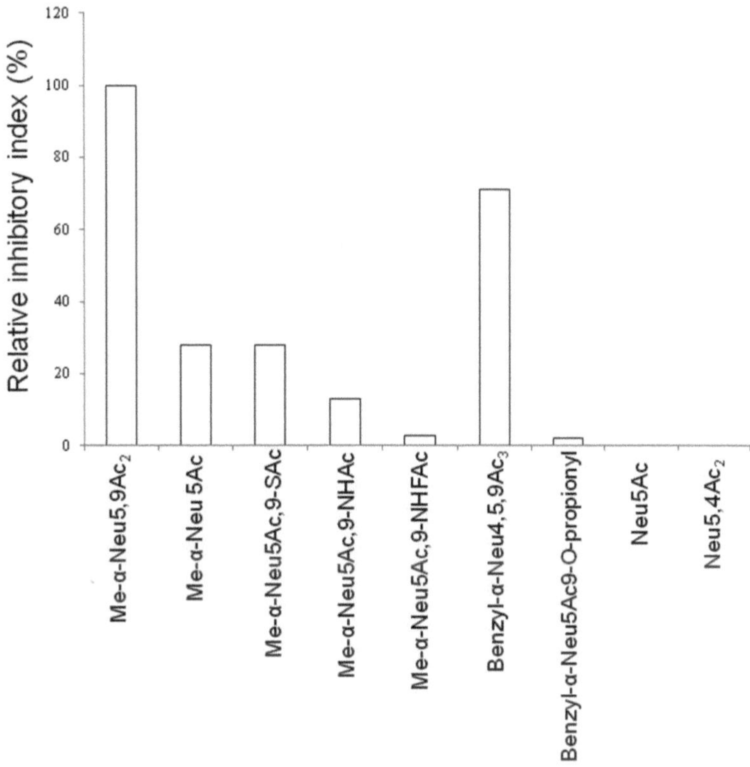

Fig. 4. Inhibition of binding of Neu5,9Ac$_2$GPs with Achatinin-H by several synthesized sialic acids. Wells of a microtiter plate are coated with Achatinin-H by incubating overnight at 4°C. Following washes, the wells are blocked with TBS–2% BSA for 2 h at cold. A constant amount of purified Neu5,9Ac$_2$GPs (1 μg/50 μL) is incubated separately with inhibitors (50 μL, 0–68 mM) in the presence of 30 mM Ca^{2+} at 4°C for 30 min. The mixture (100 μL) is incubated to Achatinin-H-coated well overnight at 4°C. Bound Neu5,9Ac$_2$GPs is detected by incubating anti-Neu5,9Ac$_2$GPs antibodies (1.0 μg), affinity-purified from sera of ALL patients, overnight at 4°C. The bound complex is detected by HRP-conjugated protein A and azino-bis-thio-sulfonic acid as substrate. The difference in binding of Neu5,9Ac$_2$GPs to Achatinin-H in the presence and absence of inhibitors reflects the specificity of Neu5,9Ac$_2$GPs towards Neu5,9Ac$_2$. This data was adopted from Pal et al. (22).

The presence of such antibodies indirectly indicates the induction of Neu5,9Ac$_2$GP on cell surface in diseased conditions (38, 40, 41). Detection of high level of corresponding anti-Neu5,9Ac$_2$GPs antibodies in serum of ALL or VL patients in comparison to normal human serum also indirectly proves the presence of a few newly induced Neu5,9Ac$_2$GPs on the cell surface in diseased conditions. Thus detection of such antibodies in patient's serum is an alternative useful method for revealing Neu5,9Ac$_2$GPs by bovine submaxillary mucin (BSM)-ELISA using BSM as coating antigen (40).

A highly sensitive fluorimetric high-performance liquid chromatography (HPLC), an invasive method, is routinely utilized to analyze *O*-AcSias. The glycosidically bound Sias are initially cleaved

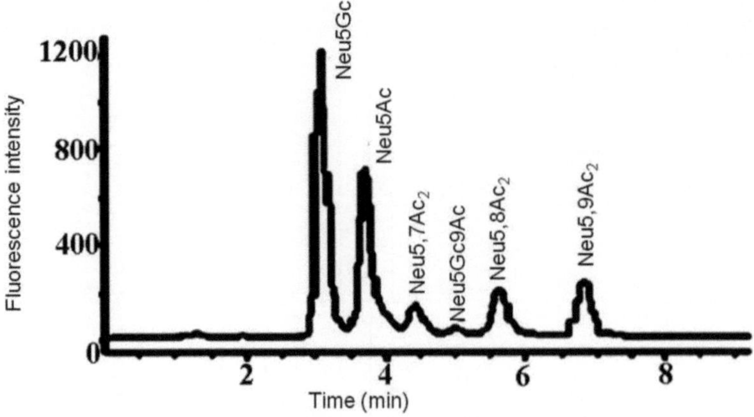

Fig. 5. Fluorimetric HPLC analysis of sialic acid purified from BSM. Sialic acids are released from BSM by the treatment of propionic acid (2 M) and are purified by anion exchange column. This purified sialic acid mixture is derivatized with DMB and analyzed as described in Subheading 3.6.1. The data was adopted from Mandal et al. (43).

by propionic acid treatment. Released Sias are purified through an anion exchange column, derivatized by 1,2-diamino-4,5-methylene-dioxybenzene (DMB), and analyzed by fluorimetric HPLC (ref. 42 and Fig. 5). The purified Sias can be identified by the gas chromatography/mass spectrometry (GC/MS, ref. 44).

Linkage-specific *O*-AcSias can be detected by indirect methods using two well-known commercially available plant lectins (15, 16) and siglecs (45–49) before and after removal of *O*-acetyl group from Sias. *Sambucus nigra agglutinin* (SNA) shows specificity for Sias in α2-6Gal/GalNAc linkage (15) whereas *Maackia amurensis agglutinin* (MAA) towards Sias in α2-3 Gal linkage (16). However these two lectins do not show any specificity towards *O*-AcSias.

Siglecs belongs to the Sia-binding immunoglobulin superfamily lectins. Their expressions are cell type specific (45). Siglec-1 prefers to bind with α2–3-linked sialyllactosamine over α2–6-linked sialyllactosamine and also to α2–8-linked Sias on glycolipids. Siglec-2 (CD22) only binds to α2–6-linked sialyllactosamines on *N*-linked glycans and is expressed exclusively on B cells. Binding preference of siglec-3a is towards α2–3-linked sialyllactosamine over α2–6-linked sialyllactosamine. Siglec-4a requires terminal α2–3-linked Sias for its binding. Siglec-5, expressed on neutrophils and monocytes, binds equally well to α2–3- and α2–6-linked sialyllactosamine. Thus siglecs, although do not bind to *O*-AcSias, show differential specificities towards different linkage-specific Sias and therefore can be used for indirect measurement of *O*-AcSias. In general, chimera of siglecs is fused with Fc domain of human immunoglobulin to produce Siglec-Fc and used for detection of *O-AcS*ias (46–49).

In general, in this approach, initially *O*-acetyl group is removed from *O*-AcSGP by treating the cells/test samples with *O*-acetyl esterase from influenza C virus (34). *O*-acetyl esterase-treated samples are subsequently used for SNA/MAA/siglecs binding separately. Therefore, increased binding of SNA/MAA/siglecs after removal of cell surface *O*-acetyl group indirectly indicates the presence of *O-AcS*ias in linkage-specific manner (45). The differences in binding with SNA and/or MAA/siglecs before and after de-*O*-acetylation of test samples can therefore be an indirect measure of *O-AcS*ias. SNA, MAA, and siglecs can thus be considered also as important tools for such purposes. Such binding is monitored by flow cytometry, dot blotting, ELISA, and Western blotting (45, 47, 49). The presence of *O-AcS*ias on *Pseudomonas aeruginosa* (49), several strains of *Leishmania donovani* (44, 47), erythrocytes (29, 50, 51), PBMC of patients suffering from VL (29, 50, 51) and ALL (21–23) has been demonstrated by several such indirect approaches.

Oxidation of Sias by mild periodic acid resulting in the formation of formaldehyde through carbon–carbon bond cleavage distinguishes between *O*-acetylated and non-*O*-AcSGPs. Free hydroxyl groups should be present for such oxidation which is prevented due to *O*-acetylation on *O*-AcSias. After removal of *O*-acetyl groups by saponification at C-8 and C-9, mild oxidation of the de-*O*-acetylated sample is done. An increase in formaldehyde production is observed in the Sia mixture and is an efficient method for quantitative determination of *O*-AcSias (52).

Therefore, after removal of *O*-acetyl group from *O*-AcSGPs, sample can be stained with mild periodic acid-Schiff (mPAS) (53–55) *which* stains only de-*O*-AcSGPs. Such staining distinguishes between *O*-acetylated and non-*O*-AcSGPs and is a useful method to identify the *O*-AcSGPs through indirect approach. Histochemical staining of *O*-AcSGPs after removal of *O*-acetyl group with NaOH is helpful for indirect demonstration of *O*-AcSGPs (52–55).

Another indirect approach for the detection of *O*-AcSias present in protein sample is acetylacetone method (43, 56). After mild oxidation of Sias, formaldehyde in combination with acetylacetone and ammonium acetate liberates a fluorogen. The formation of fluorogen has been described as a sensitive common method to determine concentration of Sia.

Several *O*-acetylated di- and mono-sialo gangliosides can also be detected by using several monoclonal antibodies (MAb) such as Jones MAb (57), UM4D4 (58), MAb U5 (59), and MAb 7H2 (60). These MAbs are generally used for immunodetection of *O*-Ac gangliosides on high-performance thin-layer chromatography (HPTLC) plates.

2. Materials

2.1. Identification of O-Acetylated Sialoglycoproteins (O-AcSGPs)

2.1.1. Purification of O-AcSias-Binding Lectin

Preparation of BSM

1. Bovine submaxillary gland (90 g).
2. 2 M acetic acid.
3. 1 M NaOH.
4. Barium acetate.
5. Methanol.
6. 0.1 M EDTA.
7. Water (cold).

Coupling of BSM with Sepharose 4B

1. BSM (100 mg).
2. 20 mL Sepharose 4B.
3. 4 mg CNBr in 6 mL dioxane.
4. Coupling buffer: 0.1 M $NaHCO_3$, 0.5 M NaCl, pH 7.2.
5. Acetate buffer: 0.1 M Na-acetate, 0.5 M NaCl, pH 4.0.
6. 0.1 M $NaHCO_3$, pH 7.2.
7. 2 M Na_2CO_3.
8. 0.2 M glycine, pH 8.0.
9. Tris buffer saline (TBS) buffer: 0.05 M Tris–HCl, 0.15 M NaCl, pH 7.2.

Source of Hemolymph

The lectin is purified from the hemolymph of freshly procured African giant land snail (*Achatina fulica*). These snails are abundantly available in Northern and Eastern part of West Bengal (Sundarban area), Assam, Arunachal Pradesh, Orissa, and Indopacific islands. They are also available in Pakistan, Bangladesh, Southern Ethiopia, southern Somalia, Northern Mozambique, Madagascar, Mauritius, Seychelles, Morocco, West Africa, Ghana, Gordonvale, and Queensland.

Snails are procured mainly during June–September in India, as this is their breeding season. Rest of the time, especially during winter months, snails go for hibernation and it is difficult to find them. Snails come out in a large number only after dark and feed themselves on paddy plants and therefore causing lots of financial harm to the farmers. Overnight they destroy paddy field.

1. Snails (size ~12–20 cm and weight ~150–600 g).
2. 2 mL syringe.
3. 26 G needles.
4. Hemolymph (~50 mL/10 snails).

5. Protease inhibitor cocktails (Calbiochem).
6. TBS/Ca^{+2}-buffer: TBS containing 0.03 M $CaCl_2$.
7. 20 mL BSM-Sepharose 4B column.
8. Citrate buffer: 0.04 M sodium citrate in TBS.

Native Gel

1. Achatinin-H (2 μg).
2. 29:1 Acrylamide:bis-acrylamide (5% gel) in 0.375 M Tris–HCl, pH 8.8, and polymerized by adding 0.05% N′N′N′N′ tetramethyl ethylenediamine (TEMED) and 0.5% ammonium persulfate.
3. Mini protein gel apparatus (BioRad).
4. Sample buffer: 0.125 M Tris–HCl, pH 6.8, 10% glycerol, and 0.001% bromophenol blue.
5. Electrode buffer: 0.025 M Tris–HCl, 0.192 M glycine, pH 8.3.
6. Staining solution: 0.2% Coomassie brilliant blue R-250 in 45% methanol, 7% acetic acid, and 48% water.
7. Destaining solution: 15% methanol, 7% acetic acid, and 78% water.

2.1.2. HA Using Red Blood Cells

1. Achatinin-H.
2. 2–10 mL syringe and 23 G needle.
3. 2 mL freshly collected blood.
4. Heparin.
5. Alsever's solution: 20.5 mg dextrose, 4.2 mg NaCl, 8 mg Na-citrate, dissolve in 1 L water, pH 7.0, store at 4°C.
6. Round-bottom 96-well plate.
7. 0.15 M NaCl.
8. 0.1 M $CaCl_2$.

2.1.3. HA-Inhibition

1. Achatinin-H.
2. 2% Suspension of erythrocyte.
3. Round-bottom 96-well plate.
4. 0.15 M NaCl.
5. 0.1 M $CaCl_2$.
6. Inhibitors: 2 mM Neu5,9Ac_2, 2 mM Neu5,4Ac_2, 2 mM Neu5Ac, 50 mg/mL galactose, 4.0 mg/mL BSM, and several sialoglycoproteins like ovine submaxillary mucin (OSM), fetuin, human chorionic gonadotropin (HCG), α1-acid glycoprotein, and transferrin along with asialo-BSM; stock solutions being 4.0 mg/mL.

2.1.4. Flow Cytometry

Preparation of Fluorescein Isothiocyanate-Achatinin-H

1. 100 μg FITC in 20 μL DMSO.
2. Labelling buffer: 0.05 M boric acid, 0.2 M NaCl, pH 9.2.
3. 1 mg Achatinin-H.
4. Dialysis buffer: 0.05 M Tris–HCl, 0.15 M NaCl, pH 7.2, 0.03 M $CaCl_2$.

Preparation of Cells

1. Lymphocytes or natural killer cells or monocytes or macrophages or dendritic cells or neutrophils or erythrocytes in diseased conditions and normal human cells for comparison. PBMC (1×10^6/100 μL), erythrocytes (5×10^6/100 μL), parasites (2×10^6/100 μL), and bacteria (1×10^7/100 μL) in culture medium containing 5% fetal calf serum (FCS) are used.
2. 0.03 M $CaCl_2$.
3. 1% Paraformaldehyde in phosphate buffer saline (PBS): 0.2 M NaH_2PO_4, 0.2 M Na_2HPO_4, 0.15 M NaCl, pH 7.2 (optional).
4. FACSCalibur BD Biosciences, San Jose, CA.

2.1.5. Lectin Blot Analysis

1. Lysate protein (75 μg/lane) from test sample (e.g., hematopoietic cells) as mentioned above.
2. 1 μg/mL Achatinin-H (HA titer 1/256, single band on native SDS-PAGE).
3. 29:1 Acrylamide:bis-acrylamide (5–12% gradient gel) in 0.375 M Tris–HCl, pH 8.8, and 2% SDS and polymerized by adding 0.05% TEMED and 0.5% ammonium persulfate.
4. Electrode buffer: 0.025 M Tris–HCl, 0.192 M glycine, pH 8.3, and 0.1% SDS.
5. Sample buffer: 2% SDS, 5% β-mercaptoethanol, 0.125 M Tris–HCl, pH 6.8, 10% glycerol, and 0.001% bromophenol blue.
6. Transfer buffer: 0.025 M Tris–HCl, 0.192 M glycine, 20% methanol.
7. Nitrocellulose membrane.
8. Nitrocellulose membrane activation buffer: 0.025 M Tris–HCl, 0.192 M glycine.
9. Anti-Achatinin-H antibodies raised in rabbit.
10. Anti-rabbit IgG conjugated with horseradish peroxidase (HRP).
11. 0.1% Tween-20 in TBS.
12. Staining solution: 15 mg diamino benzidine (DAB), 30 mL of 100 mM Tris–HCl, pH 7.2, 15 μL H_2O_2.
13. Alternatively, enhanced chemiluminescent (ECL, SuperSignal West Pico Chemiluminescent Substrate, Thermo Scientific, Rockford, USA) staining solution.

14. Protein II mini gel apparatus (Bio-Rad).
15. Semi-dry transfer apparatus (Trans-blot SD, Bio-Rad).

2.1.6. Western Blot

1. Soluble chimeric influenza C virus hemagglutinin-esterase (CHE-Fc, refs. 35, 36).
2. 1.0 mM diisopropyl fluoriphosphates.
3. 75 μg *O*-AcSGP.
4. Follow steps 3–7 in Subheading 2.1.5.
5. Anti-human IgG-HRP.
6. 0.1% Tween-20 in TBS.
7. Follow steps 11–15 in Subheading 2.1.5 for staining.

2.2. Quantitation of O-AcSias

2.2.1. RBC-ELISA

1. Achatinin-H or any suitable lectins.
2. Normal or diseased erythrocytes.
3. ELISA plate (NUNC, Denmark).
4. TBS.
5. TBS/Ca^{+2}-buffer.
6. 2% Bovine serum albumin (BSA) in TBS/Ca^{+2}.
7. 0.15 M NaCl.
8. 0.15 M NaCl containing 30 mM $CaCl_2$.
9. Water.
10. Staining solution: 1 mL of 2,7-diamino fluorene dihydrochloride (DAF, 1 mg/mL in 60% glacial acetic acid) in 1 mL H_2O_2 (30% in 0.02 M Tris–HCl and 6 M urea).
11. ELISA reader (Sartorius, Germany).

2.2.2. Scatchard Analysis

Iodination of Achatinin-H

1. Achatinin-H (10 μg/20 μL).
2. 0.5 M phosphate buffer, pH 7.5.
3. 0.5 mCi (^{125}I)Na.
4. 4 μL of 8 mg/mL chloramines T in phosphate buffer.
5. Metabisulfite (20 mg/mL).
6. 20 μL of 200 mg/mL potassium iodide in phosphate buffer.
7. Biogel P10 column (0.2 × 10 cm).
8. 0.2% Desialylated-BSA in TBS, pH 7.4.

Binding of (^{125}I)-Achatinin-H

1. (^{125}I)-Achatinin-H (specific activity 1.4×10^6 cpm/μg).
2. 1×10^6 cells in 25 μL RPMI-1640 medium containing 0.2% desialylated-BSA.
3. 0.3 mol/L $CaCl_2$.
4. 0.2% Desialylated-BSA in TBS, pH 7.4.
5. Gamma counter (Electronic Corporation of India).

2.3. Purification of Neu5,9Ac$_2$GPs

2.3.1. Coupling of Achatinin-H with Sepharose 4B

1. 2 mg Achatinin-H.
2. 3.5 mL CNBr activated Sepharose 4B (Sigma).
3. Coupling buffer: 0.1 M $NaHCO_3$, 0.5 M NaCl, pH 8.5.
4. Acetate buffer.
5. 0.2 M glycine, pH 8.0.
6. TBS/Ca^{+2}: 0.05 M Tris–HCl, 0.15 M NaCl, pH 7.2, 0.03 M $CaCl_2$.

2.3.2. Isolation of Membrane Protein

1. Cells (1×10^7), parasite (1×10^8), erythrocytes (1×10^8), expressing Neu5,9Ac$_2$GPs.
2. Homogenization buffer: 0.01 M Tris–HCl, pH 7.2, 0.001 M EDTA, protease inhibitors: 0.1 mg/mL PMSF, 10 μg/mL of each leupeptin and aprotonin.
3. Solubilization buffer: 0.02 M piperazine, 0.05 M NaCl, 0.001 M EDTA, protease inhibitor cocktail, 1% Nonidet-40.
4. Ultracentrifuge.

2.3.3. Affinity Purification

1. Achatinin-H–Sepharose4B.
2. Membrane protein.
3. Washing buffer TBS/Ca^{+2}.
4. Elution buffer: 0.04 M sodium citrate in 0.05 M Tris–HCl, pH 7.2, 0.15 M NaCl.

2.4. Characterization of O-Acetylated Sialoglycoproteins by SDS-PAGE

1. 20 μg purified *O*-AcSGPs.
2. Follow steps 3–5 in Subheading 2.1.5.
3. Fixative buffer: 50% Methanol, 12% acetic acid, 0.05% formalin, 37.95% water.
4. Wash buffer: 35% Ethanol in water.
5. Sensitizing solution: 0.02% $Na_2S_2O_3$ in water.
6. Staining solution: 0.2% $AgNO_3$, 0.076% formalin in water.
7. Developer solution: 6% Na_2CO_3, 0.05% formalin, 0.0004% $Na_2S_2O_3$.
8. Stop solution: 50% Methanol, 12% acetic acid, and 38% water.

2.5. Molecular Characterization of Purified Neu5,9Ac$_2$GPs by MALDI-TOF-TOF Analysis

1. 150 μg affinity-purified Neu5,9Ac$_2$GPs.
2. IPG strip (Bio-Rad, Hercules, CA) of pH 4–7 and 7 cm length.
3. Rehydration buffer: 7 M urea, 2 M thiourea, 2% CHAPS, 0.2% (w/v) ampholytes, 50 mM DTT, and 0.004% of bromophenol blue.

4. Equilibration buffer I: 6 M urea, 2 M thiourea, 0.375 M Tris–HCl, pH 8.8, 20% glycerol, 2% SDS, 0.005% bromophenol blue, and 2% DTT.
5. Equilibration buffer II: Identical to equilibration buffer I, but 2.5% iodoacetamide in place of 2% DTT.
6. PROTEAN IEF apparatus (Bio-Rad, Hercules, CA).
7. 0.5% Agarose.
8. Fixation buffer: 30% Methanol, 15% acetic acid, 55% water.
9. Bio-Safe Coomassie (Bio-Rad).
10. 25 mM ammonium bicarbonate buffer.
11. Digestion buffer: 25 mM ammonium bicarbonate buffer, 400 ng trypsin (Promega).
12. Destaining solution: 25 mM ammonium bicarbonate and 50% acetonitrile.
13. 50 mM Tris(2-carboxyethyl)phosphine (TCEP).
14. 100 mM Iodo-acetamide in 25 mM ammonium bicarbonate.
15. Trifluoroacetic acid.
16. α-Cyano hydroxy cinnamic acid (CHCA).
17. MALDI-TOF/TOF analyzer (Applied Biosystem, USA).

2.6. Identification of O-AcSias by Analytical Methods using Fluorimetric-HPLC

2.6.1. Release and Purification of Sias

1. Sample (100 μg of whole cell lysate protein or 50 μg purified *O*-AcSGPs).
2. 2 M propionic acid.
3. Anion exchange column (Dowex 2×8 (200–400 mesh)-$HCOO^-$).
4. 1.5 M formic acid.

2.6.2. Derivatization of Sias by DMB

1. Purified Sias.
2. DMB.
3. 0.75 M 2-mercaptoethanol.
4. 18 mM $NaHSO_3$.
5. 1.4 M acetic acid.

2.6.3. Analysis

1. DMB-derivatives of Sias.
2. Methanol.
3. Acetonitrile.
4. Reverse phase C-18 (RP-C18, 4×250 mm, 5 μm) column (Merck, Germany).

2.6.4. GC/MS

1. Peaks collected from HPLC column or purified mixture of Sias from anion exchange column.
2. Methanol.

3. Dowex H.
4. Cotton wool.
5. P_2O_5.
6. TMS-reagent: Pyridine/hexamethyl disilazane/trimethyl chlorosilane (5/1/1, v/v/v).
7. Fisons Instruments GC 8060/MD800 system (Interscience, Breda, The Netherlands), AT-1 column (30 m, 0.25 mm, Alltech, Breda, The Netherlands).

2.7. Indirect Method for Analysis of O-AcSias: Before and After Removal of O-Acetyl Group

2.7.1. Mild Periodic Acid Oxidation

Acetylacetone Method

1. Lymphocytes, natural killer cells, monocytes, macrophages, dentritic cells, neutrophils, erythrocytes, PBMC, parasites, or bacteria cells.
2. 20 μL lysate protein (1.0 mg/mL) prepared from cells before and after saponification.
3. Buffer A: 3 g/L NaCl, 0.2 g/L KCl, 1 g/L disodium hydrogen phosphate-dihydrated, 0.15 g/L sodium dihydrogen phosphate-monohydrated, 0.2 g/L potassium dihydrogen phosphate.
4. 2.5 mM sodium meta periodate in buffer A.
5. 2% Sodium arsenite in 0.5 M HCl.
6. Acetylacetone solution: 0.075 mL acetic acid, 0.05 mL acetylacetone, 3.75 g ammonium acetate in water of total final volume 25 mL.
7. 0.2 M NaOH.
8. 0.2, 0.5 M HCl.
9. 0.1 M NaCl.

mPAS Staining

1. Paraffin-embedded tissue section (5 μM) before and after saponification.
2. 5 mM periodic acid in acetate buffer.
3. Acetate buffer: 0.05 M sodium acetate, pH 5.5.
4. Schiff reagent.
5. Mayer's hematoxylin.
6. Mounting medium: PBS:glycerol 1:1 with small pinch of antiquenching reagent 1.4-diazabicyclo (2.2.2)octane (DABCO).

2.7.2. Flow Cytometric Analysis

Removal of O-Acetyl Group from Cells

1. Lymphocytes or natural killer cells or monocytes or macrophages or dentritic cells or neutrophils or erythrocytes in diseased conditions as compared to normal human cells (1×10^6/100 μL) or parasites (2×10^6/100 μL) or bacteria (1×10^7/100 μL) in 5% FCS medium.

2. 9-*O*-Acetyl hemagglutinin esterase of influenza C virus (prepared by Dr. Reinhard Vlasak, Institute of Molecular Biology, Austrian Academy of Science, Austria, ref. 34).
3. PBS.

SNA/MAA Binding

1. 5 μg/mL biotinylated SNA and MAA.
2. 0.5 μg/mL FITC-conjugated streptavidin.
3. 1% Paraformaldehyde in PBS.
4. FACSCalibur (BD Biosciences, San Jose, CA).

Siglecs Binding

Follow steps 1 and 2 in Subheading 2.7.2 (*removal of O-acetyl group from cells*).

1. PBS.
2. Siglec-Fc (prepared by Prof. Paul R. Crocker, School of Life Science, Dundee University, UK, ref. 46).
3. Protein A–biotin.
4. 0.5 μg/mL FITC-conjugated streptavidin.
5. 1% Paraformaldehyde in PBS.
6. FACSCalibur (BD Biosciences, San Jose, CA).

2.7.3. Siglecs-ELISA

FITC Labelling of Cells

1. Follow steps 1 and 2 in Subheading 2.7.2 (*removal of O-acetyl group from cells*).
2. PBS.
3. 0.1% FITC in DMSO.
4. 50 mM carbonate buffer, pH 8.0.

ELISA

1. 0.5 μg/100 μL protein A.
2. 50 mM carbonate buffer, pH 9.5.
3. PBS.
4. 2% BSA in PBS.
5. Siglec-Fc.
6. FITC-cells.
7. Fluorescent plate reader (Cyto Flour^{II}, Perkin Elmer).

2.7.4. Dot Blot

1. Follow steps 1 and 2 in Subheading 2.7.2 (*removal of O-acetyl group from cells*).
2. PBS.
3. 2% BSA in PBS.
4. Siglec-Fc.
5. Biotin-anti-human Fc specific (Cappel, St. Louis, MO).

6. Avidin–HRP (Cappel, St. Louis, MO).
7. Staining solution: As shown in step 11 in Subheading 2.1.5.

3. Methods

3.1. Identification of O-AcSGPs

The following procedure is from Murphy et al. (61, 62) with little modifications.

3.1.1. Purification of O-AcSias-Binding Lectin

Preparation of BSM

1. Collect bovine submaxillary gland (90 g) from slaughter house (see Notes 1 and 2).
2. Cut the gland into small pieces to remove the fat body on ice tray.
3. Mince in homogenator with 1.5th volume of ice-cold water.
4. Centrifuge at 12,000 × *g* for 15 min at 4°C.
5. Collect the supernatant and resuspend the pellet with same volume of ice-cold water and homogenate and centrifuge again at cold condition.
6. Pool both supernatants.
7. Bring the pH to 4.5 with 2 M acetic acid at cold condition.
8. Centrifuge at 12,000 × *g* for 15 min at 4°C.
9. Collect the supernatant and bring the pH to 6.0 with 1 M NaOH.
10. Dialyze the sample against water for 2 days with 3–4 changes per day.
11. Collect the sample and add barium acetate slowly to make final concentration 0.1 M.
12. Add distilled and chilled methanol to make it 64% by volume.
13. Precipitate formation is allowed to settle overnight at cold.
14. Centrifuge at 10,000 × *g* for 15 min at 4°C.
15. Resuspend the pellet in 80 mL 0.1 M EDTA.
16. Dialyze against water for 2 days with 4–5 changes per day.
17. Collect and centrifuge at 5,000 × *g* for 30 min at 4°C.
18. Collect the clear supernatant.
19. Store at −70°C in several aliquots.
20. Estimate protein by Lowry's method (63).
21. Estimate *O*-AcSias by acetylacetone method (see Subheading 3.7.1).

The approximate yield of BSM is about 400 mg from 90 g gland. Sia concentration is 20 μg/mg of BSM and the percentage of *O*-AcSias in BSM is about 40% (see Note 3).

Coupling of BSM with Sepharose 4B

BSM–Sepharose 4B is specifically used as an affinity matrix for purification of the Achatinin-H from hemolymph of *Achatina fulica* snail.

1. Dialyze BSM (100 mg) against coupling buffer for 1 day with three changes at cold (see Notes 4, 5).
2. Wash 20 mL of Sepharose 4B with water for five times.
3. After final washing of Sepharose 4B mix with an equal volume of 2 M Na_2CO_3.
4. Dissolve 4 g CNBr in 6 mL dioxane.
5. Add to the Sepharose 4B in beaker kept on ice.
6. Allow the reaction to continue for 15 min with gentle stirring.
7. Transfer it to the sinter glass funnel and wash extensively by 20 times bead volume of 0.1 M $NaHCO_3$, pH 8.3, inside hood, followed by a quick wash with coupling buffer, pH 7.3.
8. Quickly add dialyzed BSM to the activated Sepharose 4B and keep overnight at 4°C with very gentle stirring (see Note 6).
9. Centrifuge at 500×*g* for 5 min at 4°C to remove unbound BSM.
10. Block the uncoupled site by adding 0.2 M glycine, pH 8.0 (1 mL/100 mL of Sepharose 4B).
11. Keep at cold for 2 h.
12. Transfer it to the sinter glass funnel and alternative washing is done by coupling buffer and acetate buffer using 20 times of the bead volume at cold to remove excess glycine followed by final wash with TBS-Ca^{+2}, pH 7.2.
13. Store it in TBS-Ca^{+2} buffer, pH 7.2 (see Note 7).
14. Calculate the % of coupling by measuring the concentration of BSM before and after coupling. In general it shows above 60–70% coupling of BSM with Sepharose.

Source of Hemolymph

Based on the inhibition with BSM, Achatinin-H is purified by using BSM as an affinity matrix from the hemolymph of *Achatina fulica* (9).

1. Carefully remove shell from the snail by bone cutter without damaging inner organs.
2. Collect hemolymph in a tube by cardiac puncture using 23 G needles and 2 mL syringe.
3. Add protease inhibitor cocktails (2 μL/10 mL hemolymph) and incubate at cold for 2 h.
4. Centrifuge at 10,000×*g* at 4°C for 30 min to remove amebocytes from the bottom of the tube.

5. Dialyze supernatant against TBS/Ca^{+2}, pH 7.2 with three changes in 24 h.
6. Centrifuge again hemolymph before loading to the column.
7. Immediately use dialyzed hemolymph for purification, and do not store or freeze.
8. Equilibrate 20 mL of the BSM–Sepharose 4B column in TBS/Ca^{+2}, pH 7.2, until the OD at 280 nm reaches <0.005 (see Note 8).
9. Pass slowly 50 mL of the hemolymph into BSM–Sepharose column at least ten times and incubate overnight at 4°C.
10. Wash column with TBS/Ca^{+2}, pH 7.2, to remove unbound hemolymph till the OD at 280 nm is <0.02.
11. Add citrate buffer (column volume) and keep the column at 25°C for 2 h.
12. Elute bound Achatinin-H using citrate buffer.
13. Dialyze extensively against TBS/Ca^{+2}, pH 7.2, with 0.02% sodium azide (NaN3) with 4–5 changes per day for 3–5 days to remove trace amount of citrate buffer.
14. Store purified Achatinin-H in the same buffer at 4°C after filter sterilization (see Notes 9 and 10).
15. Protein is estimated by Lowry's method (62).
16. Yield of Achatinin-H is about 1.5 mg. It shows high HA titer against rabbit erythrocytes (see Fig. 2). Approximately 1/256 dilution of BSM (2.5 mg/mL) is required to inhibit 1/8 HA unit of Achatinin-H. It shows single band in 5% native gel (9). Achatinin-H is stable for more than 1 year, if stored at cold in aliquot after filter-sterilization through 13 mm syringe filter (0.22 micron pore size).

Native Gel

1. Prepare a 5% bis-acrylamide separating gel.
2. Mix 2 μg Achatinin-H with sample buffer.
3. Run at constant voltage (80 V) for 1–1.5 h.
4. Stain the gel with staining solution for at least 1 h.
5. Destain the gel with destaining solution and visualize.

3.1.2. HA Using Erythrocytes

This is a general method and can be used for testing any lectin. This experiment needs to perform to check the specificity of lectin towards carbohydrate. Achatinin-H shows differential agglutination with rabbit, rat, hamster, and guinea pig erythrocytes having variable amount of 9-*O*-AcSias. However, it does not agglutinate human, monkey, sheep, goat, and chicken erythrocytes, which lack these *O*-acetylated sialoglycotope although they have lot of Sias on their cell surface. Horse erythrocytes do not bind with Achatinin-H as it contains 4-*O*-AcSias (ref. 17, see Fig. 2). Rabbit erythrocytes show best hemaglutination titer, so these erythrocytes are used for routine checking of the specificity of freshly purified Achatinin-H towards 9-*O*-AcSias (9, 17). Therefore to confirm the presence of Neu5,9Ac_2GPs on RBC of patients suffering from any diseases, HA can be used using Achatinin-H as a tool.

1. Collect rabbit blood in the presence of heparin and store in Alsever's solution overnight (if needed, see Notes 11 and 12).
2. For identification of *O*-AcSias on diseased erythrocytes, collect blood from patients in the presence of heparin (21, 22).
3. Take 200 μL rabbit blood and dilute with 500 μL 0.15 M NaCl.
4. Centrifuge at 800 × *g* for 3 min.
5. Repeat washing step until supernatent looks clear (absence of red color).
6. Prepare 2% erythrocyte in 0.15 M NaCl.
7. Add 25 μL 0.15 M NaCl up to well number 12 in round-bottom 96-well plate.
8. Take 25 μL Achatinin-H and serially dilute up to well number 10.
9. Add 25 μL of 0.1 M $CaCl_2$ up to well number 11 (well number 11 serves as RBC control).
10. Add 25 μL of 2% erythrocyte into each well.
11. Mix slowly and keep it at room temperature for 30–60 min.
12. Measure the degree of agglutination by visualizing the formation of umbrella in the wells. The reciprocal of highest dilution of the lectin that produces visible agglutination is taken as the HA titer.

3.1.3. HAI

Several synthetic Neu5,9Ac_2 analogues and *O*-AcSGPs are used as HA inhibitors for confirming the specificity of lectins. HA inhibition of Achatinin-H with rabbit erythrocytes by Neu5,9Ac_2 and BSM (rich in 9-*O*-AcSiaα2-6 GalNAc content, refs. 61, 62) confirms its specificity towards Neu5,9Ac_2 linked to α2-6 GalNAc, ref. (17). Interestingly, HA shown by Achatinin-H could not be inhibited by Neu5,4Ac_2 (4-*O*-Ac-Sias), 9-*O*-acetylated ganglioside GD3 (9-OAcGD3) and OSM, fetuin, HCG, α1-acid glycoprotein, transferrin, and asialo-BSM.

1. Prepare 2% rabbit erythrocyte (alternatively RBC from normal and diseased individuals) in 0.15 M NaCl and wash as described in steps 1–3 in Subheading 3.1.2.
2. Add 25 μL of 0.15 M NaCl up to well number 12 in round-bottom 96-well plate.
3. Take a panel of synthetic sugars Neu5Ac (0–2 mM), Neu5,4Ac_2 (0–2 mM), Neu5,9Ac_2 (0–2 mM), and galactose (0–50 mg/mL, as negative control).

4. Prepare several sialoglycoproteins (0–4 mg/mL of each stock solution), e.g., BSM, asialo-BSM, OSM, fetuin, HCG, α1-acid glycoprotein, and transferrin.
5. Add each inhibitor serially diluted up to well number 10 in each row. Separate row is needed for each inhibitor.
6. Add 25 μL of 0.1 M $CaCl_2$ to each well.
7. Usually the concentration of Achatinin-H at which it shows one-fourth or one-eighth HA titer is selected for HAI.
8. Add 25 μL Achatinin-H to each well upto number 11 and leave for 30 min allowing it to bind with sugars.
9. Add 25 μL of 2% erythrocyte in each well.
10. Mix slowly by gentle shaking and incubate at room temperature for 30–60 min.
11. Measure the degree of inhibition of HA.
12. The results are expressed as the minimum concentration of the sugar required for 50% inhibition of HA. For routine HAI, use BSM (positive control) and galactose (negative control).

3.1.4. Flow Cytometry

Cell surface expression of *O*-AcSGPs is determined using FITC-Achatinin-H. This method is useful for any type of cells like erythrocytes, lymphocytes, parasites, bacteria, etc.

Preparation of FITC-Achatinin-H

1. Dialyze 1 mg/mL Achatinin-H against FITC labelling buffer with three changes per day for 2 days.
2. Dissolve FITC in 20 μL DMSO at a concentration of 5 mg/mL in the dark by covering it with aluminum-foil.
3. Add dialyzed Achatinin-H to FITC solution and keep in the dark for 2 h at cold with shaking.
4. Dialyze against dialysis buffer till the unbound FITC is removed.
5. Alternatively, Biogel P-10 column may be used to remove unbound FITC.
6. The concentration of FITC-Achatinin-H is optimized by checking its binding with normal human PBMC by FACS analysis. The concentration of FITC-Achatinin-H, which shows ~5–10% binding with normal human PBMC, is considered as ideal for testing with other cells at diseased conditions (see Note 13), as normal PBMC contain minimal amount of *O*-AcSGPs.
7. Store FITC-Achatinin-H at 4°C in the dark.

Flow Cytometric Analysis

1. Incubate cells (erythrocyte/lymphocyte/other immune cells, 1×10^6) from VL and/or ALL patients or parasites (2×10^6) or

bacteria (1×10^7) with 100 μL 5% FCS medium for 30 min at 4°C.

2. Suspend the cells in medium containing 0.03 M Ca^{+2}.
3. Add FITC-Achatinin-H (~0.04 μg, see Note 13) and keep it at 4°C for 1 h.
4. Wash the cells with TBS/Ca^{+2} and suspend them in 1% paraformaldehyde in PBS.
5. Acquire FITC-Achatinin-H-positive cells by flow cytometry (see Note 14).
6. Measure fluorescence intensity at FL1 region.
7. Analyze data by CellQuestPro software.

3.1.5. Lectin Blot Analysis

The expression of *O*-AcSGPs specifically induced on different cells in diseased condition or on parasites or bacteria can be analyzed by lectin blotting. Rabbit anti-Achatinin-H polyclonal antibody is developed by injecting Achatinin-H in rabbit using standard protocol and used as secondary antibody to detect Achatinin-H-bound Neu5,9Ac_2GPs.

1. Prepare cell lysate from erythrocytes (1×10^8) or hematopoietic cells (1×10^7) or bacteria (5×10^8) or parasite (5×10^7) by sonication (three pulses for 8 s at 100 W giving breaks for 1 min on ice) in PBS.
2. Denature 75 μg of cell lysate by boiling for 5 min in sample buffer.
3. Prepare a 5–12% gradient bis-acrylamide separating gel.
4. Run the gel at constant voltage (80 V) for 1–1.5 h.
5. Equilibrate the gel with cold transfer buffer for 15 min at 4°C.
6. Activate nitrocellulose membrane with activation buffer.
7. Transfer the proteins to the activated nitrocellulose membrane using a semidry transfer cell at constant voltage (15 V) for 30 min.
8. Block the nonspecific binding sites with 2% BSA in TBS overnight in the cold.
9. Incubate the blot with Achatinin-H in TBS/Ca^{+2} overnight at 4°C with gentle shaking.
10. Wash twice with 0.1% Tween-20 in TBS.
11. Incubate the blot with anti-Achatinin-H overnight at 4°C.
12. Wash twice with 0.1% Tween-20 in TBS.
13. Add HRP-anti-rabbit IgG in TBS for 1 h at room temperature.

14. Wash twice with 0.1% Tween-20 in TBS, followed by five washes for 5 min each, with a final wash step with TBS and develop with staining solution.

3.1.6. Western Blot

Influenza C virus hemagglutinin-esterase utilizes to hydrolyze the 9-*O*-acetyl group of Sias. However, the esterase activity irreversibly deactivated by the treatment of diisopropyl fluoriphosphate and then it can be utilized to detect 9-*O*-acetylated Sias.

1. Prepare CHE-Fc where CHE fused with the Fc portion of human IgG (35, 36).
2. Inactivate the esterase activity of CHE-Fc by treating with 1 mM diisopropyl fluoriphosphate.
3. Follow steps 4–8 in Subheading 3.1.5.
4. Incubate the blot with CHE-FcD for overnight at 4°C with gentle shaking.
5. Wash twice with 0.1% Tween-20 in TBS.
6. Incubate the blot with anti-human IgG-HRP in TBS for 1–2 h in the cold.
7. Wash twice with 0.1% Tween-20 in TBS (five washes, 5 min/wash) and finally wash with TBS.
8. Develop with staining solution.

3.2. Quantitation of O-AcSias

3.2.1. RBC-ELISA

It is a very useful noninvasive tool to detect *O*-AcSGPs on diseased erythrocytes. Normal erythrocytes do not express *O*-AcSGPs and therefore should always be used as negative control. RBC-ELISA is used routinely to detect *O*-AcSGPs on erythrocytes from patients suffering from visceral leishmaniasis (19, 20). This is the only antigen-based diagnostic available so far.

1. Coat 0.5 μg/100 μL Achatinin-H in TBS, pH 7.3, on 96-well flat-bottom ELISA plate (see Note 15).
2. Incubate overnight at 4°C.
3. Discard unbound Achatinin-H and wash three times with TBS/Ca^{+2}.
4. Block the nonspecific binding sites with 2% desialylated BSA in TBS/Ca^{+2}.
5. Incubate overnight at 4°C.
6. Collect blood from patients and normal healthy individuals.
7. Separate erythrocytes as described in Subheading 3.1.2, steps 1–3.
8. Wash with 0.15 M NaCl.
9. Add erythrocytes (4.0×10^8 cells/100 μL TBS/Ca^{+2}/well) to ELISA plate.

10. Incubate overnight at 4°C.
11. Remove unbound erythrocytes by gentle washing with 0.15 M NaCl containing 30 mM $CaCl_2$.
12. Lyze the bound cells with 50 μL water.
13. Prepare staining solution.
14. Add 100 μL/well to the lysis solution.
15. Measure absorbance at 620 nm by ELISA reader.

3.2.2. Scatchard Analysis

Iodination of Achatinin-H

1. Dialyze Achatinin-H (10 μg/μL) in phosphate buffer, pH 7.5 for 1 day with three changes/day.
2. Add 0.5 mCi (^{125}I)Na and 4 μL of 8 mg/mL chloramines T in phosphate buffer to dialyzed Achatinin-H.
3. Incubate for 1 min at 25°C.
4. Terminate the reaction by adding metabisulfite (20 mg/mL) and 20 μL of 200 mg/mL potassium iodide in phosphate buffer and raise the volume to 100 μL with phosphate buffer.
5. Apply the mixture of this solution to Biogel P10 column (0.2 × 10 cm) previously equilibrated with 0.2% desialylated-BSA in TBS, pH 7.4 to remove the free iodine.
6. (^{125}I)-Achatinin-H will come in the void volume.

Binding of (^{125}I)-Achatinin-H

1. Take (1×10^6) any cells in 25 μL RPMI-1640 medium containing 0.2% desialylated-BSA.
2. Add increasing concentration of (^{125}I)-Achatinin-H with specific activity 1.4×10^6 cpm/μg in the presence of 15 μL of 0.3 mol/L $CaCl_2$ 40 μL of 0.2% desialylated-BSA in TBS, pH 7.4.
3. Incubate for 60 min at 4°C.
4. Wash out the unbound (^{125}I)-Achatinin-H by TBS containing 0.2% desialylated-BSA.
5. Repeat the step three times.
6. Count the pellet-associated radioactivity by Gamma counter.
7. To evaluate the specific nature of binding a 50-fold excess of unlabelled Achatinin-H was added in step 2 separately.
8. Calculate the dissociation constant (K_d) and the number of binding site from Schatchard plot (28).

3.3. Purification of Neu5,9Ac$_2$GPs

Neu5,9Ac$_2$GPs from lymphoblast of childhood ALL is affinity purified using Achatinin-H as an affinity matrix. This method is also applicable for other system where *O*-AcSGPs are detected on diseased cells/systems.

3.3.1. Coupling of Achatinin-H with Sepharose 4B

1. Dialyze 2 mg Achatinin-H against coupling buffer for a day with three changes.
2. Take 3.5 mL commercially available CNBr-activated Sepharose 4B and soak in 1.0 mM HCl for 15 min at room temperature.
3. Spin at 500 × *g* at 4°C for 10 min.
4. Wash the beads in 10 mL coupling buffer.
5. Immediately add dialyzed Achatinin-H to the activated Sepharose 4B and keep at 4°C with very gentle stirring.
6. Centrifuge at 500 × *g* for 5 min at 4°C and keep unbound Achatinin-H for protein estimation.
7. Block the uncouple site of Sepharose 4B by adding 100 μL of 0.2 M glycine, pH 8.0.
8. Keep at cold for 2–4 h.
9. Centrifuge at 500 × *g* for 5 min in 4°C.
10. Wash with coupling buffer and acetate buffer alternatively with 20 times of the bead volume to remove excess glycine.
11. Store in TBS/Ca^{+2} at 4°C.
12. Calculate the % of coupling by measuring the concentration of Achatinin-H before and after conjugation. It should show above ~80–85% coupling.

3.3.2. Isolation of Membrane Protein

1. Lymphoblasts are separated from peripheral blood or bone marrow by Ficoll density gradient centrifugation at 500 × *g* for 45 min. Any cell type (bacteria or parasite or any diseased cells) can be similarly processed.
2. Homogenate cells (1×10^7) in homogenization buffer by sonication (three pulses for 8 s at 100 W) for 1 min on ice.
3. Centrifuge at 800 × *g* for 10 min.
4. Collect supernatant and centrifuge at 100,000 × *g* for 1 h at 4°C.
5. Suspend the pellet in solubilization buffer and stir for 45 min at 4°C.
6. Centrifuge at 100,000 × *g* for 1 h.
7. Take the supernatant and dialyze against TBS/Ca^{+2} (pH should always between 7–7.2).
8. Estimate protein concentration by Lowry's method (63).

3.3.3. Affinity Purification

1. Load 800 μg of membrane protein on the Achatinin-H–Sepharose 4B column pre-equilibrated with TBS/Ca^{+2}, pH 7.2, at 4°C.
2. For maximum binding of Neu5,9Ac_2GPs present in membrane fraction, protein mixture is passed into column for several times at a very low speed and incubate overnight at cold.

3. Wash column with TBS/Ca^{+2} to remove unbound protein.
4. Elute lectin-bound Neu5,9Ac_2GPs from the column with elution buffer (Fig. 3).
5. Estimate the protein concentration by Lowry's method (63).

3.4. Characterization of O-Acetylated Sialoglycoproteins

3.4.1. SDS-PAGE

1. Fill the gel (protein II mini gel) trunk with electrode buffer.
2. Denature the purified protein by boiling for 5 min in sample buffer.
3. Prepare the resolving 5–12% gradient gel and follow steps 3 and 4 in Subheading 3.1.5.
4. Fix the gel in fixative buffer for 2 h.
5. Wash the gel with wash buffer for 20 min and repeat three times.
6. Sensitize the gel with sensitizing solution for 2 min.
7. Wash with water for 5 min and repeat for three times.
8. Stain the gel with silver staining solution for 20 min.
9. Wash the gel with water for 1 min two times.
10. Develop the gel with developing solution until color appears.
11. Stop the reaction with stop solution for 5 min.

3.5. Molecular Characterization of Purified Neu5,9Ac2GPs

3.5.1. MALDI-TOF-TOF Analysis

1. Precipitate 150 μg purified *O*-AcSGPs to remove all the interfering materials by using 2D-cleanup kit according to the manufacturer's protocol.
2. Solubilize in rehydration buffer by mild shaking for 30 min.
3. Rehydrate IPG strip with the sample by passive rehydration for 18 h.
4. Focus IPG strip using PROTEAN IEF apparatus following manufacturer's protocol with minute modification of voltage gradient.
5. Equilibrate the focused IPG strip 30 min each by using equilibration buffer I and buffer II.
6. Place IPG strip on top of a gradient polyacrylamide gel.
7. Seal with 0.5% agarose in electrophoresis buffer and electrophoreses with Bio-Rad gel apparatus for second dimension (80 V for 2 h).
8. Fix the gel with fixation buffer and stain with Bio-safe Coomassie.
9. Cut Coomassie-stained spots from the 2D-SDS-PAGE gel.
10. Incubate overnight at 37°C in 25 mM ammonium bicarbonate buffer using 400 ng trypsin per sample according to the manufacturer's protocol of in-gel tryptic digestion kit.
11. Destain the bands with destaining solution at 37°C for 30 min with shaking.

12. Reduce proteins with 50 mM TCEP and 25 mM ammonium bicarbonate at 60°C for 10 min.
13. After cooling, add 100 mM iodo-acetamide in 25 mM ammonium bicarbonate and further incubate in the dark for 1 h.
14. Wash gel pieces consecutively with destaining solution followed by acetonitrile and air-dry.
15. Rehydrate the gel pieces with trypsin containing 25 mM ammonium bicarbonate digestion buffer and incubate overnight at 37°C with shaking.
16. Dry the solutions and dissolve in acetonitrile and trifluoroacetic acid.
17. Spot the samples directly on MALDI plate using CHCA as a matrix.
18. Do sequence analysis of selected tryptic fragments with an Ultraflex III MALDI-TOF-TOF mass spectrometer. Acquire spectra over the *m/z* range of 800–4,000 Da.
19. Do peptide mass fingerprint (PMF) analysis and MS/MS ion searches with MASCOT.

3.6. Identification of O-AcSias by Analytical Methods by Fluorimetric-HPLC

This is a direct method to identify *O*-AcSias. In this method glycosidically linked Sias are released and purified by anion exchange chromatography. The purified Sias are analyzed by fluorimetric-HPLC and GC/MS.

3.6.1. Release and Purification of Sias

1. Incubate 100 μg of lysate proteins prepared from any cells such as lymphocytes or natural killer cells or monocytes or macrophages or dendritic or neutrophil or erythrocytes or parasites or bacteria, etc. with 2 M propionic acid.
2. Keep it in an 80°C water bath for 4 h.
3. Centrifuge at 20,000 × *g* for 30 min at 4°C.
4. Take the supernatant and lyophilize the sample.
5. Dissolve the sample in minimum volume of water (500–1,000 μL).
6. Load the sample on anion exchange column (1 cm × 20 cm) Dowex 2 × 8 (200–400 mesh)-$HCOO^-$.
7. Wash with three column volumes of water.
8. Elute the Sias with three column volume of 1.5 M formic acid (see Note 16).
9. Lyophilize the sample.

3.6.2. DMB Derivative of Sias

1. Prepare 7.0 mM DMB solution by dissolving DMB in 1.4 M acetic acid containing 0.75 M 2-mercaptoethanol and 18 mM $NaHSO_3$ (see Note 17).

2. Add 200 μL DMB solutions in 200 μL Sia-containing solution.
3. Heat at 50°C for 2.5 h in the dark.
4. Stop the reaction by chilling on ice.

3.6.3. Analysis

1. Equilibrate the reverse-phase (RP) C-18 column with methanol/acetonitrile/water (7/9/84).
2. Load 20 μL of DMB derivative to the column with the flow rate 1 mL/min.
3. Measure fluorescence intensity at 373 nm excitation and 448 nm emissions (see Note 18, Fig. 5).

3.6.4. GC/MS

1. Dissolve lyophilized Sias in 0.5 mL dry methanol.
2. Add Dowex H in 80 μL methanol.
3. Filter the samples over cotton wool to remove the Dowex.
4. Treat the sample with diazomethane in ether for 5 min at 25°C.
5. Evaporate the solution using a stream of nitrogen and dried over P_2O_5.
6. Dissolve the residue in 6 mL TMS-reagent.
7. Incubate for 2 h at 25°C.
8. Analyze 3 μL samples by GC/MS.
9. Use Fisons Instruments GC 8060/MD800, an AT-1 column with the following temperature program: 220°C for 25 min, 6°C/min to 300°C, 6 min; the injector temperature should be 230°C; and the detection is done by electron impact MS with a mass range of 150 ± 800 *m/z*.

3.7. Analysis of O-AcSias by Indirect Method Before and After Removal of O-Acetyl Group

This is an indirect method to identify and quantitate *O*-AcSias. The presence of *O*-AcSGPs on the cell surface or tissue section or cell lysate can be demonstrated by removing *O*-acetylation by saponification. After hydrolysis of *O*-acetyl group, Sias are identified. Now these Sias are oxidized by periodic acid to form aldehyde. This aldehyde can be detected by acetylacetone method and mPAS staining.

3.7.1. Mild Periodic Acid Oxidation

Acetylacetone Method

1. Take 20 μL of Sias containing lysate protein (1 mg/mL) prepared from blood cells (lymphocytes or natural killer cells or monocytes or macrophages or dendritic cells or neutrophils or erythrocytes) or parasites or bacteria in buffer A with total volume of 100 μL.
2. Add 200 μL of 0.1 M NaCl.

3. In another set, add 100 μL of 0.2 M NaOH to the samples, keep 45 min on ice for removal of *O*-acetyl group (saponification), and neutralize with 0.2 M HCl.
4. Oxidize both the sample with 200 μL 2.5 mM sodium meta periodate in buffer A at 4°C for 15 min in the dark.
5. Centrifuge the sample at 1,000 × *g* for 2 min in the cold.
6. Take 400 μL supernatant and mix with 2% sodium arsenite in 0.5 M HCl and acetylacetone solution.
7. Incubate at 60°C for 15 min in the dark.
8. Add 1.7 mL of water.
9. Measure the relative fluorescence intensity at 410 nm excitation and 510 nm emissions.
10. Determine the percentage of *O*-AcSias by subtracting the relative fluorescence intensity of un-substituted Sia from that obtained after saponification.
11. Use 1, 2, 4, and 8 μg pure Neu5Ac as standard.

mPAS Staining

1. Prepare 5 μM thin paraffin-embedded tissue section.
2. Deparaffinize and rehydrate to water.
3. Saponify the section with 0.5% potassium hydroxide in 70% ethanol for 30 min at room temperature.
4. Wash with water 2–3 times.
5. Oxidize in 5 mM periodic acid solution in acetate buffer for 15 min.
6. Wash extensively with distilled water.
7. Place in Schiff reagent for 15 min (sections become light pink color during this step).
8. Wash in lukewarm tap water for 5 min (immediately sections turn dark pink color).
9. Counterstain in Mayer's hematoxylin for 1 min.
10. Wash in tap water for 5 min.
11. Dehydrate and coverslip using a synthetic mounting medium.

3.7.2. Flow Cytometric Analysis

The presence of *O*-AcSGPs on the cell surface can be demonstrated by removing *O*-acetylation by using esterase activity of 9-*O*-acetyl-specific hemagglutinin esterase of influenza C virus. After hydrolysis of *O*-acetyl group, Sias are identified by SNA, MAA, and siglecs binding.

Removal of *O*-Acetyl Group from Cells

1. Treat PBMC (1×10^6) or parasite (1×10^7) or bacteria (1×10^7) with 100 μL recombinant 9-*O*-acetyl hemagglutinin esterase (100 U/mL) of influenza C virus.

2. Incubate at 37°C for 3 h.
3. Wash the cells three times with PBS.

SNA/MAA Binding

1. Take untreated and esterase-treated cells separately.
2. Add 5 μL biotinylated SNA and MAA (5 μg/mL) in cell suspension separately.
3. Incubate at 4°C for 30 min.
4. Wash the cells three times with PBS.
5. Add 5 μL FITC-streptavidin (0.5 μg/mL) in cell suspension.
6. Incubate at 4°C for 30 min.
7. Wash the cells three times with PBS.
8. Fix the cells in 1% paraformaldehyde in PBS.
9. Acquire SNA- and/or MAA-positive cells by flow cytometry.
10. Check fluorescence intensity at FL1 region.
11. Analyze the data by CellQuestPro.
12. Difference in SNA- and or MAA-positive cells before and after esterase treatment is a measure of *O*-AcSias.

Siglecs Binding

1. Take untreated and esterase-treated cells (1×10^6) or parasite (1×10^7) or bacteria (1×10^7)/100 μL separately.
2. Add 0.05 μg siglecs-Fc per test to 100 μL cells.
3. Incubate at 4°C for 60 min.
4. Wash the cells three times with PBS.
5. Add 0.02 μg biotin-conjugated Protein A per test and incubate at 4°C for 30 min.
6. Wash cells three times with PBS.
7. Add 5 μL FITC-streptavidin (0.5 μg/mL) in 100 μL cell suspensions.
8. Incubate at 4°C for 30 min.
9. Wash the cells three times with PBS.
10. Fix the cells in 1% paraformaldehyde in PBS.
11. Examine by flow cytometry to check fluorescence intensity at FL1 region.
12. Analyze by CellQuestPro software.
13. Difference in siglec-positive cells before and after esterase treatment is a measure of *O*-AcSias.

3.7.3. Siglecs-ELISA

FITC Labelling of Cells

1. Take untreated and esterase-treated cells (1×10^7) or parasite (1×10^7) or bacteria (1×10^7) separately.
2. Dissolve 0.1% FITC in DMSO.
3. Take the cells in carbonate buffer and add 20 μL of FITC solution.
4. Incubate for 1 h at 37°C.
5. Wash extensively with PBS to remove unbound FITC.

ELISA

1. Coat the well of 96-well plates with 0.5 μg protein A in 100 μL carbonate buffer overnight at 4°C.
2. Block with 2% BSA in PBS.
3. Add 100 μL siglec-Fc (0.05 μg/μL) and incubate for 3 h at 25°C.
4. Wash with PBS three times.
5. Add esterase-treated FITC-labelled cells in diseased conditions as compared to human cells (1×10^6) from normal individuals/ parasites (2×10^6)/heat-killed bacteria (1×10^7).
6. Wash with PBS four times to remove the unbound cells.
7. Measure the fluorescence intensity (excitation, 488 nm; emission, 530 nm) using a Cyto Flour^{II} fluorescent plate reader.
8. Difference in readings before and after esterase treatment is a measure of *O*-AcSias.

3.7.4. Siglecs Dot Blot

1. Apply 20 μL of untreated and esterase-treated cells onto nitrocellulose membrane separately.
2. Leave for air dry.
3. Block the membrane with 3% BSA in PBS.
4. Overlay with 50 μL siglec-Fc (1 μg/μL) and incubate for 1 h at 25°C.
5. Wash with PBS three times.
6. Add biotin-anti-human Fc specific (1:3,000) and incubate for 1 h at 25°C.
7. Wash with PBS three times.
8. Add avidin–HRP conjugate (1:10,000) and incubate for 1 h at 25°C.
9. Wash with PBS three times.
10. Develop with staining solution.
11. Quantify by densitometric analysis using Master Totallab Software, version 1.11.
12. Difference in readings before and after esterase treatment is a measure of *O*-AcSias.

4. Notes

1. Always store bovine submaxillary glands soon after collection at −70°C to avoid destruction of *O*-Ac group.
2. Use gland immediately after collection (if possible).
3. Do not store glands for longer time (more than 2 months) even at −70°C; otherwise the yield of BSM having OAcSias will be less.
4. Special care is needed in handling BSM to prevent the loss of *O*-acetyl groups at alkaline pH and higher temperature. To bring BSM (20 mL) from −70 to 4°C, always use ice and it takes ~4 h to thaw 20 mL BSM.
5. Coupling pH of BSM conjugation to Sepharose 4B has to be strictly between 7.0 and 7.2.
6. This step has to be performed quickly; otherwise reactive group of the Sepharose hydrolyzes at alkaline pH.
7. Always store the BSM column at 4°C.
8. Same BSM-Sepharose column for purification of Achatinin-H from hemolymph should not be used after 3–4 time applications due to destruction of *O*-acetyl group present in BSM by temperature shock during purification as well as many proteases present in hemolymph.
9. Always store Achatinin-H at 4°C with 0.2% azide and filter sterile in aliquots.
10. Never freeze Achatinin-H as it is a high-molecular-weight glycoprotein; it tends to aggregate and biological activity is drastically reduced.
11. Do not store washed erythrocyte for >24 h for HA and/or HAI experiment.
12. Always store unwashed erythrocyte at 4°C in Alsever's solution but not more than 24 h.
13. Always optimize the concentration of FITC-Achatinin-H before use. Typically the concentration at which it shows ~5–10% binding with normal human PBMC by FACS is considered as ideal for testing other cells at diseased condition.
14. Always use cell-specific anti-CD antibodies to prove the presence of *O*-AcSias on specific cells in double-color flow cytometry analysis. Anti-glycophorin A, anti-CD10/CD-19, anti-CD3/CD4/CD8, antibodies are commonly used to detect erythrocytes, B and T lymphocytes.
15. Before using Achatinin-H, always check its *O*-AcSia binding property by HA, HAI with every batch of purification. The purity is checked by 5% native PAGE.

16. Always keep the collection tube on ice during purification of Sias from anion exchange column.
17. Always store 200 μL DMB solution in several aliquots at −20°C.
18. Due to unavailability of *O*-AcSias commercially, *O*-AcSias purified from BSM by fluorimetric-HPLC can be used as standard.

Acknowledgement

This work received financial support from the Council of Scientific and Industrial Research (CSIR, TLP-006, IAP-001, NWP-004, HCP-004), Department of Biotechnology (GAP-235), Indian Council of Medical Research (GAP-266), New Delhi, India. CM acknowledges support from Sir J.C. Bose National Fellowship, Department of Science and Technology, Govt. of India.

References

1. Schauer R (2000) Achievements and challenges of sialic acid research. Glycoconj J 17:485–499
2. Klein A, Roussel P (1998) O-Acetylation of sialic acids. Biochimie 80:49–57
3. Angata T, Varki A (2002) Chemical diversity in the sialic acids and related α-keto acids: an evolutionary perspective. Chem Rev 102:439–469
4. Schauer R (2004) Sialic acids: fascinating sugars in higher animals and man. Zoology 107:49–64
5. Kamerling JP, Schauer R, Shukla AK, Stoll S, VanHalbeek H, Vliegenthart JF (1987) Migration of O-acetyl groups in N, O-acetylneuraminic acids. Eur J Biochem 162:601–607
6. Vandamme-Feldhaus V, Schauer R (1998) Characterization of the enzymatic 7-O-acetylation of sialic acids and evidence for enzymatic O-acetyl migration from C-7 to C-9 in bovine submandibular gland. J Biochem (Tokyo) 124:111–121
7. Ghosh S, Bandhyopadhyay S, Pal S, Das B, Bhattacharya DK, Mandal C (2005) Increased interferon gamma production by peripheral blood mononuclear cells in response to stimulation of overexpressed disease-specific 9-*O*-acetylated sialoglycoconjugates in children suffering from acute lymphoblastic leukaemia. Br J Haematol 128:35–41
8. Ghosh S, Bandyopadhyay S, Mukherjee K, Mallick A, Pal S, Mandal C, Bhattacharya DK, Mandal C (2007) *O*-Acetylation of sialic acids is required for the survival of lymphoblasts in childhood acute lymphoblastic leukemia (ALL). Glycoconj J 24:17–24
9. Sen G, Mandal C (1995) A unique specificity of the binding site of Achatinin-H, a sialic acid binding lectin from *Achatina fulica*. Carbohydr Res 268:115–125
10. Fragkiadakis GA, Stratakis EK (1997) The lectin from the crustacean *Liocarcinus depurator* recognizes *O*-acetyl sialic acids. Comp Biochem Physiol 117:545–552
11. Ravindranath MH, Higa HH, Cooper EL, Paulson JC (1985) Purification and characterization of an *O*-acetyl sialic acid-specific lectin from a marine crab *Cancer antennarius*. J Biol Chem 260:8850–8856
12. Ravindranath MH, Morton DL, Irie RF (1989) An epitope common to gangliosides 0-acetyl-GD3 and GD3 recognized by antibodies in melanoma patients after active specific immunotherapy. Cancer Res 49:3891–3897
13. Denis M, Palatty PD, Bai NR, Suriya SJ (2003) Purification and characterization of a sialic acid specific lectin from the hemolymph of the freshwater crab *Paratelphusa jacquemontii*. Eur J Biochem 270:4348–4355
14. Ahmed H, Gabius HJ (1989) Purification and properties of a Ca2+-independent sialic acid-binding lectin from human placenta with preferential affinity to O-acetylsialic acids. J Biol Chem 264:18673–18678

15. Van Damme EJM, Bane A, Rouge P, Van Leuven F, Peumans WJ (1996) The NeuAc(a-2,6)Gal/GaINAc-binding lectin from elderberry *(Sanbucus nigra)* bark, a type-2 ribosome-inactivating pmtein with an unusual specificity and structure. Eur J Biochem 235:128–137
16. Kaku H, Mon Y, Goldstein IJ, Shibuya N (1993) Monomeric, monovalent derivative of Maackio arnurensis leukoagglutinin. Preparation and application to the study of cell surface glycoconjugates by flow cytometry. J Biol Chem 268:13237–13241
17. Mandal C, Basu S (1987) A unique specificity of a sialic acid binding lectin Achatinin-H, from the hemolymph of a *Achatina fulica* snail. Biochem Biophys Res Commun 148:795–801
18. Mandal C, Mandal C (1990) Sialic acid binding lectins—a review. Cell Mol Life Sci 46:433–441
19. Chava AK, Chatterjee M, Sundar S, Mandal C (2002) Development of an assay for quantification of linkage-specific *O*-acetylated sialoglycans on erythrocytes; its application in Indian visceral leishmaniasis. J Immunol Methods 270:1–10
20. Chava AK, Chatterjee M, Sharma V, Sundar S, Mandal C (2004) Variable degree of alternative complement pathway–mediated hemolysis in Indian visceral leishmaniasis induced by differential expression of 9-*O*–acetylated sialoglycans. J Infect Dis 189:1257–1264
21. Pal S, Ghosh S, Bandhyopadhyay S, Mandal C, Bandhyopadhyay S, Bhattacharya DK, Mandal C (2004) Differential expression of 9-*O*-acetylated sialoglycoconjugates on leukemic blasts: a potential tool for long-term monitoring of children with acute lymphoblastic leukaemia. Int J Cancer 111:270–277
22. Pal S, Ghosh S, Mandal C, Kohla G, Brossmer R, Isecke R, Merling A, Schauer R, Schwartz-Albiez R, Bhattacharya DK, Mandal C (2004) Purification and characterization of 9-*O*-acetylated sialoglycoproteins from leukemic cells and their potential as immunological tool for monitoring childhood acute lymphoblastic leukaemia. Glycobiology 14:859–870
23. Sinha D, Mandal C, Bhattacharya DK (1999) A novel method for prognostic evaluation of childhood acute lymphoblastic leukemia. Leukemia 13:309–312
24. Sinha D, Mandal C, Bhattacharya DK (1999) Identification of 9-O acetyl sialoglycoconjugates (9-OAcSGs) as biomarkers in childhood acute lymphoblastic leukemia using a lectin, AchatininH, as a probe. Leukemia 13:119–125
25. Chowdhury S, Bandyopadhyay S, Mandal C, Chandra S, Mandal C (2008) Flow-cytometric monitoring of disease-associated expression of 9-O-acetylated sialoglycoproteins in combination with known CD antigens, as an index for MRD in children with acute lymphoblastic leukaemia: a two-year longitudinal follow-up study. BMC Cancer 8:40
26. Sinha D, Mandal C, Bhattacharya DK (1999) A colorimetric assay to evaluate the chemotherapeutic response of children with Acute Lymphoblastic Leukemia (ALL) employing achatinin: a 9-O acetylated sialic acid binding lectin. Leukemia Res 23:803–809
27. Sinha D, Mandal C, Bhattacharya DK (1999) Development of a simple, blood based lymphoproliferation assay to assess the clinical status of patients with acute lymphoblastic leukemia. Leukemia Res 23:433–439
28. Scatchard G (1949) The attraction of protein for small molecules and ions. Ann NY Acad Sci 51:660–672
29. Ghoshal A, Mukhopadhyay S, Demine R, Forgber M, Jarmalavicius S, Saha B, Sundar S, Walden P, Mandal C, Mandal C (2009) Detection and characterization of a sialoglycosylated bacterial ABC-type phosphate transporter protein from patients with visceral leishmaniasis. Glycoconj J 26:675–689
30. Harms G, Reuter G, Corfield AP, Schauer R (1996) Binding specificity of influenza C-virus to variably O-acetylated glycoconjugates and its use for histochemical detection of N-acetyl-9-O-acetylneuraminic acid in mammalian tissues. Glycoconj J 13:621–630
31. Hubl U, Ishida H, Kiso M, Hasegawa A, Schauer R (2000) Studies on the specificity and sensitivity of the influenza C virus binding assay for 9-O-acetylated sialic acids and its application to human melanomas. J Biochem (Tokyo) 127:1021–1031
32. Herrler G, Rott R, Klenk HD, Muller HP, Shukla AK, Schauer R (1985) The receptor-destroying enzyme of influenza C virus neuraminate-O-acetylesterase. EMBO J 4:1503–1506
33. Rogers GN, Herrler G, Paulson JC, Klenk HD (1986) Influenza C virus uses 9-O-acetyl-N-acetylneuraminic acid as a high affinity receptor determinant for attachment to cells. J Biol Chem 261:5947–5951
34. Vlasak R, Krystal M, Nacht M, Palese P (1987) The influenza C virus glycoprotein (HE) exibits receptor-binding (hemagglutinin and receptor-destoying (esterase)). Virology 160:419–425
35. Klein A, Krishna M, Varki NM, Varki A (1994) 9-*O*-Acetylated sialic acids have widespread but selective expression: analysis using a chimeric dual-function probe derived from

influenza C hemagglutinin-esterase. Proc Natl Acad Sci U S A 91:7782–7786

36. Shi WX, Chammas R, Varki A (1996) Regulation of sialic acid 9-*O*-acetylation during the growth and differentiation of murine erythroleukemia cells. J Biol Chem 271: 31517–31525
37. Siebert HC, von der Lieth CW, Dong X, Reuter G, Schauer R, Gabius HJ, Vliegenthart JF (1996) Molecular dynamics-derived conformation and intramolecular interaction analysis of the N-acetyl-9-O-acetylneuraminic acid-containing ganglioside GD1a and NMR-based analysis of its binding to a human polyclonal immunoglobulin G fraction with selectivity for O-acetylated sialic acids. Glycobiology 6:561–572
38. Bandyopadhyay S, Bhattacharyya A, Sen AK, Das T, Sa G, Bhattacharya DK, Mandal C (2005) Over expressed IgG2 antibodies against *O*-acetylated sialoglycoconjugates incapable of proper effector functioning in childhood acute lymphoblastic leukemia (ALL). Int Immunol 17:177–191
39. Bandyopadhyay S, Chatterjee M, Pal S, Waller RF, Sundar S, McConville MJ, Mandal C (2004) Purification, characterization of O-acetylated sialoglycoconjugates-specific IgM and development of an enzyme-linked immunosorbent assay for diagnosis and follow-up of Indian visceral leishmaniasis patients. Diagn Microbiol Infect Dis 50:15–24
40. Pal S, Chatterjee M, Bhattacharya DK, Bandhyopadhyay S, Mandal C (2000) Identification and purification of cytolytic antibodies directed against *O*-acetylated sialic acid in childhood acute lymphoblastic leukemia. Glycobiology 10:539–549
41. Pal S, Chatterjee M, Bhattacharya DK, Bandhyopadhyay S, Mandal C, Mandal C (2001) O-acetyl sialic acid specific IgM as a diagnostic marker in childhood acute lymphoblastic leukaemia. Glycoconj J 18:529–537
42. Hara S, Yamaguchi M, Takemori Y, Furuhata K, Ogura H, Nakamura M (1989) Determination of mono-O-acetylated N-acetylneuraminic acids in human and rat sera by fluorimetric highperformance liquid chromatography. Anal Biochem 179:162–166
43. Mandal C, Srinivasan GV, Chowdhury S, Chandra S, Mandal C, Schauer R, Mandal C (2009) High level of sialate-*O*-acetyltransferase activity in lymphoblasts of childhood acute lymphoblastic leukaemia (ALL): enzyme characterization and correlation with disease status. Glycoconj J 26:57–73
44. Ghoshal A, Gerwig GJ, Kamerling JP, Mandal C (2010) Sialic acids in different Leishmania sp., its correlation with nitric oxide resistance and host responses. Glycobiology 20:553–566
45. Varki A (1997) Sialic acids as ligands in recognition phenomena. FASEB J 11:248–255
46. Crocker PR, Varki A (2001) Siglecs, sialic acids and innate immunity. Trends Immunol 22:337–342
47. Chatterjee M, Chava AK, Kohla G, Pal S, Merling A, Hinderlich S, Unger U, Trasser P, Gerwig GJ, Kamerling JP, Vlasak R, Crocker PR, Schauer R, Schwartz-Albiez R, Mandal C (2003) Identification and characterization of adsorbed serum sialoglycans on *Leishmania donovani* promastigotes. Glycobiology 13:351–361
48. Brinkman-Van der Linden ECM, Varki A (2000) New Aspects of Siglec Binding Specificities, Including the Significance of Fucosylation and of the Sialyl-Tn Epitope. J Biol Chem 275:8625–8632
49. Khatua B, Ghoshal A, Bhattacharya K, Mandal C, Saha B, Crocker PR, Mandal C (2010) Sialic acids acquired by *Pseudomonas aeruginosa* are involved in reduced complement deposition and siglec mediated host-cell recognition. FEBS Lett 584:555–561
50. Bandyopadhyay S, Chatterjee M, Sundar S, Mandal C (2004) Identification of 9-*O*-acetylated sialoglycans on peripheral blood mononuclear cells in Indian visceral leishmaniasis. Glycoconj J 20:531–536
51. Ghoshal A, Mukhopadhyay S, Saha B, Mandal C (2009) 9-*O*-Acetylated sialoglycoproteins: Important immunomodulators in Indian visceral leishmaniasis. Clin Vaccine Immunol 18:889–898
52. Haverkamp J, Schauer R, Wember M, Kamerling JP, Vliegenthart JF (1975) Synthesis of 9-*O*-acetyl- and 4,9-di-*O*-acetyl derivatives of the methyl ester of N-acetyl-beta-D-neuraminic acid methylglycoside. Their use as models in periodate oxidation studies. Hoppe Seylers Z Physiol Chem 356:1575–1583
53. Veh RW, Meessen D, Kuntz D, May B (1982) A new method for histochemical demonstration of side chain substituted sialic acids. In: Malt RA, Williamson RCN (eds) Colonic carcinogenesis. MTP, Lancaster, pp 355–365
54. Jass JR, Sugihara K, Love SB (1988) Basis of sialic acid heterogeneity in ulcerative colitis. J Clin Pathol 41:388–392
55. Campbell F, Puller CE, Williams GT, Williams ED (1994) Human colonic stem cell mutation frequency with and without irradiation. J Pathol 174:175–182
56. Reuter G, Schauer R (1984) Determination of sialic acids. Methods Enzymol 230:168–199

57. Blum AS, Barnstable CJ (1987) *O*-acetylation of a cell-surface carbohydrate creates discrete molecular patterns during neural development. Proc Natl Acad Sci U S A 84:8716–8720
58. Kniep B, Peter-Katalinic J, Flegel W, Northoff H, Rieber EP (1992) CDw 60 antibodies bind to acetylated forms of ganglioside GD3. Biochem Biophys Res Commun 187:1343–1349
59. Kniep B, Claus C, Peter-Katalinic J, Monner DA, Dippold W, Nimtz M (1995) 7-*O*-acetyl-GD3 in human T-lymphocytes is detected by a specific T-cell-activating monoclonal antibody. J Biol Chem 270:30173–30180
60. Cerato E, Birkle S, Portoukalian J, Mezazigh A, Chatal JF, Aubry J (1997) Variable region gene segments of nine monoclonal antibodies specific to disialogangliosides (GD2, GD3) and their O-acetylated derivatives. Hybridoma 16:307–316
61. Murphy WH, Gottschalk A (1961) Studies on mucoproteins. VII. The linkage of the prosthetic group to aspartic and glutamic acid residues in bovine submaxillary gland mucoprotein. Biochim Biophys Acta 52:349–360
62. Bertolini M, Pigman W (1970) The existence of oligosaccharides in bovine and ovine submaxillary mucins. Carbohydr Res 14:53–63
63. Lowry OH, Rosebrough NJ, Farr AL, Randall RJ (1951) Protein measurement with the Folin phenol reagent. J Biol Chem 193: 265–275

Chapter 7

HPLC-Based Quantification of In Vitro N-Terminal Acetylation

Rune H. Evjenth, Petra Van Damme, Kris Gevaert, and Thomas Arnesen

Abstract

Protein N-terminal acetylation is a widespread modification in eukaryotes catalyzed by N-terminal acetyltransferases (NATs). The various NATs and their specific substrate specificities and catalytic mechanisms are far from fully understood. We here describe an in vitro method based on reverse-phase HPLC to quantitatively measure in vitro acetylation of NAT oligopeptide substrates, enabling the determination of NAT specificity as well as kinetic parameters.

Key words: N-terminal acetylation, NAT, Reversed-phase high-pressure liquid chromatography (RP-HPLC), Oligopeptide, Acetyl CoA

1. Introduction

N-terminal acetylation is a highly abundant protein modification in eukaryotes (1). Several different N-terminal acetyltransferases (NATs) catalyze the transfer of an acetyl group from acetyl coenzyme A (acetyl-CoA) to the α-amino group of the N-terminal amino acid residue. NatA–NatE are conserved from *S. cerevisiae* to humans (2), while higher eukaryotes also express NatF as well as additional NAT isoforms (3). These NATs catalyze N-terminal acetylation of specific subsets of substrates largely depending on the sequence of the N-terminal amino acid residues. However, there is still a great need for detailed substrate investigations as the specificity profiles of many of these enzymes are not (fully) explored and for many species, NATs have not been studied at all. Furthermore, eukaryotic NATs have not been studied thoroughly in terms of enzyme kinetics. We here present a nonradioactive method that was recently employed to reveal a novel and unexpected

Sandra B. Hake and Christian J. Janzen (eds.), *Protein Acetylation: Methods and Protocols*, Methods in Molecular Biology, vol. 981, DOI 10.1007/978-1-62703-305-3_7, © Springer Science+Business Media, LLC 2013

substrate specificity of the major Naa10p (NatA) enzyme towards acidic actin N-termini (4). Recombinantly produced (*E. coli*) and endogenously immunoprecipitated enzyme preparations can be used as enzyme input in combination with synthetic oligopeptides and acetyl-CoA for the in vitro acetylation assay. A reverse-phase high-performance liquid chromatography (HPLC) readout of products and substrates is used for quantification of the reaction. As all NATs use two substrates in their catalysis (acetyl-CoA and oligopeptides), their activity will lead to the formation of two products (CoA and N-acetylated oligopeptides). From an enzyme-kinetic point of view, it is very beneficial to have the opportunity to monitor the consumption of both substrates and the formation of both products. The HPLC method presented is currently the only one available, allowing for such detailed analysis and as such was recently used to determine K_M and k_{cat} kinetic parameters, besides revealing the first catalytic mechanism of a eukaryotic NAT (refs. 5, 6, Evjenth et al., submitted). This method can profile NAT complexes as well as individual NAT catalytic subunits quantitatively in vitro (4), and is thus a useful tool for studying NAT enzymes.

2. Materials

The acetylation reaction requires a thermomixer (for instance Eppendorf Thermomixer comfort). All samples are analyzed on a Shimadzu Prominence HPLC system, operated by the software LCSolution version 1.21. The main hardware configuration on the Shimadzu Prominence HPLC system is a solvent delivery module (LC-20AB), a photodiode array detector (SPD-M20A), and an autosampler (SIL-20AC).

1. Acetylation buffer: 1 mM dithiothreitol (DTT), 50 mM Tris–HCl, pH 8.5, 800 μM EDTA, and 10% glycerol (v/v). The buffer is stored at 4°C without DTT and DTT is freshly added to the buffer immediately before use.
2. Acetyl-CoA: Acetyl-CoA lithium salt, dissolved in H_2O.
3. Oligopeptides: 24-Mers synthesized to at least 90% purity, dissolved in H_2O.
4. Reverse-phase HPLC column: 250/3 Nucleosil 100-3 C18 HD (Macherey-Nagle).
5. HPLC washing solvent (solvent A): 5% Acetonitrile (ACN) (v/v) and 0.1% trifluoroacetic acid (TFA) (v/v).
6. HPLC elution solvent (solvent B): 90% ACN (v/v) and 0.1% TFA (v/v).

3. Methods

3.1. Assay for Determining Enzyme Kinetic Parameters

When determining the enzyme turnover number (k_{cat}) for a NAT or a catalytic subunit of a NAT complex, the enzyme is initially purified as described in ref. 4. After determining the protein concentration by measuring the absorbance at 280 nm and/or performing a standard Bradford assay, the enzyme is used for the in vitro NAT assay. The purified enzyme is mixed with acetylation buffer, oligopeptides, acetyl-CoA, and enzyme in the specified order (see Note 1). A typical setup contains a mixture volume of 60 μL or more per sample (i.e., for instance a total volume of approximately 350 μL if five time points are to be analyzed requiring 60 μL aliquots per time point) with the following component concentrations to give initial rate conditions: oligopeptides (200 μM), acetyl-CoA (500 μM), and enzyme (for recombinant NATs, typically within a range of 30–100 nM).

1. Immediately after adding the enzyme, quickly transfer the samples to a Thermomixer, preset to 37°C, and mix for 30 s with 1,100 rpm shaking (see Note 2).
2. At specific time points (for instance at 0, 10, 20, 30, and 60 min), collect 60 μL aliquots and stop the enzyme activity by lowering the pH (adding 5 μL of 10% TFA, v/v). At this point the samples may be either analyzed directly on the HPLC system or stored at -20°C.
3. Before proceeding to RP-HPLC-based analyses, thaw all samples and centrifuge at 17,000 × g for 5 min to remove any precipitate. Transfer 60 μL of the supernatant to HPLC-tubes and place in the autosampler rack at 4°C (see Note 3).

3.2. Performing the Reverse-Phase HPLC Analysis

1. Before starting the sample analysis, the elution gradient needs to be determined, that is, finding the best mixing ratio between solvent A and solvent B. A standard gradient (Fig. 1) is used as starting point, and is adjusted depending on the capacity to separate different oligopeptide forms. One method to assess the gradient separation capacity is to inject pure oligopeptide substrate that has not been in contact with any enzyme. The resulting oligopeptide absorbance peak should elute in the area of the gradient being indicated with a grey box in Fig. 1. Since the goal of this method is to quantify the acetylation rate, it is useful to have the same oligopeptide also as an N-terminally acetylated variant (synthetically made). Mix them in a predefined ratio (see Note 4), and you can easily assess if the enzyme substrates and products are sufficiently separated.
2. The RP-HPLC gradient, as the principle of reverse-phase HPLC, is difference in the interaction between the molecules

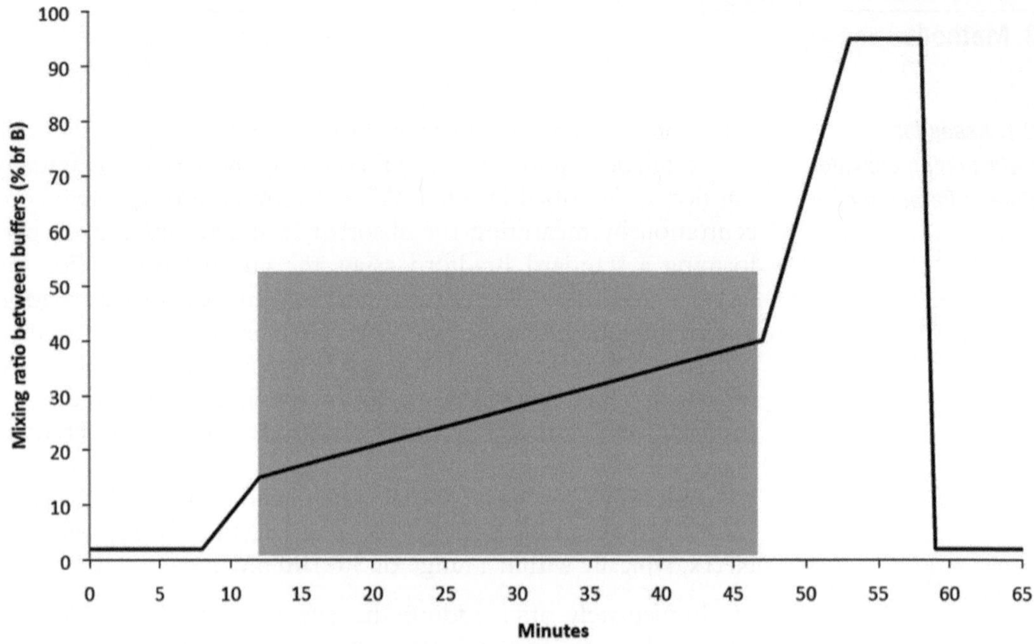

Fig. 1. Example of a gradient used to separate acetylated oligopeptides from non-modified forms, and separation of acetyl-CoA from CoA. The *grey box* indicates where in the gradient the separation of oligopeptides is optimal.

being analyzed and the column matrix; the gradient should contain a step that cleans the column. In this step, the level of solvent B should go up to 95 % to ensure thorough cleaning of the column to generate maximum binding capacity for the next sample. Maintain this step for at least 4–5 min, and follow up by a re-equilibrating step to restore the binding conditions in the column (Fig. 1).

3. During the HPLC oligopeptide separation, the absorbance values are continuously recorded from 200–700 nm in 2-nm increments. This allows for identifying molecules based on specific absorbance information. For instance, acetyl-CoA and CoA have absorbance maxima around 260 nm due to the ATP nucleotide structure. Acetyl-CoA and CoA, in addition to oligopeptides and acetylated oligopeptides, are significantly separated with this gradient (Figs. 2 and 3). Therefore, when plotting the elution profiles recorded at 260 nm, one can interpret the enzyme reaction based on the consumption of acetyl-CoA and corresponding formation of CoA.

   Having the opportunity to monitor the formation of both products and the consumption of both substrates allows for a superior control to maintain enzymatic steady-state conditions. If the concentration of the reaction products accumulates too much (>15% of substrates consumed), it could lead to loss of

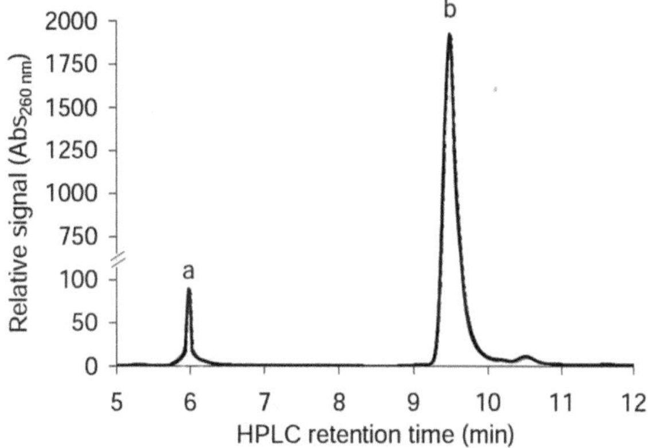

Fig. 2. With the reported gradient, acetyl-CoA (**b**) is efficiently separated from CoA (**a**).

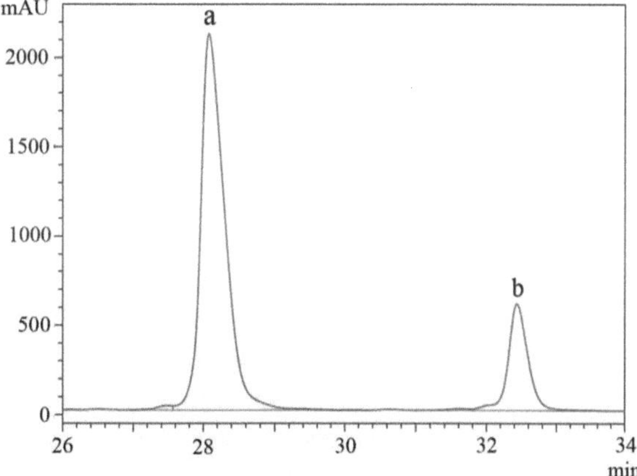

Fig. 3. The separation of unmodified oligopeptides (**a**) from the acetylated forms (**b**).

steady-state conditions and product inhibition, generating falsely low enzyme rates (7). Therefore, it is advised to specifically monitor the amount of both substrates and both products after reactions complete.

4. Calculating the level of acetylation: The elution profiles of the oligopeptides are recorded by measuring their absorbance at 215 nm (see Note 5). After oligopeptide separation, the absorbance peaks corresponding to oligopeptide substrates (non-acetylated oligopeptides) and oligopeptide products (acetylated oligopeptides) are automatically integrated by the HPLC software (Fig. 3). The area ratio of the oligopeptide products and substrates is used to calculate the actual acetylation velocity. The amount of acetylated oligopeptides is determined by

multiplying the total amount of oligopeptides with the ratio between area b and the total area $(a+b)$ using the following equation, $[\text{oligopeptides}] * \frac{b}{a+b}$ (Fig. 3, and see Note 6).

4. Notes

1. Pipetting normally starts with the largest volume of the reaction mixture, the buffer. The other components are then added to the buffer to ensure that they are mixed. To maintain good buffer conditions during the enzyme reactions, the volumes of the other components should never exceed 15% of the total volume (60 μL), as it would lead to dilution of the buffer (50 mM Tris–HCl, pH 8.5) and potentially cause problems due to wrong pH or ionic strength. The buffer conditions must be tested and optimized for each enzyme.
2. To avoid prolonged time for the samples to reach 37°C, the acetylation buffer and acetyl-CoA should be at room temperature when starting mixing components.
3. Based on our experience, the first sample analyzed by RP-HPLC very often gives low signal or a poor signal-to-noise ratio (for all columns, not only for new ones). To avoid loss of a sample, it could be a good idea to put an extra control sample in position 1.
4. The reason for using two different ratios of oligopeptide substrates and acetylated oligopeptide products is to easily distinguish them in the elution profile. If the oligopeptide product (N-terminally acetylated) is injected at half the concentration of the other, the smallest absorbance peak represents the acetylated oligopeptide form and vice versa. Further, this serves as a control for where the acetylated oligopeptides are expected to elute. If, when analyzing the enzyme samples, significant elution shifts (>1 min) are observed as compared to what is expected, this might indicate that more than one acetyl-moiety is being transferred to the oligopeptide substrates, which in most cases involves lysine acetylation. The number of acetyl groups being transferred must of course be verified by other techniques, like mass spectrometry. The same procedure can be used to identify the elution profile of acetyl-CoA and CoA. One other thing that has shown to improve the separation power is to increase the concentration of TFA in the HPLC solutions. Increasing the concentration of TFA from 0.1 to 0.4% made the N-terminal histone H3 oligopeptides elute later, so that the different acetylated peptide forms could be separated (empirically verified).
5. The absorbance wavelength used to quantify the oligopeptides (215 nm) appears to generate very good sensitivity, but also

Table 1
The correlation between increases in oligopeptide concentrations and resulting increases in absorbance peak areas

Conc. of oligopeptides (μM)	Ratio of theoretical oligopeptide concentration increase*	Ratio of measured absorbance area of oligopeptide**	Observed absorption peak area (*a* + *b*)
25	0.05	0.04	8.00E+06
50	0.10	0.12	2.40E+07
75	0.15	0.16	3.20E+07
150	0.30	0.32	6.40E+07
300	0.60	0.63	1.26E+08
500	1	1	2.00E+08

*The theoretical oligopeptide concentration increase is calculated by dividing each concentration on the highest value. For example, 300/500, 150/500, 75/500, etc.

**Observed absorbance area of oligopeptides is based on dividing each area on the largest area. For example, (1.26E+08)/(2.00E+08), (6.40E+07)/(2.00E+08), (3.20E+07)/(2.00E+08), etc.

shows peaks from buffer compounds, so the reaction buffer should be analyzed in a separate sample to identify the peaks derived from the buffer compounds.

6. In addition to quantifying the amount of acetylation, the absorbance peaks also serves as a control for monitoring the amount of oligopeptides being added. In a time velocity experiment, the amount of added oligopeptides is constant; thus the sum of area "*a*" and area "*b*" (Fig. 3) normally should be constant. In experiments designed to determine a K_M value, the concentration of substrate increases for each sample. In those experiments, the total area ($a + b$) should increase in a similar way as the substrate concentration (see Table 1 for an example). If the two ratio values deviate too much, it could indicate poor peak integration (occurs most often at low substrate concentrations), inaccurate pipetting when adding the oligopeptide substrate, or HPLC injection errors.

Acknowledgements

P.V.D. is a postdoctoral fellow of the research foundation—Flanders (FWO-Vlaanderen). K.G. acknowledges support of research grants from the Fund for Scientific Research—Flanders (Belgium) (project numbers G.0042.07 and G.0440.10), the Concerted

Research Actions (project BOF07/GOA/012) from the Ghent University, and the Inter University Attraction Poles (IUAP06). T.A. is supported by the Norwegian Research Council (Grant 197136), and the Norwegian Cancer Society.

References

1. Arnesen T, Van Damme P, Polevoda B, Helsens K, Evjenth R, Colaert N et al (2009) Proteomics analyses reveal the evolutionary conservation and divergence of N-terminal acetyltransferases from yeast and humans. Proc Natl Acad Sci U S A 106:8157–8162
2. Polevoda B, Arnesen T, Sherman F (2009) A synopsis of eukaryotic Nalpha-terminal acetyltransferases: nomenclature, subunits and substrates. BMC Proc 3(Suppl 6):S2
3. Van Damme P, Hole K, Pimenta-Marques A, Helsens K, Vandekerckhove J, Martinho RG et al (2011) NatF contributes to an evolutionary shift in protein N-terminal acetylation and is important for normal chromosome segregation. PLoS Genet 7(7):e1002169
4. Van Damme P, Evjenth R, Foyn H, Demeyer K, De Bock PJ, Lillehaug JR et al (2011) Proteome-derived peptide libraries allow detailed analysis of the substrate specificities of N{alpha}-acetyltransferases and point to hNaa10p as the posttranslational actin N{alpha}-acetyltransferase. Mol Cell Proteomics 10:M110.004580
5. Evjenth R, Hole K, Karlsen OA, Ziegler M, Arnesen T, Lillehaug JR (2009) Human Naa50p (Nat5/San) displays both protein N alpha- and N epsilon-acetyltransferase activity. J Biol Chem 284:31122–31129
6. Evjenth R, Hole K, Ziegler M, Lillehaug JR (2009) Application of reverse-phase HPLC to quantify oligopeptide acetylation eliminates interference from unspecific acetyl CoA hydrolysis. BMC Proc 3(Suppl 6):S5
7. Berndsen CE, Denu JM (2005) Assays for mechanistic investigations of protein/histone acetyltransferases. Methods 36:321–331

Chapter 8

Separation and Purification of Multiply Acetylated Proteins Using Cation-Exchange Chromatography

Romeo Papazyan and Sean D. Taverna

Abstract

High-performance liquid chromatography (HPLC) is extremely useful for the study of proteins and the characterization of their posttranslational modifications. Here we describe a method that utilizes cation-exchange HPLC to separate multiply acetylated histone H3 species on the basis of their charge and hydrophilicity. This high-resolution method allows for the separation of histone H3 species that differ by as few as one acetyl group, and is compatible with subsequent analysis by a variety of techniques, including mass spectrometry and western blotting.

Key words: Cation-exchange chromatography, PolyCAT A, Acetylation, HPLC, Histone H3, Acetic anhydride

1. Introduction

Dynamic protein acetylation is known to impart key functional changes in biological systems, in particular, acetylation of eukaryotic histone proteins helps govern transcriptional competency of the associated DNA. It is increasingly appreciated that acetylation can reside at multiple sites on the same histone polypeptide, and can even occur together with other small covalent posttranslational modifications (PTMs) like methylation and phosphorylation. Specific combinations of such PTMs likely "code" for a variety of integrative chromatin-templated processes. However, the mechanistic basis of this regulation is poorly understood, in part because detection of multiple PTM states is complicated. For example, antibody-based recognition of PTMs is often dependent on epitope accessibility, which can vary depending on the modification state of

Sandra B. Hake and Christian J. Janzen (eds.), *Protein Acetylation: Methods and Protocols*, Methods in Molecular Biology, vol. 981, DOI 10.1007/978-1-62703-305-3_8, © Springer Science+Business Media, LLC 2013

surrounding residues. Fortunately, mass spectrometry methodologies have recently emerged that permit relatively unbiased PTM analysis on long stretches of amino acids or even intact proteins, although the complexity of PTM states usually necessitates extensive front-end purification to be suitable for such approaches (1, 2).

While gel-based systems including acid–urea gel electrophoresis can be used to resolve acetylated (and phosphorylated) protein isoforms, the amount of material purified is relatively small and is in a form that is not compatible with limited proteolytic treatment necessary to utilize the mass spectrometry approaches referred to above. These deficiencies are remedied by cation-exchange HPLC. Under certain conditions, cation-exchange columns, like polyCAT A, permit separation of protein isoforms by exploiting their hydrophilic and electrostatic differences (3–5). For example, protein acetylation decreases positive charge and hydrophilicity, which contribute to the superior resolving power of polyCAT A chromatography. Furthermore, this chromatographic approach can also be performed on a preparative scale, yielding amounts of protein that are compatible with subsequent analysis by mass spectrometry and western blotting (6).

Below we describe our approach to separate and purify distinctly acetylated isoforms of histone H3 from the ciliated protozoan *Tetrahymena thermophila* by using polyCAT A-based chromatography. We start with highly purified histone H3 from *Tetrahymena* which contains a complex mixture of acetylation states ranging from no acetylation to five (and possibly more) acetylated lysines. We treat this purified H3 with acetic anhydride (Ac_2O) to generate hyperacetylated isoforms in vitro. The hyperacetylated and endogenous H3 samples are independently applied to the polyCAT A column and eluted over a steep salt gradient. The concentrations of salt required to elute hyper- and endogenously acetylated H3 isoforms are used to demarcate elution conditions for all possible acetylated H3 isoforms. Endogenous H3 is then reapplied to the column and eluted using the truncated range of salt concentrations and an extended run time, thus achieving maximal resolution among acetyl isoforms. Lastly, samples are desalted by reversed phase-high-performance liquid chromatography (HPLC) (RP-HPLC) and stored for subsequent analysis by mass spectrometry and western blots. Although we resolve histone H3 acetyl states in this protocol, this methodology can be adapted for many other acetylated proteins as well.

2. Materials

All commercially available solutions and reagents should be of analytical grade. Water used to prepare solutions should be ultrapure and highly deionized. Additionally, any solution made for HPLC should be purified using at least a 0.45-μm filter and degassed prior to injection into the column.

2.1. *In Vitro* Acetylation of Proteins

1. Acetic anhydride (Ac_2O).
2. Methanol (MeOH).
3. 2× Reaction buffer: 100 mM ammonium bicarbonate (NH_4HCO_3, m.w. = 79.05). Dissolve 79 mg in 10 mL of water and do not adjust the pH (see Note 1).
4. RP-HPLC-purified protein dissolved in water (~2 mg/mL) (see Note 2).
5. Microcentrifuge tubes.
6. Speed-vac centrifuge.

2.2. Desalting by Reversed Phase-HPLC

1. HPLC pump and fraction collector.
2. C8 column (220×4.6 mm, RP-300, #0711-0059, Perkin Elmer, Waltham, MA, USA).
3. Acetonitrile (CH_3CN), HPLC grade.
4. Trifluoroacetic acid (TFA).
5. Graduated cylinders, 1 and 100 mL.
6. Magnetic stirrer, stir bars, and 1 L suction flasks.
7. Kontes vacuum filtration assembly (#XX1504700, Fisher Scientific, Fair Lawn, NJ, USA).
8. Durapore PVDF membrane filters 0.45 μm (#HVLP047000, Millipore, Billerica, MA, USA).
9. 1 mL Hamilton syringe equipped with blunt-end needle point.

2.3. PolyCAT A Chromatography Buffers

1. Phosphate buffer: Stock solution is made by mixing appropriate amounts of NaH_2PO_4 (sodium phosphate monobasic, solution 1) and Na_2HPO_4 (sodium phosphate dibasic, solution 2) to obtain the final buffer at pH 7. Start by making 1 M stocks of each solution. Dissolve 6.9 g of monobasic NaH_2PO_4 H_2O (m.w. = 138) in water to a final volume of 50 mL (solution 1) and 7.1 g of dibasic Na_2HPO_4 (m.w. = 142) in water to a final volume of 50 mL (solution 2). Shake both solutions vigorously to help dissolve. While stirring 50 mL of solution 2, adjust to pH 7 by slowly (1 mL at a time) adding solution 1. Store the 1 M phosphate buffer, pH 7 at room temperature.

2. Urea ($CO(NH_2)_2$).
3. Dithiothreitol (DTT): 1 M solution dissolved in water (see Note 3).
4. Sodium chloride (NaCl).
5. Bio-Rex MSZ 501 (D) deionizing resin (#142-7425, BioRAD, Hercules, CA, USA).
6. Gravity column (#732-1010, BioRAD, Hercules, CA, USA).
7. Two 500 mL graduated cylinders.
8. Magnetic stirrer, stir bar, and 1 L flask.

2.4. PolyCAT A Column Chromatography Components

1. HPLC pump and fraction collector.
2. PolyCAT A column (200 × 4.6 mm, 5 μm, 1,000 Å, PolyLC, Columbia, MD, USA) (see Notes 4 and 5).
3. 1 mL Hamilton syringe equipped with blunt-end needle point.
4. Dot blot apparatus (#170-6545, BioRAD, Hercules, CA, USA).
5. PVDF membrane.
6. Necessary reagents and instructions for immunodetection are described in ref. 6.

3. Methods

3.1. In Vitro Acetylation of Proteins

1. Make 50 μL of acetylation buffer by mixing 12.5 μL of acetic anhydride with 37.5 μL of methanol (see Note 6).
2. In a separate microcentrifuge tube, mix 10 μL 2× reaction buffer and 10 μL of protein (2 mg/mL) (see Note 7).
3. Combine the two solutions (70 μL final volume) and briefly mix by gentle tapping.
4. Incubate at room temperature for 30 min.
5. Stop the reaction by drying with a speed-vac centrifuge.
6. Store dried samples at −20°C.
7. The extent of acetylation can be tested by mass spectrometry, immunoblot assays, or acid–urea gels after desalting the samples following the procedures in Subheading 3.2 (Fig. 1 and see Note 8).

3.2. Desalting by Reversed Phase-HPLC

1. Prepare mobile phase for RP-HPLC: 1 L solution A—5% acetonitrile, 95% water, and 0.1% TFA; and 1 L solution B—90% acetonitrile, 10% water, and 0.1% TFA (see Note 9).
2. Vacuum filter both chromatography solutions using a 0.45 μm PVDF filter (see Note 10).

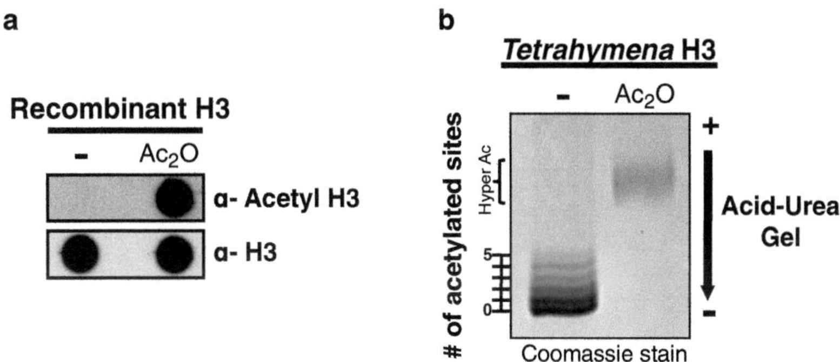

Fig. 1. In vitro acetylation of proteins using acetic anhydride. (**a**) Unmodified, recombinant histone H3 was treated with acetic anhydride (Ac_2O) and analyzed by dot blot immunodetection using antibodies specific to acetylated histone H3 or total H3. (**b**) The extent of in vitro acetylation of *Tetrahymena* histone H3 was measured by acid–urea gels, which combine size- and charge-based protein separation. The untreated H3 sample contains six bands with each band containing the indicated number of acetyl marks. Treatment with acetic anhydride generates hyperacetylated H3, which is represented by a significant upward shift in acid–urea gels.

3. Transfer solutions into clean 1 L suction flasks with stir bar and stir at medium speed for 10 min under vacuum to degas each buffer.
4. Equilibrate the C8 column with 10 mL of solution A at a flow rate of 0.8 mL/min.
5. Take dried samples from Subheading 3.1 and resuspend in 200 μL of solution A by vortexing.
6. Spin down >15,000 × *g* for 5 min to pellet any particulates.
7. Load sample in Hamilton syringe, inject into the HPLC machine, and run the chromatography method at 0.8 mL/min according to Table 1.
8. Record absorbance at 214 nm and collect relevant fractions (see Note 11).
9. Dry down fractions by speed-vac centrifugation.
10. Once dried, the samples can be stored at −20°C.

3.3. Preparation of the Mobile Phase for PolyCAT A Chromatography

These solutions are made fresh before use and discarded at the end of each day. A flowchart for the procedures in this section and the final concentration of the components for the mobile phases are provided in Fig. 2.

1. In a 1 L flask, dissolve 192.2 g of urea (m.w. = 60.06) in 230 mL of water. Stir at room temperature until the urea completely dissolves (see Note 12).
2. Remove ions from the urea solution by adding 1–5 g of deionizing resin. Deionize for 1 h while stirring. Add more resin as needed (see Note 13).
3. Remove the resin by passing the urea solution through a gravity column. Collect the flow through in a 500 mL graduated cylinder.

Table 1
Protocol for desalting by RP-HPLC

Gradient	% B	Volume (mL)	Time (min)
Isocratic	0	4	5
Linear	0→100	24	30
Isocratic	100	4	5
Linear	100→0	1.6	2
Isocratic	0	4	5

Use a flow rate of 0.8 mL/min for all steps

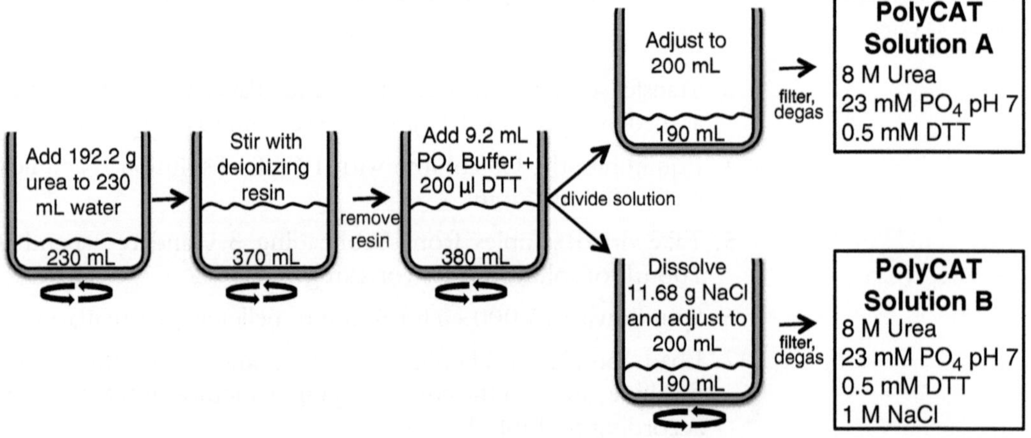

Fig. 2. Flowchart of mobile-phase preparation for PolyCAT A chromatography. Circulating *arrows* indicate that stirring is required at that step.

4. While stirring on a magnetic plate add 9.2 mL of 1 M phosphate buffer to the flow through.
5. Add 200 μL of 1 M DTT.
6. Adjust the volume to 380 mL with water.
7. Split the solution evenly into two graduated cylinders (190 mL in each cylinder). In the first cylinder add 10 mL water and label this "solution A" for polyCAT chromatography.
8. While stirring the second cylinder dissolve 11.68 g of NaCl (m.w. = 58.44) to obtain a final concentration of 1 M NaCl and label this as solution "B" for polyCAT chromatography. Adjust the final volume of solution B to 200 mL.
9. Filter and degas both buffers following the instructions above.

Table 2
Protocol for polyCAT A chromatography

Gradient	%B	Volume (mL)	Time (min)
Isocratic	0	4	10
Linear	0→100	36	90
Isocratic	100	4	10
Linear	100→0	0.8	2
Isocratic	0	4	10

Use a flow rate of 0.4 mL/min for all steps

3.4. PolyCAT A Chromatography

1. Equilibrate the column with at least 10 mL polyCAT solution A or until absorbance at 230 nm stabilizes. Use a flow rate of 0.4 mL/min for this step and all steps in the chromatography protocol (see Note 5).
2. While equilibrating, resuspend dried and desalted sample from Subheading 3.2 in 200 μL polyCAT solution A by vortexing for several seconds (see Note 14).
3. Pellet any particulates in the tube by centrifuging >15,000 × g for 5 min. Immediately after centrifugation, load the sample into a Hamilton syringe fitted with an appropriate blunt needle point designed for injections into HPLC machines.
4. Inject the sample into the HPLC machine and follow the chromatography protocol outlined in Table 2. Detect the presence of protein by constantly measuring absorbance at 230 nm (see Note 15). Collect fractions every 2 min (0.8 μL fraction volume) during the first 110 min of the run (see Note 16). Retention time increases as the number of acetylated sites on the protein decreases. For example, hyperacetylated proteins will elute before hypoacetylated species.
5. To obtain an optimal range of elution conditions repeat procedures 1–4 above using RP-HPLC-purified protein samples that are not treated with Ac_2O. Compare all fractions from the two polyCAT A chromatography runs by using a dot blot apparatus (load ~10% of each fraction) followed by immunodetection (Fig. 3a). The optimal salt concentration range will encompass fractions that include hyperacetylated (Ac_2O treated) and endogenously acetylated species.

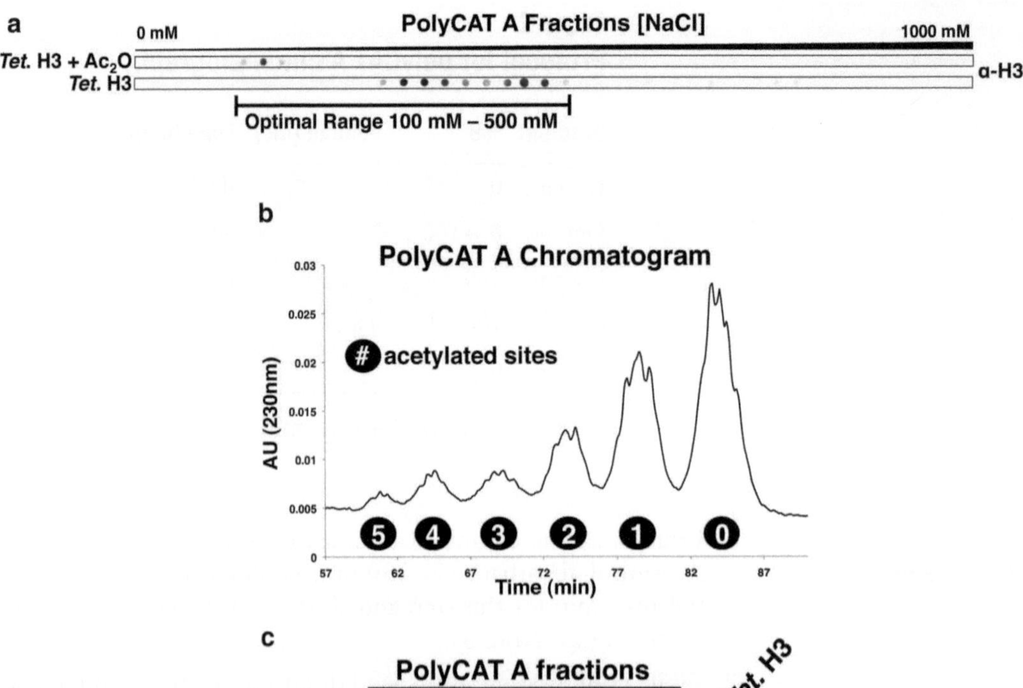

Fig. 3. Separation of acetylated proteins using polyCAT A chromatography. (**a**) Dot blot analysis comparing *Tetrahymena* H3 species that were either hyperacetylated in vitro using Ac_2O or endogenously acetylated. These samples represent the extremes of acetylation that may exist in a biologically relevant sample and serve as ideal starting material to optimize polyCAT A chromatography conditions. (**b**) Chromatogram of an optimized polyCAT A run using 75 μg of endogenously acetylated *Tetrahymena* histone H3. Solutions described in Fig. 2 were used with the following chromatography protocol at a rate of 0.4 mL/min: inject sample; 0% B for 10 min; 0–10% B for 10 min; 10% B for 15 min; 10–50% B for 65 min; 50–100% B for 10 min; 100% B for 10 min; 100–0% B for 5 min; and 0% B for 10 min. Sub-peaks within each major acetylation peak likely represent differently methylated H3 species. (**c**) Peaks 0–4 from (**b**) were desalted and analyzed by acid–urea electrophoresis and Coomassie staining.

6. Inject more sample into the polyCAT A column and adjust the elution gradient by increasing the time of elution within the newly defined elution range, as described in Fig. 3b (see Note 17). Successful optimization of elution conditions will yield individual peaks on the chromatogram that differ by one acetyl group store fraction at −80°C (Fig. 3b and see Note 18).
7. Desalt each relevant fraction by thawing it individually and injecting it onto an RP-HPLC column using the desalting procedures above followed by drying to completion using a speed-vac centrifuge. Dried fractions may be stored in −80°C.
8. Desalted samples may be analyzed by gel electrophoresis or mass spectrometry (Fig. 3c and ref. 6).

4. Notes

1. Due to the volatility of ammonium bicarbonate we recommend making fresh solution each time before use.
2. We obtain purified histone H3 by following the nuclear preparation, acid extraction, and RP-HPLC purification protocols described in Shechter et al. (7). During nuclear preparation, we try to preserve endogenous protein acetylation by including 10 mM butyric acid, an HDAC inhibitor, in buffers.
3. To make 1 M stock solution, dissolve 1.54 g of DTT (m.w. = 154.25) in water to a final volume of 10 mL. Filter, divide to 1 mL aliquots, and store in −20°C.
4. The specifications of the column will vary depending on the mass and properties of the protein being studied. Contact the manufacturer for additional help with selecting the proper column.
5. PolyCAT A column performance is significantly enhanced when prewashed with 40 mM EDTA-2Na overnight using a slow flow rate (0.25 mL/min). Do not adjust the pH of this solution and remember to filter purify the solution prior to passing it through the column. After the EDTA prewash, wash the column with 10 mL filtered water and proceed to equilibration with "polyCAT solution A."
6. The in vitro acetylation protocol was adapted from ref. 8.
7. This procedure is used to acetylate 20 μg of protein. If more acetylated protein is desired adjust the volume of the final reaction as needed and keep the same ratio between the acetylation and reaction buffers.
8. In addition to acetylating lysines, this treatment also yields acetylated N-termini.
9. Prepare appropriate volumes of acetonitrile and water in separate graduated cylinders prior to mixing. Upon mixing the actual volume (~980 mL) will be less than the expected volume (1,000 mL) due to entropy changes. Carefully add TFA in the end and proceed to filtration.
10. This critical step is required for all solutions used in HPLC. Removal of particulates extends the life of the column and minimizes clogging during a run.
11. In general, acetylation or other small PTMs do not dramatically affect the column retention time of proteins, so hypoacetylated proteins will elute in the same fractions as hyperacetylated proteins using RP-HPLC. If the elution time of the sample is not known, collect fractions over the entirety of the initial run. After determining the elution time of the sample, the fraction collection window can be narrowed.

12. Do not heat the urea solution to dissolve because heating promotes the breakdown of urea into ions that lead to chemical modification (carbamylation) of the proteins being studied.
13. Deionizing the urea solution is critical to eliminate carbamylation of proteins and to facilitate the interaction of the protein with the column. This resin contains a colorimetric indicator that turns gold or clear when saturated with ions. Continue adding resin until some of the resin remains blue (1–5 g).
14. Samples injected into the column should be free of salts. When possible, we recommend drying the protein sample using a speed-vac centrifuge and resuspending it in polyCAT solution A prior to column injection.
15. When possible, absorbance at 230 nm should be recorded for polyCAT A chromatography of histones, because urea saturates the 214 nm signal.
16. This is not necessary for subsequent runs, but for the initial run we recommend collecting all fractions to help determine the extent of binding and retention in the column.
17. While we narrowed the range of salt concentrations to optimize elution conditions, the samples were still resuspended in polyCAT A "solution A" containing no salt. The details of the optimized polyCAT A chromatography protocol are highlighted in the caption for Fig. 3b. In general, samples should be injected onto the column using a solution that lacks salt. The presence of salt will disrupt the retention time and resolution of the analyte.
18. Following polyCAT A chromatography, we highly recommend washing the column with filtered water (>20 mL) followed by a total system clean of the HPLC, especially if the same HPLC machine will be used for subsequent desalting steps by RP-HPLC. Mixing the mobile phases from each type of chromatography may cause precipitation of salts and significant damage to the column or HPLC pump.

Acknowledgements

We thank T. Gilbert and other members of the Taverna laboratory for comments on this protocol. This work was supported by NIH grant RO190035489.

References

1. Garcia BA, Shabanowitz J, Hunt DF (2007) Characterization of histones and their post-translational modifications by mass spectrometry. Curr Opin Chem Biol 11:66–73
2. Thomas CE, Kelleher NL, Mizzen CA (2006) Mass spectrometric characterization of human H3: a bird's eye view. J Prot Res 5:240–247
3. Lindner HH (2008) Analysis of histones, histone variants, and their post-translationally modified forms. Electrophoresis 29:2516–2532
4. Zhang K, Siino JS, Jones PR, Yau PM, Bradbury EM (2004) A mass spectrometric "Western blot" to evaluate the correlations between histone methylation and histone acetylation. Proteomics 4:3765–3775
5. Alpert AJ (1983) Cation-exchange high-performance liquid chromatography of proteins on poly(aspartic acid)-silica. J Chromatogr 266:23–37
6. Taverna SD, Ueberheide BM, Liu L, Tackett AJ, Diaz RL, Shabanowitz J, Chait BT, Hunt DF, Allis CD (2006) Long-distance combinatorial linkage between methylation and acetylation on histone H3 termini. Proc Natl Acad Sci U S A 104:2086–2091
7. Shechter D, Dormann HL, Allis CD, Hake SB (2007) Extraction, purification and analysis of histones. Nat Protoc 2:1445–1457
8. Acetylation of proteins and peptides (2011) http://www.ionsource.com/Card/acetylation/mono0003.htm. Accessed 7 March 2011

Chapter 9

In-Gel N-Acetylation for the Quantification of the Degree of Protein In Vivo N-Terminal Acetylation

Petra Van Damme, Thomas Arnesen, Bart Ruttens, and Kris Gevaert

Abstract

Maturation of protein N-termini occurs in all kingdoms of life, with major protein modifications being proteolytic processing (e.g., removal of initiator methionines) and N-terminal acetylation. The functional consequences of these modifications are only known for a few substrates, and techniques to study such modifications have begun to emerge only recently. We here report on a method enabling targeted, mass spectrometry based analysis of protein N-termini from polyacrylamide gel-separated proteins. In our method, stable isotope incorporation by in-gel N-acetylation of free primary amines permits calculating the extent of in vivo N-terminal acetylation, proven to reveal crucial information with reference to N-terminal protein biology.

Key words: N-terminal acetylation, Co-translational protein modification, Protein N-terminus, Quantitative mass spectrometry, In-gel labelling, N-terminal acetyltransferase, Strong cation exchange

1. Introduction

Our understanding of the biology and impact of N-terminal protein modifications, which affect nearly all protein species from the moment of ribosome-emergence, has recently been extended following the development of various proteomics strategies aimed at specific isolation and characterization of N-terminal parts of proteins. Identifying protein N-termini and their associated modifications indeed reveals important information about the mature (processed) status of proteins. Processes affecting protein N-termini include proteolysis events such as initiator methionine and signal peptide removal by co-translationally acting methionine aminopeptidases (MetAPs) and signal peptidases respectively, besides other N-terminal protein modifications. N-terminal

Sandra B. Hake and Christian J. Janzen (eds.), *Protein Acetylation: Methods and Protocols*, Methods in Molecular Biology, vol. 981, DOI 10.1007/978-1-62703-305-3_9, © Springer Science+Business Media, LLC 2013

acetylation, the most ubiquitous N-terminal modification, is catalyzed by several, conserved N-terminal acetyltransferases (NATs) that transfer the acetyl moiety from acetyl coenzyme A to the alpha-amino group of a protein N-terminus (1).

Evolutionary convergences and differences observed in various species-specific N-acetylomes have only recently been revealed by so-called positional proteomics screens (2–5). However, protein-specific analyses opposed to these proteome-wide screens are typically needed to follow up on candidate substrates. Of note is that only very few N-terminal acetylation epitope-specific antibodies exist; for example the N-terminal acetylation specific antibodies used in the study of Hwang et al. (6), and therefore, we here report on a method enabling targeted analysis of protein N-termini, focusing on a single protein. Stable isotope incorporation by in-gel N-acetylation of primary amines of SDS-PAGE separated proteins was achieved using variants of *N*-acetoxysuccinimides that contain different stable isotopes. These introduce an MS-distinguishable mass tag that allows MS-based quantification. Trypsin digestion of such in-gel modified proteins generates Arg-C type of peptides as trypsin is unable to cleave at the C-terminal side of the modified lysines. Depending on the original sample complexity, the resulting peptide mixture may further be pre-enriched for protein terminal peptides by strong cation exchange chromatography (SCX) at pH 3.0 (7); at this pH only the internal peptides, carrying one extra positive charge by their primary alpha-amine, are retained, and thus, terminal peptides are recovered in the SCX flow-through fraction. Figure 1a shows the overall workflow of an in-gel N-acetylation based analysis for studying protein N-terminal acetylation. This method was successfully used to demonstrate the functional implication of the chaperone-like HYPK protein, a newly identified interactor of the human NatA complex and Huntingtin, in steering the overall degree of NatA mediated N-terminal acetylation (8). In this study and upon knockdown of HYPK, the N-terminus of a C-terminally tagged, immunoprecipitated form of the hNatA substrate, PCNP (PEST proteolytic signal-containing nuclear protein), clearly showed a decrease in the degree of in vivo N^{α}-acetylation, analogous to the decrease observed when knocking down the catalytic subunit of the NatA complex (8).

Overall, our method thus allows for the determination of the extent of in vivo N^{α}-acetylation of a specific protein without the necessity or requirement of other (complementary) methodologies. The representative MS spectra shown in Fig. 1b clearly demonstrate the use of the stable encoded variants of *N*-acetoxysuccinimides for distinguishing the light (in vivo N^{α}-acetylated) from the heavy (in vitro N^{α}-acetylated) form of a protein N-terminus. Furthermore, this strategy can be combined with metabolic labelling such as stable isotope labelling with amino acids in cell culture (SILAC) (9), enabling differential analysis.

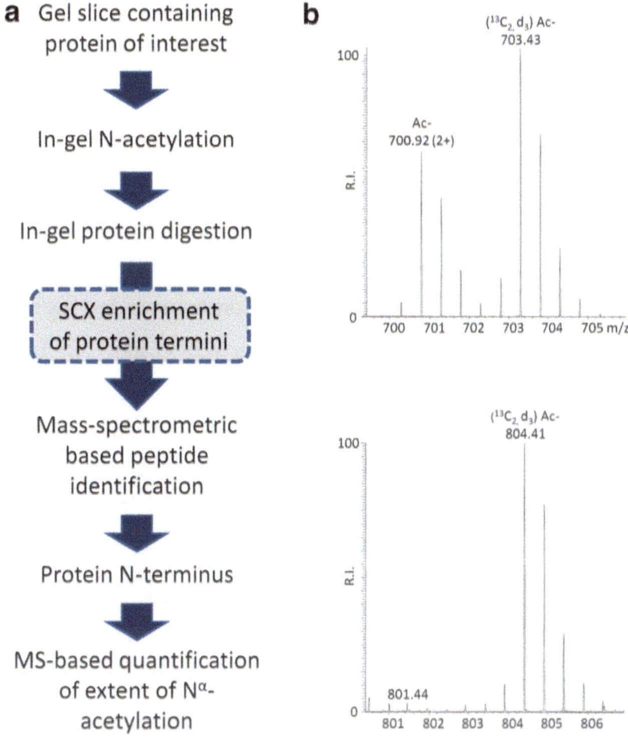

Fig. 1. Flowchart for assessing the degree of in vivo N-terminal acetylation and representative MS spectra of N-termini from yeast proteins and their degree of in vivo N-terminal acetylation. (**a**) Flowchart of the in-gel N-acetylation assay for quantification of protein in vivo N-terminal acetylation. (**b**) The MS spectrum shown in the upper panel originates from the Ala-starting N-terminal peptide (doubly charged precursor) of the yeast actin patches distal protein 1 ([2]AFLNIFKQKR[12]) (Swiss-Prot accession P38281). Two isotopic envelopes are distinguished; those of the acetylated and $^{13}C_2$, d_3-acetylated forms, indicative for the fact that this N-terminus is only partially (39%) in vivo Nα-acetylated. The lower panel displays a MS spectrum of the completely Nα-free MK-starting protein N-terminus [1]MKLDSGIYSEAQR[13] from the stationary phase gene 1 protein (Swiss-Prot accession P50088). The percentage of N-terminal acetylation was calculated using the MS-isotope pattern calculator (http://prospector.ucsf.edu).

2. Materials

2.1. Reagents

1. Acetic acid-$^{13}C_2$, d_4 (Sigma-Aldrich).
2. Aceticanhydride-d_6 (Sigma-Aldrich).
3. Acetone (Sigma-Aldrich).
4. Acetonitrile, HPLC-grade (Biosolve).
5. Acetylation buffer: 100 mM sodium phosphate pH 8.0.
6. Ammonium bicarbonate (Sigma-Aldrich).
7. Dichloromethane (Acros).

8. Digestion buffer: 50 mM ammonium bicarbonate in water–acetonitrile, 90:10 (v/v).
9. Diphenyl phosphoryl chloride (Sigma-Aldrich).
10. Ethyl acetate (Acros).
11. Formic acid (Biosolve) (caution: causes severe burns).
12. Hydrogen peroxide, 30% (w/w) in water (Sigma-Aldrich).
13. Hydroxylamine, 50% (w/w) in water (Sigma-Aldrich).
14. *N*-Hydroxysuccinimide (Sigma-Aldrich).
15. Nitrogen or argon gas (Air Liquide).
16. SCX solvent: 10 mM sodium phosphate pH 3 in water–acetonitrile, 50:50 (v/v).
17. Sequencing-grade modified trypsin (Promega Corp).
18. Sodium bicarbonate (Sigma-Aldrich).
19. Sodium sulfate (Acros).
20. Trifluoroacetic acid, peptide synthesis grade (Biosolve).
21. Tris(2-carboxyethyl)phosphine hydrochloride (TCEP, Pierce).
22. Trypsin digestion buffer: 2.5 μl trypsin stock solution (= 0.5 μg trypsin) in 100 μl digestion buffer.
23. Trypsin stock solution: 20 μg sequencing-grade modified trypsin in 100 μl digestion buffer.
24. Water, HPLC grade (purified with a Milli-Q system from Millipore) (see Note 1).

2.2. Equipment

1. (Ultra-) high-resolution and mass-accuracy mass-spectrometer coupled to nano-HPLC.
2. 1.5 ml enzyme-free microcentrifuge tubes (Eppendorf Biopur®).
3. 10 μl or 100 μl C18 reversed phase sorbent containing pipette tips.
4. 10 μl or 100 μl SCX pipette tips (OMIX®, Agilent).
5. Air-blow ionizer.
6. Magnetic stirrer with heater (Heidolph MR-Hei-standard).
7. Mascot (Distiller) software (Matrix Science).
8. Microcentrifuge.
9. Oil bath for magnetic stirrer with heater.
10. Oil pump (Edwards E2M28).
11. Rotary evaporator (Heidolph Laborota 4003 control).
12. Scalpel or razor blade for gel-slicing.
13. Silica gel, pre-packed flash chromatography cartridge (Interchim, PF-30SIHP/40G).

14. Solvent pump (Analogix BSR).
15. SpeedVac concentrator.
16. Thermomixer.
17. TLC plates (Macherey-Nagel ALUGRAM SIL G/UV$_{254}$).
18. Vortexer.

2.3. Reagent Setup

Gels: Standard SDS-PAGE gels and 2D-gels can be used. Gels should be stained with standard protocols for MS compatibility using dyes such as Coomassie brilliant blue or Sypro Ruby, or other MS-compatible staining protocols.

2.4. Equipment Setup

Mass spectrometer and HPLC system: Examples of high-resolution mass spectrometers include an LTQ-Orbitrap XL hybrid mass spectrometer (Thermo Electron Corp.) coupled to a Dionex Ultimate 3000 HPLC.

3. Methods

3.1. Synthesis of N-Aceoxy succinimides

3.1.1. Synthesis of N-(Acetoxy-d3)Succinimide

1. Mix acetic anhydride-d_6 (2.8 g, 25.89 mmol) and *N*-hydroxysuccinimide (NHS) (1.0 g, 8.69 mmol) in a round-bottom flask containing a magnetic stir bar.
2. Stir overnight at room temperature under inert atmosphere.
3. Concentrate the mixture with a rotary evaporator and dry the solid white residue thoroughly.
4. Remove unreacted NHS by quick chromatography over silica gel using dichloromethane as eluent.
5. Pool the fractions containing the product (check by TLC, see also Note 2) and remove the solvent using a rotary evaporator.
6. Dry the white solid at high vacuum (oil pump) (see Note 3).

3.1.2. Synthesis of N-(Acetoxy-$^{13}C_2$, d_3) Succinimide

1. Add acetic acid-$^{13}C_2$, d_4 (286 μl, 5.0 mmol) to a stirred mixture of NHS (633 mg, 5.5 mmol), sodium bicarbonate (1.680 g, 20 mmol), and acetone (15 ml).
2. Add a solution of diphenyl phosphoryl chloride (1.24 ml, 6.0 mmol) in acetone (5 ml) dropwise over 5 min.
3. Stir 24 h at room temperature under inert atmosphere.
4. Stir 3 h at 50°C under inert atmosphere (see Note 4).
5. Concentrate the mixture with a rotary evaporator and dry the solid white residue thoroughly.
6. Titrate the residue with dichloromethane (3 × 25 ml) and dry the combined organic phases over sodium sulfate (15 min).

7. Filter off the drying agent and concentrate the filtrate with the rotary evaporator until a white solid is obtained.
8. Purify by quick chromatography over silica gel with 2% ethyl acetate using dichloromethane as eluent.
9. Pool the fractions containing the product (check by TLC) and remove the solvent with the rotary evaporator (see Note 2).
10. Dry the white solid at high vacuum (oil pump) (see Note 3).

3.2. Determination of the Degree of In Vivo Protein N-Terminal Acetylation

3.2.1. In-Gel N-Acetylation and Protein Digestion

1. Perform 1D-SDS-PAGE separation.
2. Perform MS-compatible in-gel protein staining.
3. Excise stained protein bands (see Note 5).
4. Place the gel piece(s) in separate 1.5 ml microcentrifuge tubes and treat all samples identically and in parallel.
5. Wash the gel pieces with 200 µl (or the volume required to completely submerge the gel pieces) water, water–acetonitrile 50:50 (v/v) and acetonitrile (in this order) by shaking the samples at room temperature in a Thermomixer at 600 rpm for 10 min (see Note 6).
6. Dry the gel samples in a SpeedVac concentrator.
7. Prepare a *N*-(acetoxy-d_3)succinimide or *N*-(acetoxy-$^{13}C_2$, d_3)succinimide solution by dissolving 1 mg of the *N*-acetoxysuccinimides in 10 µl of acetonitrile and thereafter add 90 µl 100 mM sodium phosphate (pH = 8.0) (N-acetylation buffer).
8. Perform in-gel N-acetylation of free protein amines by re-swelling the gel pieces in 100 µl of the freshly prepared N-acetylation buffer (or the volume required to completely submerge the gel pieces).
9. Vortex the sample(s) for 1 min.
10. Incubate for 1 h at 30°C.
11. Wash the gel pieces as in step 5.
12. Repeat steps 6–10 once to assure as quantitative N-acetylation as possible.
13. Wash the gel pieces twice with 200 µl of 50 mM ammonium bicarbonate (pH = 8.0) for 10 min to remove and quench excess of reagent. Add 2 µl water–hydroxylamine 50:50 (v/v) to the final wash solution (see Note 7).
14. Incubate for 20 min at room temperature.
15. Remove the solution and wash the gel pieces with 200 µl of 0.1% TFA for 10 min at 30°C (see Note 8).
16. Perform methionine oxidation by adding hydrogen peroxide to a final concentration of 0.5% in 0.1% TFA and incubate for 30 min at 30°C (see Notes 8 and 9).

17. Wash the gel pieces for 10 min with 200 μl acetonitrile.
18. Dry the gel pieces in a SpeedVac concentrator.
19. Re-swell the gel samples in 10–20 μl trypsin-digestion buffer for 10 min (see Note 10).
20. Add digestion buffer until the gel pieces are completely immersed.
21. Incubate samples overnight at 37°C.

3.2.2. Peptide Extraction

1. Spin down the gel pieces by centrifugation at room temperature at 1,000×*g* for 1 min and transfer the peptide mixture (supernatant) to microcentrifuge tubes.
2. Extract any remaining peptides from the gel parts by adding 100 μl water–acetonitrile, 50:50 (v/v), with incubation at room temperature for 10 min at 600 rpm in a Thermomixer.
3. Repeat step 1 and add the peptide mixture to the first peptide mixture.
4. Proceed either directly with MS analysis or (optionally) perform terminal peptide enrichment (see Note 11).

3.2.3. Solid Phase Extraction

1. Partially dry the peptide mixture in a SpeedVac concentrator to obtain a final volume between 10 and 100 μl (see Note 12).
2. In order to achieve acceptable binding conditions (pH<4) for C18 solid phase extraction (SPE), add TFA to the sample to reach a final concentration of 0.1% TFA (see Note 13).
3. Perform solid phase extraction of peptides using C18 reversed phase sorbent containing pipette tips (we recommend using pipette tips from ZipTip™ (Millipore) or OMIX® (Agilent)) using the manufacturer's instructions.
4. Condition the tip by aspirating the maximum pipette tip volume of 0.1% TFA in water–acetonitrile, 50:50 (v/v). Discard the solvent (see Notes 12 and 14).
5. Equilibrate the tip by washing three times with the maximum pipette tip volume of 0.1% TFA in water–acetonitrile, 98:2 (v/v).
6. Dispense and aspirate the sample during 3–5 cycles for maximum efficiency. Up to ten cycles may be used for efficient binding.
7. Wash the tip three times with the maximum pipette tip volume of 0.1% TFA in water–acetonitrile, 98:2 (v/v).
8. Elute peptides with the maximum pipette tip volume of water–acetonitrile, 30:70 (v/v).
9. Dry the samples in a SpeedVac concentrator.

3.2.4. SCX-Based Terminal Peptide Enrichment

1. Dissolve peptides in 100 μl of SCX solvent.
2. Perform strong cation exchange (SCX) of peptides using SCX-resin containing pipette tips (we recommend using OMIX® SCX-pipette tips (Agilent)) using the manufacturer's instructions.
3. Condition and equilibrate the SCX tip by aspirating the maximum pipette tip volume of SCX solvent. Discard the solvent and repeat three times.
4. Dispense and aspirate the sample for 3–5 cycles for maximum efficiency. Up to ten cycles may be used for efficient binding.
5. Discard the SCX-tip with the bound peptide material.
6. Keep the unbound peptide material which contains alpha-amino blocked and terminal peptides.

3.2.5. Sample Preparation for MS Analysis

1. Dry the samples in the SpeedVac and redissolve in 20 μl of 2 mM TCEP in 2% acetonitrile (see Note 15).
2. Transfer to LC–MS/MS vials (samples can be stored for several weeks at –20°C).
3. Proceed with LC–MS/MS analysis.

3.2.6. General Info on LC–MS/MS Based Peptide Identification

Since we were unable to observe a noteworthy influence on peptide retention when comparing N-terminal acetylated versus (d_3)- or ($^{13}C_2$, d_3)-acetylated peptides on reverse-phase HPLC columns, mass spectrometers can simply operate in a data-dependent LC–MS/MS modus for identification and quantification. The resulting MS/MS-spectra are typically searched against a nonredundant protein sequence database such as UNIPROT or NCBInr using Mascot or another search engine. (d_3)- or ($^{13}C_2$, d_3)-acetylation of lysine side-chains (+45.029395 or +47.036105 (both monoistopic masses)) and methionine oxidation to methionine-sulfoxide are set as fixed modifications. Variable modifications include (d_3)- or ($^{13}C_2$, d_3)-acetylation, and acetylation of alpha-N-termini and pyroglutamate formation of N-terminal glutamine (see also Table 1). Endoproteinase Arg-C/P (Arg-C specificity with arginine-proline cleavage allowed) should be set as the protease, allowing for no missed cleavages. The mass tolerance on the precursor ion is typically set to 10 ppm, and on fragment ions to 0.5 Da. The identification confidence level is usually set at 99% for spectra generated on an LTQ Orbitrap mass spectrometer.

3.2.7. Quantification of the Degree of N-Terminal Acetylation

Following identification of the N-terminal peptide(s) of interest, one can manually calculate the degree of N-terminal acetylation or, alternatively, make use of quantification software packages such as Mascot Distiller. For manual determination, the *m/z* value, charge state and retention time of the identified peptide are used to assign the matching MS-spectrum. The peptide's isotopic

Table 1
Recommended parameters for searching protein sequence databases with MS/MS spectra of peptides

Fixed modifications	Variable modifications
$^{13}C_2$, d_3-acetylation (K) or d_3-acetylation (K)	Acetylation (N-term)
Oxidation (Met)	$^{13}C_2$, d_3-acetylation (N-term) or d_3-acetylation (N-term)
	Pyroglutamic acid (N-terminal Q)

Since the in-gel N-acetylation strategy leads to additional modifications, an overview of recommended and essential settings of amino acid modifications at the peptide level is given

pattern is reconstructed by summing the peptide ion profile over the entire peptide elution profile. The peptide sequence is used to calculate the theoretical isotope peak distribution using the MS-Isotope pattern calculator (http://prospector.ucsf.edu).

In the case of ($^{13}C_2$, d_3)-N$^{\alpha}$-acetylation; for both possible peptide variants; i.e., the in vivo N-acetylated (peak at m/z) and in vitro $^{13}C_2$, d_3-N-acetylated (peak at $m/z+5$ Da), the predicted intensity of the fifth contributing isotope (isotopic pattern of contributing $+m/z$ 5 Da peak) was subtracted from the measured intensity of all corresponding monoisotopic peaks of the other isotopic clusters in order to correct for peak contributions by overlapping isotopic envelopes.

In the case of (d_3)-N$^{\alpha}$-acetylation; for both possible peptide variants; i.e., the in vivo N-acetylated (peak at m/z) and in vitro d_3-N-acetylated (peak at $m/z+3$ Da), the predicted intensity of the third and following contributing isotope peaks (isotopic pattern of contributing $m/z+3$ Da peak) was subtracted from the measured intensity of all corresponding monoisotopic peaks of the other isotopic clusters in order to correct for peak contributions by overlapping isotopic envelopes.

When using Mascot Distiller, the extent of N-terminal acetylation was calculated after extracting the corresponding peak intensities (extracted from the resulting .rov-files).

4. Notes

1. The water used should have a resistivity of 18.2 MΩ cm and a TOC (total organic carbon) of 2 ppb or less.
2. Thin layer chromatography or TLC is most illustrative when NHS and diphenyl phosphoryl *N*-hydroxysuccinimide are co-spotted as references. Note that NHS esters decompose

on silica gel and that in a series of spots, for example when several column fractions are being analyzed, the first spots will show partial on-plate decomposition. This is also the reason purification on silica gel should be done quickly. When the purification is done with a flash chromatography system equipped with a UV detector, TLC analysis of the collected fractions is not necessary.

3. When properly dried and pure, NHS esters can be stored for several months under argon at –20°C. Before opening the tube make sure it reaches room temperature to avoid moisture condensation leading to hydrolysis of the reagent. The purity of the reagent can be checked by Nuclear Magnetic Resonance (NMR) spectroscopy.
4. Diphenyl phosphoryl *N*-hydroxysuccinimide is formed upon reaction between diphenyl phosphoryl chloride and NHS. This intermediate product can still react with the carboxylic acid in the reaction mixture but is considerably less reactive than diphenyl phosphoryl chloride. Increasing the temperature to 50°C for 3 h drastically improves the yield of NHS ester by accelerating the reaction between diphenyl phosphoryl *N*-hydroxysuccinimide and the carboxylic acid.
5. When working with gels, one should try to work as clean as possible. To avoid keratin contamination, one may work under deionizing air conditions by the use of an air-blow ionizer.
6. Unless otherwise indicated, all incubation and washing steps are performed by shaking the samples in a Thermomixer at 600 rpm.
7. The use of hydroxylamine will reverse (partial) O- and S-acetylation on Ser, Thr, or Tyr and Cys residues, respectively. Only perform this step if the peptide of interest contains Cys-, Ser-, Tyr-, or Thr-residues, or if its sequence is unknown.
8. Only perform steps 15 and 16 if the N-terminal peptide of interest contains methionine or if its sequence is unknown.
9. A uniform oxidation of methionine to methionine-sulfoxide can be achieved by the use of hydrogen peroxide. This modification prevents accidental hydrophilic shifts of methionyl peptides when performing a chromatographic separation and as such prevents the introduction of peptide variants (oxidized versus non-oxidized). The final concentration and incubation time should be respected since prolonged incubation leads to uncontrolled oxidation of methionine to methionine-sulfone, and to oxidation of the side-chain of other amino acids such as cysteine and tryptophan.
10. Other sequence-grade proteases—such as chymotrypsin and endoproteinase Glu-C—may be used for in-gel digestion in order to create an MS-detectable and identifiable peptide.

Alternatively, trypsin digestion may precede in vitro in-gel protein N-acetylation if a Lys-ending peptide-species is of interest (tryptic digest of in-gel N-acetylated proteins, only results in the creation of Arg-ending peptides). In such a case in vitro peptide N-acetylation follows the in-gel digestion, carried out in a non amine-reactive and enzyme-compatible buffer (typically we make use of 10 mM triethylammoniumbicarbonate (TEA) buffer (pH = 8.0)). When used at the peptide level, hydroxylamine may be inactivated by acidification with formic acid (to reach a final concentration of 0.5% formic acid). Thereafter, proceed directly with MS analysis since the optional terminal peptide enrichment by use of SCX cannot be applied in such cases.

11. Dependent on the sample complexity and/or the sensitivity of the mass spectrometer used, one may either proceed immediately to LC–MS/MS or further enrich the N-terminal peptide of interest by performing SCX.
12. The volumes depend on the pipette tips used for subsequent SPE extraction and SCX fractionation. 100 μl volume tips are ideal for peptide mixture in volumes larger than 10 μl where very high capacity is required (up to 80 μg of initial protein material), while for proteolytic digests up to 10 μl in volume, 10 μl volume C18 tips with an intermediate capacity and elution volume may be used.
13. SPE extraction is preferred to prepare the sample for subsequent SCX-based fractionation to assure a complete removal of SCX-interfering cations.
14. Always set the pipettor to the maximum pipette tip volume (10 or 100 μl) and secure the pipette tip tightly to the end of the pipettor for optimum aspiration and tip-to-pipettor seal.
15. Samples are dissolved in 2 mM TCEP.HCl (the stock solution has a pH of 2.8) to reduce cysteine residues. Since TCEP is active over a wide pH-range (10), including lower acidic pH, no pH adjustment of this solution is needed since an acidic environment preserves the reduced status of sulfhydryl functions.

Acknowledgments

P.V.D. is a Postdoctoral Fellow of the Research Foundation—Flanders (FWO-Vlaanderen). T.A. is supported by the Norwegian Research Council (Grant 197136) and the Norwegian Cancer Society. K.G. acknowledges support of research grants from the Fund for Scientific Research—Flanders (Belgium) (project numbers G.0042.07 and G.0440.10), the Concerted Research Actions (project BOF07/GOA/012) from the Ghent University and the Inter University Attraction Poles (IUAP06).

References

1. Polevoda B, Arnesen T, Sherman F (2009) A synopsis of eukaryotic Nalpha-terminal acetyltransferases: nomenclature, subunits and substrates. BMC Proc 3(Suppl 6):S2
2. Aivaliotis M, Gevaert K, Falb M, Tebbe A, Konstantinidis K, Bisle B, Klein C, Martens L, Staes A, Timmerman E et al (2007) Large-scale identification of N-terminal peptides in the halophilic archaea *Halobacterium salinarum* and *Natronomonas pharaonis*. J Proteome Res 6:2195–2204
3. Arnesen T, Van Damme P, Polevoda B, Helsens K, Evjenth R, Colaert N, Varhaug JE, Vandekerckhove J, Lillehaug JR, Sherman F et al (2009) Proteomics analyses reveal the evolutionary conservation and divergence of N-terminal acetyltransferases from yeast and humans. Proc Natl Acad Sci USA 106:8157–8162
4. Goetze S, Qeli E, Mosimann C, Staes A, Gerrits B, Roschitzki B, Mohanty S, Niederer EM, Laczko E, Timmerman E et al (2009) Identification and functional characterization of N-terminally acetylated proteins in Drosophila melanogaster. PLoS Biol 7:e1000236
5. Helbig AO, Rosati S, Pijnappel PW, van Breukelen B, Timmers MH, Mohammed S, Slijper M, Heck AJ (2010) Perturbation of the yeast N-acetyltransferase NatB induces elevation of protein phosphorylation levels. BMC Genomics 11:685
6. Hwang CS, Shemorry A, Varshavsky A (2010) N-terminal acetylation of cellular proteins creates specific degradation signals. Science 327:973–977
7. Staes A, Van Damme P, Helsens K, Demol H, Vandekerckhove J, Gevaert K (2008) Improved recovery of proteome-informative, protein N-terminal peptides by combined fractional diagonal chromatography (COFRADIC). Proteomics 8:1362–1370
8. Arnesen T, Starheim KK, Van Damme P, Evjenth R, Dinh H, Betts MJ, Ryningen A, Vandekerckhove J, Gevaert K, Anderson D (2010) The chaperone-like protein HYPK acts together with NatA in cotranslational N-terminal acetylation and prevention of Huntingtin aggregation. Mol Cell Biol 30: 1898–1909
9. Ong SE, Blagoev B, Kratchmarova I, Kristensen DB, Steen H, Pandey A, Mann M (2002) Stable isotope labeling by amino acids in cell culture, SILAC, as a simple and accurate approach to expression proteomics. Mol Cell Proteomics 1:376–386
10. Han JC, Han GY (1994) A procedure for quantitative determination of tris(2-carboxyethyl)phosphine, an odorless reducing agent more stable and effective than dithiothreitol. Anal Biochem 220:5–10

Chapter 10

Computational Prediction of Lysine Acetylation Proteome-Wide

Amrita Basu

Abstract

Several studies have contributed to our knowledge of the enzymology underlying acetylation, including focused efforts to understand the molecular mechanism of substrate recognition by several acetyltransferases; however, conventional experiments to determine intrinsic features of substrate site specificity have proven challenging. In this chapter, I describe in detail a computational method that involves clustering analysis of protein sequences to predict protein acetylation based on the sequence characteristics of acetylated lysines within histones. This method illustrates that sequence composition has predictive power on datasets of acetylation marks, and can be used to predict other posttranslational modifications such as methylation and phosphorylation. Later in this chapter, other recent methods to predict lysine acetylation are described and together, these approaches combined with more traditional experimental methods, can be useful for identifying acetylated substrates proteome-wide.

Key words: Prediction acetylation, Hierarchical clustering, Sequence motif acetylation

1. Introduction

Modifying enzymes that "write" a posttranslational modification (PTM), enzymes that "erase" the PTM, and molecules that interact specifically with the PTM or "readers," are linked to enzyme activity in gene expression regulation, but how do these molecules recognize their epitopes, and what dictates their recognition machinery? Limited biochemical and structural studies of lysine acetyltransferases (KATs) and bromodomains coupled to histones display some sequence preferences (1, 2); however, a rigorous and thorough analysis is still needed to pinpoint specific sequence motifs or patterns on the substrate that help establish appropriate acetyl-lysine-dependent interactions with chromatin.

Sandra B. Hake and Christian J. Janzen (eds.), *Protein Acetylation: Methods and Protocols*, Methods in Molecular Biology, vol. 981, DOI 10.1007/978-1-62703-305-3_10,

Primary sequence context surrounding a target lysine could be one major factor dictating these types of recognition; however, this factor is still relatively understudied. Few studies so far have aimed at looking at the sequence context of a lysine in trying to establish a global view of the acetylome. Structural analyses of KATs coupled to histone tail peptides have been the subject of intense study. KAT enzymes catalyze the transfer of an acetyl group from the cofactor acetyl-Coenzyme A (acetyl-CoA) to the ε-amine of a substrate lysine side chain. A large number of KATs have now been identified and characterized. Studies on the divergent histone KAT enzymes Gcn5/PCAF, Esa1, and Hat1 have provided insights into the underlying mechanism of acetylation by KAT proteins (3, 4). These three histone KAT enzymes contain a conserved core domain that plays a role in binding the cofactor acetyl-CoA in catalysis. More recently, biochemical and structural studies of the metazoan-specific p300/CBP (human homolog of Gcn5) and fungal-specific Rtt109 histone KATs have provided a new understanding into the evolutionary and ancestral relationship between histone KATs and their divergent catalytic and substrate-binding properties (5).

Co-crystallized histone KATs coupled to peptides also provide insight regarding substrate recognition. Positively charged residues are typically present within three to four amino acid residues (about 10 Å) upstream or downstream of the acetylated lysine residues of known p300 (a human homolog of Gcn5)/CBP substrates (5). Moreover, the X-ray crystal structure of p300, in complex with a bi-substrate inhibitor, Lys-CoA reveals the preference for nearby basic residues, such as a lysine in the +2 and −2 positions (5). Specificity for a random-coil structure containing a G-K-X-P recognition sequence on the histone substrate is revealed the when the Gcn5 crystal structure is coupled to the H3K1-20 peptide (2, 6). Critical contacts between Gcn5 and H3K14 are displayed; the glycines (G13), and proline (P15) preceding and following the lysine make the tightest contacts with Gcn5 and suggest residues that are specific to Gcn5 recognition. These studies demonstrate that residues in the core domain of the KAT together with proximal residues surrounding acetyl-lysines on the histone substrate can help achieve substrate specificity (2).

To a lesser extent, nonhistone protein acetylation has also been implicated in a wide variety of biological processes, such as DNA binding or the stabilization of multi-subunit complexes (7–9). One of the most famous nonhistone proteins that was discovered as acetylated was p53, where pioneering studies by Roeder and colleagues demonstrated that p53 acetylation was critical for the regulation of its binding to the DNA (10). Since then, KATs, such as p300, have been shown to acetylate multiple other nonhistone transcription-related proteins. Additional studies have identified additional acetylated proteins, such as HIV1 (11), an HIV protein whose acetylation is required for its viral integration.

Acetylation of alpha tubulin was also recently shown to be critical for the formation of cortical neurons in the developing mouse brain (12). Intriguingly, several transcription factors, such as E2F1, and MYC implicated in cancer pathways, have also been shown to be acetylated (8, 13). Conventional experiments, such as mutagenesis of potential acetylated lysines, acetylation-specific antibodies, metabolic labeling, mass spectrometry (MS), and in vitro histone acetyltransferase assays have typically been used in order to identify acetylated lysines in substrate proteins. More recently, large-scale proteomic studies have emerged as a result of high-throughput technology. In one study, the KAT NuA4 (the essential nucleosome acetyltransferase of H4) was incubated with yeast proteome microarrays in the presence of radioactive acetyl-CoA. Many non-chromatin substrates of complex were identified and validated, including acetylation (Lys19 and Lys514) of phosphoenolpyruvate carboxykinase (Pck1p). Additionally, advanced proteomic tools have enabled identification of several hundred acetylation sites in approximately two hundred proteins, using samples derived from HeLa cells, mouse liver and bacteria (14). In addition to regulators of chromatin-based cellular processes, non-nuclear proteins with diverse functions were also identified. Most strikingly, acetylated lysines were found in more than 20% of mitochondrial proteins, including many metabolic enzymes (14). Another high resolution mass spectrometry study published recently revealed that lysine acetylation preferentially targets large macromolecular complexes involved in diverse cellular processes, such as chromatin remodeling, cell cycle, splicing, nuclear transport, and actin nucleation. The study also reveals that acetylation substrates had enriched residues flanking the target lysine depending on whether the protein resided in the nuclear, cytoplasm, or mitochondria compartment in the cell (15).

A limited number of studies suggest that there may be sequence recognition target(s) for certain KATs (16, 17). A putative "rule" for lysine selection in a primary sequence by a KAT has been proposed previously by examining the N-terminal tail of histones (17). For example, TIP60 (a mammalian KAT), recognizes specific glycine-lysine G-K patterns in human proteins in vitro and in vivo (16). Additionally, the acetylation of Rch1 (a nuclear importin factor) is severely inhibited when a glycine adjacent to the modified lysine is mutated, supporting the view that G-K is part of a recognition motif for acetylation (18). Moreover, the KAT CLOCK1 acetylates H3 lysine 14, which bears a similar sequence environment to the acetylated lysine of the CLOCK-mediated substrate BMAL1 (19). The term "histone mimic" was recently put forward to describe short stretches in nonhistone proteins that closely resemble histone sequences containing PTMs (20). Observations such as these provide early indications of specific "rules" that can potentially define enzyme recognition of target substrates. More recently, several computational methods have been used to investigate such substrate specificities more rapidly. The success of some of these methods can be attributed to multiple research groups that have generated large proteome-wide datasets of lysine acetylation sites. For example, Kim et al. reported the first proteomic survey of acetylation, identifying 388 acetylation sites in 195 proteins among proteins derived from HeLa cells and mouse liver mitochondria (14). In addition to regulators of chromatin-based cellular processes, nonnuclear localized proteins with diverse functions were identified. On a larger scale, Choudhary et al. utilized high-resolution mass spectrometry to identify 3,600 lysine acetylation sites on 1,750 proteins and quantified acetylation changes in response to the deacetylase inhibitors suberoylanilide hydroxamic acid and MS-275 (15). These are examples of large datasets that have been used as validation or training sets, and have been used in methods I describe in this chapter. I carefully describe a particular method in the next section, followed by other prediction methods that have been described in literature.

2. Materials

To computationally predict lysine acetylation no experimental materials are needed.

1. A computer with 2.5 GHz CPU or above and 32 GB RAM or above.
2. High-speed Internet connection.
3. Protein sequence of interest.

3. Methods

3.1. Unsupervised Learning: Hierarchical Clustering

Hierarchical clustering is a method that can be used to learn how acetylated lysine-embedded sequences can group together by their primary amino acid sequence. To begin, sequences from histone proteins were used as a training set because of the wealth of information known about their PTM patterns and well-developed purification and analytical detection methods and focused on the major human core histones bearing a total of 56 lysines (H2A: 13, H2B: 19, H3: 13, H4: 11).

Parameters were selected that could influence the ability to predict acetylation sites on histones by making a series of assumptions. The detailed methodology is as follows:

1. Compile short stretches of amino acids N- and C-terminal of all 56 lysines in histone proteins. Since structural studies of published KAT domains coupled with peptide substrates typically do not exceed 14–20 amino acids in length (3), utilize a sliding window of a maximum number of 20 residues flanking each lysine. Residues most proximal to the lysine are given the highest weight, assuming that these residues are most important for enzyme recognition, as several studies have shown (4, 21).
2. Use standard Blast sequence alignment parameters including gap penalty, extension, insertion, and deletion scores. For lysines in the extreme N- and C-terminal region, such as H3K4 or H2AK129, normalize the raw alignment score based on the length of the sequence. Note that both orientations of the protein sequence (N-terminal to C-terminal or vice versa) are weighted equally. For sequences with lysines located in close proximity to each other, such as H3K36 and H3K37, restrict the alignment matrix so that these sequences do not receive an alignment score. The weight is added to the raw alignment score such that the Ts = Raw score + $\sum$wi of each variable, where Ts = total score, and wi = weight of each residue inversely proportional to distance. This restriction prevents the training set to be overrepresented with sequences from overlapping fragments of the same protein.
3. Hierarchical cluster core histone lysines based on the sequences surrounding each of these given lysines, where a standard BLAST algorithm is used. All 56 histone core sequences are aligned to each other creating a matrix of pairwise alignment scores; generating a hierarchical tree of histone sequences. Sequence alignment scores were computed by performing BLAST local alignments using the NCBI BLAST 2.0 server. A standard BLOSOM62 evolutionary substitution matrix was applied (22). The hierarchical clustering works in an iterative process with the sequence alignment score representing the metric value: it begins with each protein sequence as a singleton cluster; during each iteration, it finds two clusters with the lowest metric value, then joins these two clusters into a new cluster, and updates the metric value between this new cluster and all others (see text for details). An average alignment score is calculated when there are multiple leaves under a node, thereby assigning a single metric value to each node. As an example, H2AK127, a lysine not observed as acetylated, and H4K12, an acetylated lysine, clustered tightly via sequence alignment. As a result, H2AK127 was predicted to be acetylated via this approach. In contrast, H2BK27 clustered with H2BK11 more weakly, and thus, their shared node was positioned on the left side of the threshold line. Hence, this lysine H2BK27 was not predicted as acetylated.

4. Classify each lysine into one of two categories based on its acetylation status reported in literature: "validated" or "not observed."
5. Visually categorize each of the 56 lysines is performed by color-coding the tree based on the acetylation status of each lysine.
6. To assess how robust the clustering was and how well it could actually predict lysine acetylation, all 56 lysines are used to and performed a Leave One out cross-Validation (LOV) (23) by iteratively excluding one lysine from a training set. The hierarchical tree is reconstructed with the remaining 55 lysines, and incorporated the excluded single lysine observation as test data. For each set and combination of predefined parameters (stated above) and in a single run, LOV analysis is used to examine the predictive power on all 56 lysines to discover which set of parameters best optimized classification power. If two lysines were in overlapping fragments of the same protein, both of these lysines are excluded from the training set when either lysine is a test case. Each test lysine (56) is traversed through the training tree to find which subgroup of sequences my target sequence formed the tightest cluster with.
7. Perform Receiving Operating Curve (ROC) analysis on the test dataset where the statistics measure used was the Area Under Curve (AUC). An AUC of one represents a perfect prediction and an AUC of 0.5 random predictions. Each point on a single curve of the ROC plot can be calculated by measuring the false positive versus true positive rate of the performance on all 56 lysines for a given parameter(s) under a cutoff alignment score. If the "test lysine" clusters within a group of "validated" acetylated lysines above the cutoff score, the lysine is predicted to be acetylated. Conversely, if the test lysine clusters within a group of "not observed" lysines above the alignment score, the lysine is predicted as not acetylated. The default status of the lysine when it did not fall into the above criteria was "not acetylated." To test the significance of an ROC score, the above procedure can be applied to 1,000 random permutations of the labels of the observed and not observed lysines, and the median AUC can be measured to establish whether the observed AUC is statistically significant ($p<0.001$).
8. In vivo validation of the computational prediction method is the final step in this analysis. To test whether these predicted lysines are acetylated in vivo, an MS-based approach to examine histone peptides from human cell lines that were asynchronously growing and treated without any HDAC inhibitors can be performed. HDAC inhibitors can be excluded so that the non-hyper levels of acetylation in the asynchronously growing cells can be captured. Further detail on the experimental method can be found in Chapters 1–3.
9. To identify which amino acids play a critical role in acetylation site selection, preferences for certain amino acids near the target acetylated lysines in the validation datasets can be analyzed.

Notably, when the surrounding residues (six residues to the left and right) are examined of a "validated" acetylated lysine versus a "not observed" one in human histone and nonhistone protein, an enrichment can be calculated for each group of residues, i.e., small residues, lysines, and phosphorylatable residues. To test whether the observed enrichment of G, K, and S is statistically significant, frequency of these residues flanking a lysine in the entire human proteome is computed. Hypergeometric test can be used to measure the statistical relevance of this observation (see Formulas).

Using this histone-trained method, two independent validation datasets of acetylated lysines were used. Using the Kim et al. dataset (14), PredMod achieves a sensitivity of 0.60, and specificity of 0.91 at an optimal window length of six. A second dataset, where a list of 32 proteins containing 1,378 lysines with 73 of these reported in literature to be acetylated in vivo and/or in vitro was compiled. With PredMod, a sensitivity of 0.58, and specificity of 0.91 is achieved also at an optimal window length of six.

Both test datasets exhibited a decrease in performance when larger numbers of residues N- and C-terminal to the target lysine were used such as eight residues flanking the lysine, suggesting that KATs may recognize a smaller and defined set of residues. Overall, results from both approaches revealed that selected parameters for histones were also valid for the prediction of acetylated nonhistone substrates using a ROC analysis approach.

3.2. Supervised Methods for Computational Prediction of Acetylation

Additional methods have been used to predict lysine acetylation proteome-wide. Schwartz et al. describe a *motif-x* program to determine lysine acetylation motifs in human (24). These motifs were then scanned against proteomic sequence data using a newly developed tool called *scan-x* to globally predict other potential modification sites within these organisms. Ten-fold cross-validation was used to determine the sensitivity and minimum specificity for each set of predictions (24). Like the Basu et al. study, this study also suggests that there is a preference for glycine and lysine in the residues immediately surrounding the acetylation site as well as aromatic residues at the +1 position. They use a ten-fold cross-validation strategy, where the negative data was obtained by selecting all of those residue-specific sites in the appropriate proteome not found in the positive training data, and the positive set is X. *motif-x* was run on each of the ten training sets and extracted motifs were scanned against the mutually exclusive test sets to make predictions. For each cross-validation run, the sensitivity and specificity of the method was calculated according to the following formulas (see Formulas). Average values of sensitivity and specificity at each threshold were then calculated based on the data obtained from the ten cross-validation sets, and ROC curves. In the case of lysine acetylation 16 and 18% sensitivity is achieved at 95% specificity.

More recently, supervised analyses methods such as support vector machines have been used to predict lysine acetylation. Gnad et al. use the Choudhary et al. dataset (15), where the in vivo acetylation sites with their surrounding sequences are established as the positive set (25). To create a negative set of the same size, the same size of randomly selected lysines from identified human peptides that have not been found to be acetylated according to the MAPU database (26). Several common kernels were investigated such as the Gaussian radial basis function (RBF), linear, polynomial and sigmoid functions, and SVM was trained on the basis of two to eight amino acids surrounding the site to the N- and C-terminus as described (27). For each kernel and for each surrounding sequence length, parameters were optimized C and γ by varying them from 2^{-10} to 2^{10} in multiplicative steps of two and chose the best combination of both parameters. Having selected the best performing kernel function and surrounding sequence size, the accuracy of the optimal model was finally determined by using the test set. On the test set, at the score cutoff yielding 77% precision and 77% recall in the training set, the predictor generated 78% true positives and 78% true negatives, verifying the expected performance. In contrast to a single support vector machine classifier, Xu et al. report using an ensemble of support vector machine classifiers which uses a nine residue window flanking the lysine and using a position weighted matrix method (28). Here, they extract experimentally validated acetylation sites from the Swiss-Prot database, which contains 282 proteins covering 766 acetylated lysines. At the core of the algorithm is an ensemble of SVM classifiers which is a collection of SVM classifiers each trained on a subset of the training set. Prediction of ensemble of SVMs is computed from prediction of an individual SVM. EnsemblPail achieves a 78.84% sensitivity and 95.12% specificity. Other methods incorporate the standard Bayesian Discriminant Method (BDM) (29) which has been employed in the software program PAIL. At a window length of 9 and using various cross validation methods, a sensitivity between 40 and 60% is achieved, and a specificity of 83–96% is achieved. Overall, these prediction programs have achieved reasonable specificities, but there are a few reasons why a prediction algorithm for posttranslational modifications may achieve a lower specificity than other types prediction algorithms (see Notes 1–4).

3.3. Prediction Tools Currently Available

PredMod is available on the server http://www.cs.cornell.edu/w8/~amrita/predmod.html. It enables users to input any protein sequence and the output reveals a list of likely acetylation sites in the protein with a level of confidence. Another acetylation predictor is available online via the PHOSIDA database (http://www.phosida.com). It enables Web users to predict acetylation sites within any input protein sequence using a specified precision–recall cutoff. LysAcet (30), not described in this chapter is found at http://www.biosino.org/LysAcet/. Ensemble Pail can be accessed at http://app.aporc.org/EnsemblePail/, and PAIL is at http://bdmpail.biocuckoo.org.

3.4. Numerical Formulas

Hypergeometric probability calculation: Pr = in human proteome; K = number of times the particular residue is seen flanking in each position in human proteome; n = total number of lysines in each independent validation dataset; m = number of times the particular residue is seen flanking in each position in validation dataset.

$$\text{Specificity} = \text{TP} / \text{TP} + \text{FN}$$

$$\text{Sensitivity} = \text{TN} / \text{TN} + \text{FP}$$

4. Notes

There are a few reasons why a prediction algorithm for posttranslational modifications may achieve a lower specificity than other types of predictions.

1. The MS approach has limited detection and sensitivity capabilities and cannot recover peptides that are acetylated at only low levels.
2. Lysines are modified in discrete environmental conditions, cell cycle stages, cell types, and thus are undetectable in the cell extracts used.
3. Acetylation may be inhibited by adjacent PTMs, and therefore, the responsible KAT might be prevented from binding to or accessing its target site.
4. Acetylation is a transient, dynamic modification, and thus mass spectrometry results may depend on a time-dependent acetylation state whose kinetic properties have not been adequately captured by our experimental parameters.

References

1. Mujtaba S, Zeng L, Zhou MM (2007) Structure and acetyl-lysine recognition of the bromodomain. Oncogene 26:5521–5527
2. Liu X, Wang L, Zhao K, Thompson PR, Hwang Y, Marmorstein R et al (2008) The structural basis of protein acetylation by the p300/CBP transcriptional coactivator. Nature 451:846–850
3. Marmorstein R (2001) Structure of histone acetyltransferases. J Mol Biol 311:433–444
4. Marmorstein R, Roth SY (2001) Histone acetyltransferases: function, structure, and catalysis. Curr Opin Genet Dev 11:155–161
5. Wang L, Tang Y, Cole PA, Marmorstein R (2008) Structure and chemistry of the p300/CBP and Rtt109 histone acetyltransferases: implications for histone acetyltransferase evolution and function. Curr Opin Struct Biol 18: 741–747

6. Rojas JR, Trievel RC, Zhou J, Mo Y, Li X, Berger SL et al (1999) Structure of Tetrahymena GCN5 bound to coenzyme A and a histone H3 peptide. Nature 401:93–98
7. Glozak MA, Sengupta N, Zhang X, Seto E (2005) Acetylation and deacetylation of non-histone proteins. Gene 363:15–23
8. Sterner DE, Berger SL (2000) Acetylation of histones and transcription-related factors. Microbiol Mol Biol Rev 64:435–459
9. Yang XJ, Seto E (2008) Lysine acetylation: codified crosstalk with other posttranslational modifications. Mol Cell 31:449–461
10. Gu W, Roeder RG (1997) Activation of p53 sequence-specific DNA binding by acetylation of the p53 C-terminal domain. Cell 90:595–606
11. Cereseto A, Manganaro L, Gutierrez MI, Terreni M, Fittipaldi A, Lusic M et al (2005) Acetylation of HIV-1 integrase by p300 regulates viral integration. EMBO J 24:3070–3081
12. Wynshaw-Boris A (2009) Elongator bridges tubulin acetylation and neuronal migration. Cell 136:393–394
13. Ozaki T, Okoshi R, Sang M, Kubo N, Nakagawara A (2009) Acetylation status of E2F-1 has an important role in the regulation of E2F-1-mediated transactivation of tumor suppressor p73. Biochem Biophys Res Commun 386:207–211
14. Kim SC, Sprung R, Chen Y, Xu Y, Ball H, Pei J et al (2006) Substrate and functional diversity of lysine acetylation revealed by a proteomics survey. Mol Cell 23:607–618
15. Choudhary C, Kumar C, Gnad F, Nielsen ML, Rehman M, Walther TC et al (2009) Lysine acetylation targets protein complexes and co-regulates major cellular functions. Science 325:834–840
16. Kimura A, Horikoshi M (1998) How do histone acetyltransferases select lysine residues in core histones? FEBS Lett 431:131–133
17. Kimura A, Horikoshi M (1998) Tip60 acetylates six lysines of a specific class in core histones in vitro. Genes Cells 3:789–800
18. Bannister AJ, Miska EA, Gorlich D, Kouzarides T (2000) Acetylation of importin-alpha nuclear import factors by CBP/p300. Curr Biol 10: 467–470
19. Hirayama J, Sahar S, Grimaldi B, Tamaru T, Takamatsu K, Nakahata Y et al (2007) CLOCK-mediated acetylation of BMAL1 controls circadian function. Nature 450: 1086–1090
20. Sampath SC, Marazzi I, Yap KL, Krutchinsky AN, Mecklenbrauker I, Viale A et al (2007) Methylation of a histone mimic within the histone methyltransferase G9a regulates protein complex assembly. Mol Cell 27:596–608
21. Marmorstein R (2001) Structure and function of histone acetyltransferases. Cell Mol Life Sci 58:693–703
22. Eddy SR (2004) Where did the BLOSUM62 alignment score matrix come from? Nat Biotechnol 22:1035–1036
23. Cooper GF, Aliferis CF, Ambrosino R, Aronis J, Buchanan BG, Caruana R et al (1997) An evaluation of machine-learning methods for predicting pneumonia mortality. Artif Intell Med 9:107–138
24. Schwartz D, Chou MF, Church GM (2009) Predicting protein post-translational modifications using meta-analysis of proteome scale data sets. Mol Cell Proteomics 8: 365–379
25. Gnad F, Ren S, Choudhary C, Cox J, Mann M (2010) Predicting post-translational lysine acetylation using support vector machines. Bioinformatics 26:1666–1668
26. Gnad F, de Godoy LM, Cox J, Neuhauser N, Ren S, Olsen JV et al (2009) High-accuracy identification and bioinformatic analysis of in vivo protein phosphorylation sites in yeast. Proteomics 9:4642–4652
27. Gnad F, Ren S, Cox J, Olsen JV, Macek B, Oroshi M et al (2007) PHOSIDA (phosphorylation site database): management, structural and evolutionary investigation, and prediction of phosphosites. Genome Biol 8:R250
28. Chang WC, Lee TY, Shien DM, Hsu JB, Horng JT, Hsu PC et al (2009) Incorporating support vector machine for identifying protein tyrosine sulfation sites. J Comput Chem 30: 2526–2537
29. Li A, Xue Y, Jin C, Wang M, Yao X (2006) Prediction of Nepsilon-acetylation on internal lysines implemented in Bayesian Discriminant Method. Biochem Biophys Res Commun 350:818–824
30. Xu Y, Wang XB, Ding J, Wu LY, Deng NY (2010) Lysine acetylation sites prediction using an ensemble of support vector machine classifiers. J Theor Biol 264:130–135

Chapter 11

Generation and Characterization of Pan-Specific Anti-acetyllysine Antibody

Wei Xu and Shimin Zhao

Abstract

Pan-specific anti-acetylated lysine antibody is a valuable tool to detect, validate, and quantify protein lysine acetylation. Compared to site-specific acetyllysine antibodies that are not readily available in most cases, polyclonal pan-specific acetyllysine antibodies can be generated with synthetic antigen of a carrier protein that has its lysine residues chemically acetylated. Here, we describe protocols of synthesizing acetylated ovalbumin (OVA), immunizing rabbits, and characterizing pan-specific polyclonal anti-acetylated lysine antibodies.

Key words: Acetyllysine, Chemical modification, Antigen, Polyclonal antibody, Affinity, Specificity

1. Introduction

Proteomic approaches had revealed that protein ε-lysine acetylation (referred as lysine acetylation hereafter) is a cell-wide posttranslational modification (PTM) that, in addition to its well-established roles in transcriptional regulation through modifying histones and transcription factors (1, 2), may have global impact on cellular physiology (3, 5). Notably, lysine acetylation is also an evolutionary conserved PTM that exists from bacteria to human (6, 7), making it possibly a fundamental PTM in all lives. Functional studies carried out recently had revealed that lysine acetylation has defined roles in regulation of metabolic enzymes and controlling of reactive oxygen species (ROS) (5, 8, 9). In bacteria, lysine acetylation had been found to exert extensive regulatory functions in metabolic regulation and chemotaxis controlling (6, 10). The versatile

Sandra B. Hake and Christian J. Janzen (eds.), *Protein Acetylation: Methods and Protocols*, Methods in Molecular Biology, vol. 981, DOI 10.1007/978-1-62703-305-3_11, © Springer Science+Business Media, LLC 2013

roles of lysine acetylation have made scientists to speculate that protein lysine acetylation can rival protein phosphorylation (11).

Despite extensive efforts had been made, detection of protein lysine acetylation remains challenging and is still one of the bottlenecks of protein lysine acetylation studies. Among currently available methods, mass spectrometry serves as a useful tool to detect protein lysine acetylation (see Chapters 1–3). Matrix-assisted laser desorption ionization/time of flight MS (MLDI–TOF-MS) is used to identify acetylated peptides from a pool of peptides, but its utilization is usually limited by interference of high-abundance non-acetylated peptides and lack of quantification; tandem mass spectrometry followed by liquid chromatography (LC–MS/MS) has better resolution than MLDI–TOF-MS, but preparation of good samples remains challenging for this method and the assay is still suffered by interferences of non-acetylated peptides. Besides, all MS-based analysis requires expensive instruments and is time-consuming. Simple biochemical analysis of protein lysine acetylation has yet developed largely because of acetyl group's lack of utilizable properties such as electric charge and unique steric structures. Antibody-based Western blotting analysis is convenient and widely employed. However, the quality of anti-acetyllysine antibody is the key for a successful detection of acetylated proteins.

Generation of monoclonal acetyllysine antibody results in limited success so far, at least partially due to either the small size of acetyl group (42 Da) or lack of unique steric conformation of acetyllysine group. However, people have successfully generated pan-specific anti-acetylated lysine antibodies by using either conjugated or synthetic acetyllysine antigens (3, 12). Because polyclonal acetyllysine antibodies consist of a pool of antibodies that recognize acetylated lysine residues flanked by various sequences, they exert higher affinity to acetylated proteins/peptides than that of monoclonal antibodies and sufficient specificity to detect acetyllysine in most cases. Moreover, polyclonal anti-acetyllysine antibodies made from chemically modified carrier proteins are particularly useful to make cell-wide survey for acetylated proteins. The reason is, again, that these antibodies are of reasonable high affinities to acetylated peptides and they can recognize acetyllysine with various flanking sequences (3–5).

In this chapter, we describe protocols of polyclonal pan-specific anti-acetylated lysine antibody generation and characterization. We chemically synthesize acetylated carrier proteins, such as ovalbumin (OVA) or bovine serum albumin (BSA), by acetylating these proteins under mild conditions to obtain intact carrier proteins. The synthesized acetylated OVA were then employed to immunize rabbits. The specificity of the resulted antibodies was verified by analyzing whether the binding of antibody to acetylated proteins can be completed by acetylated BSA, and the affinity of the antibody was analyzed by testing if the antibody is able to enrich endogenous acetylated peptides.

2. Materials

Prepare all solutions using Milli-Q water. Use analytical grade reagents for chemical modification solutions and other buffers. Prepare and store all chemical modification needed reagents and SDS gel electrophoresis buffers at room temperature (unless indicated otherwise). Store Western blot solutions containing antibody or protein at 4°C. Diligently follow all waste disposal regulations when disposing waste materials.

2.1. Synthesis of Antigen

1. Na_2CO_3 solution: 0.1 M Na_2CO_3. Weigh 2.12 g Na_2CO_3 and dissolve it in 200 mL water at room temperature (**see** Note 1).
2. Saturated Tris base: Weigh 150 g Tris base and dissolve it in 200 mL water at room temperature (see Note 2).

2.2. Immunization of Rabbits

1. PBS buffer: 9.1 mM Na_2HPO_4, 1.7 mM NaH_2PO_4, 0.15 M NaCl, pH 7.4.
2. Freund's adjuvant, complete: Paraffin oil, mannide monooleate, and heat-killed *Mycobacterium butyricum* (see Note 3).
3. Freund's adjuvant, incomplete: Paraffin oil, mannide monooleate (see Note 3).
4. New Zealand white rabbits (see Note 4).

2.3. ELISA Components

1. Carbonate buffer: Na_2CO_3, $NaHCO_3$, NaN_3. Make the buffer by dissolving 1.59 g Na_2CO_3, 2.93 g $NaHCO_3$, and 2 mL 10% NaN_3 in 1 L H_2O, adjust pH to 9.5.
2. Wash buffer: PBS buffer with 0.02% Thimerosal and 0.05% Tween-20.
3. Blocking buffer: PBS with 0.1% BSA and 0.02% Thimerosal.

2.4. SDS Gel Electrophoresis Components

1. Gel solution: 30% acrylamide/bisacrylamide solution (29.2:0.8 acrylamide:bisacrylamide). Make this solution by adding 250 g acrylamide into 400 mL H_2O. Add 1.67 g bisacrylamide and stir at room temperature overnight. Make up to 500 mL with water. Filter the solution with 3 M filter paper, and store it at room temperature (20–25°C) in a dark container protected from light (see Note 5).
2. Running buffer (Tris–glycine buffer): Make the buffer by dissolving 3.02 g Tris base, 18.8 g glycine, and 1 g SDS in 1 L water.
3. 5× SDS loading buffer (10% SDS buffer): Make the buffer by adding 50% glycerol, 20% β-mercaptoethanol, 10% SDS, and 0.0012% bromophenol blue into 0.25 M Tris–HCl solution to final concentrations (w/v) as indicated, adjust to pH 6.8.

4. Ammonium persulfate solution: 10% (w/v) ammonium persulfate. Use freshly prepared solution (see Note 6).
5. Tris buffer for gel preparation: 1.5 M Tris solution, adjust to pH 8.8 before use.
6. Coomassie blue staining solution: 0.25% (w/v) Coomassie brilliant blue G-250, 45% (v/v) methanol, 10% (v/v) acetic acid. Make it by solubilizing brilliant blue G-250 powder in methanol, followed by adding acetic acid and H_2O to desired volume. Filter the solution by 3 M filter paper before use.

2.5. Immunoblotting Components

1. Gel transfer buffer (Tris–glycine buffer): Make this buffer by dissolving 4.55 g Tris base and 21.6 g glycine in 500 mL deionized water; add methanol and water to obtain a final concentration of 20% (v/v) methanol and a total volume of 1 L. This buffer can be stored at room temperature until it is used.
2. Peptone-blocking buffer: 0.1 M Tris–HCl, pH 7.5, 100 mM NaCl, 1% peptone, 10% Tween-20 (see Note 7).
3. Tris-buffered saline and Tween-20 (TBST) buffer: 0.1 M Tris–HCl, pH 7.5, 100 mM NaCl, 0.5% Tween-20.
4. Peptone–antibody buffer: 0.1 M Tris–HCl, pH 7.5, 100 mM NaCl, 0.1% peptone, 0.5% Tween-20.
5. Saline: 0.9% (w/v) sodium chloride.
6. Amersham ECL plus Western blotting detection reagent.
7. Nitrocellulose membranes.
8. Bio-Rad wet gel transfer apparatus.

2.6. Verification of Anti-acetyllysine Antibody

1. TSA stock solution: 500 μM trichostatin A. Dissolve trichostatin A in absolute ethanol, store at −20°C.
2. Nicotinamide stock solution: 2.5 M nicotinamide. Dissolve nicotinamide in water, store at −20°C.
3. (TMB) horseradish peroxidase solution (commercially available).
4. Nonidet-P40 (NP-40) buffer: 6 mM Na_2HPO_4, 4 mM NaH_2PO_4, 1% Nonidet-P40, 150 mM NaCl, 2 mM EDTA, 50 mM NaF, 4 μg/mL leupeptin.
5. 12% (w/v) SDS gel: 3.3 mL ddH_2O, 4.0 mL 30% acrylamide, 2.5 mL 1.5 M Tris, pH 8.8, 0.1 mL 10% SDS, 0.1 mL 10% ammonium persulfate, 0.004 mL TEMED.
6. TBST buffer (PBS–Tween-20 buffer): Dissolve 8 g NaCl, 2.37 g Tris–HCl, and 0.6 g Tris base and raise volume to 1 L, adjust to pH 7.6 and add 0.1% (v/v) Tween-20.
7. NH_4HCO_3 buffer: 50 mM NH_4HCO_3, pH 8.0.
8. NETNA buffer: 50 mM Tris–HCl, pH 8.0, 100 mM NaCl, 1 mM EDTA, 0.5% NP-40, 10% acetonitrile.
9. Elution buffer: 0.1% trifluoric acid, 50% acetonitrile (v/v).

3. Methods

Carry out all procedures at room temperature unless otherwise specified. All the animals should be housed in a pathogen-free facility and given a standard diet and water ad libitum. Rules of care and use of experimental animals should be obeyed in this protocol.

3.1. Preparation of Acetyllysine Antigen

1. Add 20 mg OVA into 20 mL of 0.1 M Na_2CO_3; stir at room temperature until OVA is completely dissolved.
2. Slowly add 100 μL of acetic anhydride (drop by drop) into OVA solution while the solution is constantly stirred (see Note 8).
3. Slowly add 400 μL of pyridine (drop by drop) into solution while the solution is stirred (see Note 9).
4. Allow the reaction proceed at room temperature (25–30°C) for 4–5 h.
5. Add 400 μL of 1 M Tris base into the reaction mixture, and stir for 5–10 min to stop the reaction.
6. Load 2 mL of reaction mixture each time to a 10 mL Sephadex G-25 gel filtration column, and remove organic reagents by running the reaction mixture through the column with 50 mM Tris buffer in an AKTA-FPLC system; monitor optical density at 280 nm (O.D. 280) to collect the eluted protein (see Note 10).
7. Repeat step 6 until all the reaction mixture is separated.
8. Combine fractions of acetylated OVA from repeated separation operations.
9. Vacuum-dry acetylated OVA overnight in a concentrator (Eppendorf 5301) (see Note 11).

3.2. Verification of Acetylated Antigen

1. Prepare 10% SDS gel solution by mixing 4.0 mL H_2O, 3.3 mL gel solution, 2.5 mL 1.5 M Tris, pH 8.8, and 0.1 mL 10% SDS; add 0.1 mL 10% ammonium persulfate and 4 μL TEMED immediately before pour the solution into gel-making cassette. Wash gel wells with running buffer before sample loading.
2. Dissolve both acetylated and unacetylated OVA in 1× SDS loading buffer; boil the solution by putting samples in heat block for 5 min.
3. Run unacetylated and acetylated OVA in 10% SDS gel, stain gel in Coomassie blue staining solution, and check molecular weight gain of acetylated OVA to verify sufficient acetylation of OVA (Fig. 1).

3.3. Immunization of Rabbits

1. Dissolve antigen in saline to a final concentration of 0.5 mg/mL on day 0.

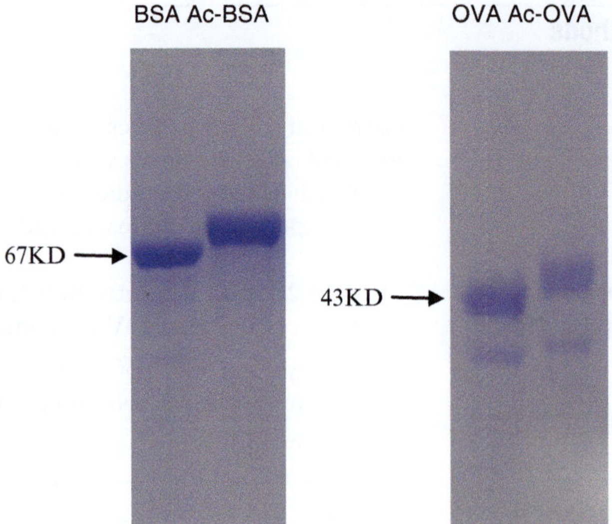

Fig. 1. Verification of chemically acetylated proteins. BSA, acetylated BSA (Ac-BSA), OVA, and acetylated OVA (Ac-OVA) were resolved in 10% SDS PAGE and stained with Coomassie brilliant blue.

2. Add equal volume of complete Freund's adjuvant to the antigen solution; make antigen by thorough mixing of adjuvant and the antigen solution. Prepared antigen should contain 200–400 μg acetylated OVA/mL.
3. Inject 1 mL of prepared antigen into a New Zealand rabbit subcutaneously in multiple places. Two rabbits should be injected for each antigen.
4. Cage rabbit singly and make sure each rabbit has free access to food and water until next injection.
5. Prepare boost antigen by mixing antigen solution with incomplete Freund's adjuvant (1:1, v/v). Prepared boost antigen should contain 100–200 μg acetylated OVA per mL.
6. On any day during 28–30 days after first inject, inject 1 mL prepared boost antigen into previously immunized New Zealand rabbit subcutaneously in multiple places.
7. Cage rabbit singly and make sure each rabbit has free access to food and water until next injection.
8. On any day during 28–30 days after second inject, inject 1 mL prepared boost antigen into previously double immunized New Zealand rabbit subcutaneously in multiple places to further boost immunization.
9. 10–15 days after the second boost immunization, obtain the blood sample (1 mL) from rabbit ear, and determine antibody titer by indirect ELISA (see next section).
10. On any day during 28–30 days after third inject, inject 1 mL prepared boost antigen into previously triple immunized New

Zealand rabbit subcutaneously in multiple places to final boost immunization (see Note 12).

11. 10–15 days after final immunization, sacrifice the rabbits by bleeding from the carotid arteries. A proper operation will result in ~50 mL blood per rabbit.
12. Put collected blood in a 37°C incubator for 2 h.
13. Transfer blood from 37 to 4°C, and maintain at 4°C overnight (see Note 13).
14. Centrifuge blood at 6,000 × *g* at 4°C for 20 min.
15. Carefully collect antiserum and discard pellets.
16. Determine antibody titer by indirect ELISA (see next section).

3.4. Determine Antibody Titer by Indirect ELISA

1. Dissolve acetylated OVA in carbonate solution, pH 9.6 to a final concentration of 2–5 μg/mL.
2. Add 50 μL of acetylated OVA solution into wells of an ELISA plate.
3. Incubate the ELISA plate at 4°C overnight to allow acetylated OVA coating (see Note 14).
4. Wash the wells with PBST buffer five times; remove PBST by vacuum each time.
5. Add 200 μL of TBST with 5% (w/v) nonfat milk to each well to block the uncoated area at 37°C in an incubator for at least 2 h (see Note 15).
6. Remove nonfat milk completely, and wash wells with PBST briefly.
7. Prepare a serial dilution of antiserum with TBST (1:1,000 to 1:500,000; 4–6 dilutions).
8. Add 50 μL each of diluted antiserum to ELISA plate wells. Triplicated tests were performed for each dilution.
9. Incubate ELISA plate at 37°C for 1 h to allow antibody–antigen formation.
10. Remove ELISA plate from incubator, and wash wells with TBST five times.
11. Remove TBST by vacuum completely.
12. Add 50 μL 3,3′,5,5′-tetramethylbenzidine (TMB) horseradish peroxidase solution to each well.
13. Leave the ELISA plate at room temperature for 3–5 min to allow color development (see Note 16).
14. Add 20 μL of 2 M sulfuric acid to each well to stop the reaction (see Note 17).
15. Measure color intensity with plate reader at 450 and 630 nm.
16. Antibody titer was estimated by determining highest detectable antibody dilution (lowest antibody concentration).

3.5. Characterization of Pan-Specific Anti-acetylated Lysine Antibody

1. Grow HEK293T cells in Dulbecco's Modified Eagle's Medium (DMEM) supplemented with 10% calf serum to ~80% confluence.
2. Wash cells with PBS one time; briefly trypsinize cells in 2 mL PBS, and collect cells by centrifuge at 1,000 × *g* for 2 min.
3. Split cells equally into two culture plates, culture cells in parallel, one for experiments and one for control, with DMEM supplemented with 10% calf serum.
4. Add TSA stock solution to the experiment plate to a final TSA concentration of 5 μM 18 h before harvesting cells.
5. Add nicotinamide stock solution to the experiment plate to a final nicotinamide concentration of 5 mM 6 h before harvesting cells.
6. Remove culture medium from both control and experiment plates, and wash cells gently by PBST.
7. Collect cells by scraping cells in PBST; use 5–10 mL PBST for each plate.
8. Centrifuge-collected cells at 1,000 × *g* for 5 min to harvest cells; discard supernatant.
9. Lyse cells using NP-40 buffer. For cells collected from each culture plate, add 1.5 mL NP-40 buffer and vortex 1–2 min.
10. Centrifuge cell lysates at 12,000 × *g* at 4°C for 15 min.
11. Retain the supernatant by carefully removing supernatant from centrifuge tube, and discard the pellet.
12. Determine total protein concentration in each of the supernatant using Bradford reagent.
13. Take 0.5 mL supernatant to make SDS sample, and verify specificity of antibody (start from step 14) and 0.5 mL supernatant to make digested peptides to verify antibody affinity (start from step 22).
14. Run equal amount (~200–300 μg) of the experiment and the control samples on a 12% (w/v) SDS gel.
15. Transfer proteins from the gel to a nitrocellulose membrane using a Bio-Rad wet transferring apparatus.
16. Block the membrane with peptone buffer for 3–4 h at room temperature with gentle shaking.
17. Wash the membrane twice with TBST buffer; continue experiments with either of the following conditions (steps 18 and 19).
18. Incubate the membrane with pan-anti-acetyllysine antibody (titer 1:100,000) diluted at 1:500 (v/v) in 10 mL of peptone–antibody buffer, shaking at room temperature for 4–6 h (see Note 18).

19. Incubate the membrane with pan-anti-acetyllysine antibody (titer 1:100,000) diluted at 1:500 (v/v) in 10 mL of peptone–antibody buffer with 2 mg/mL chemically acetylated BSA, shaking at room temperature for 4–6 h (see Note 18).
20. Wash the membrane three times with TBST buffer.
21. Incubate the membrane with secondary antibody diluted at 1:1,000 (v/v) in 10 mL peptone–antibody buffer, shaking at room temperature for 2–3 h.
22. Wash the membrane three times with TBST buffer.
23. Apply 50–200 μL Amersham© ECL Plus to the membrane, keep it in the dark for 5 min, and detect signal with GE Typhoon Trio. Alternatively, signals can be detected by exposure membrane to films.
24. Analyze the specificity of pan-specific anti-acetylated lysine antibody by determining if TSA and NAM treatment efficiently increases acetylation levels of histones and other proteins in the lysates (Fig. 2a) and if the acetylation signal of tubulin can be efficiently competed by acetylated BSA (Fig. 2b).
25. Add 5–10 mL prechilled acetone to the supernatant; vortex and place the solution at −80°C for 30 min (see Note 19).
26. Centrifuge at 1,000×*g* at 4°C for 10 min to collect the proteins (see Note 20).
27. Air-dry protein pellet.
28. Dissolve the protein pellet in 5 mL of NH_4HCO_3 buffer and sonicate for 5 min at room temperature to increase solubility (see Note 21).
29. Add sequencing-grade trypsin to the redissolved proteins at a 1:50 (trypsin:total protein) ratio; incubate the reaction at 37°C overnight.
30. Add additional trypsin to the tube at a ratio of 1:100 (trypsin:total protein) and incubate at 37°C for additional 3 h.
31. Heat the sample at 99°C for 5 min to inactivate trypsin (see Note 22).
32. Cool down samples to room temperature; centrifuge at 12,000×*g* for 10 min; retain the supernatant and discard the pellet.
33. Vacuum dry the supernatant and redissolve the pellet in 1 mL of deionized water.
34. Repeat step 30 three to four times (see Note 23).
35. Redissolve the peptide pellet in 1.5 mL of NETNA binding buffer (see Note 24).
36. Conjugate the pan-anti-acetyllysine antibody to protein A beads by adding 200–500 μL antiserum to 1 mL protein A beads.

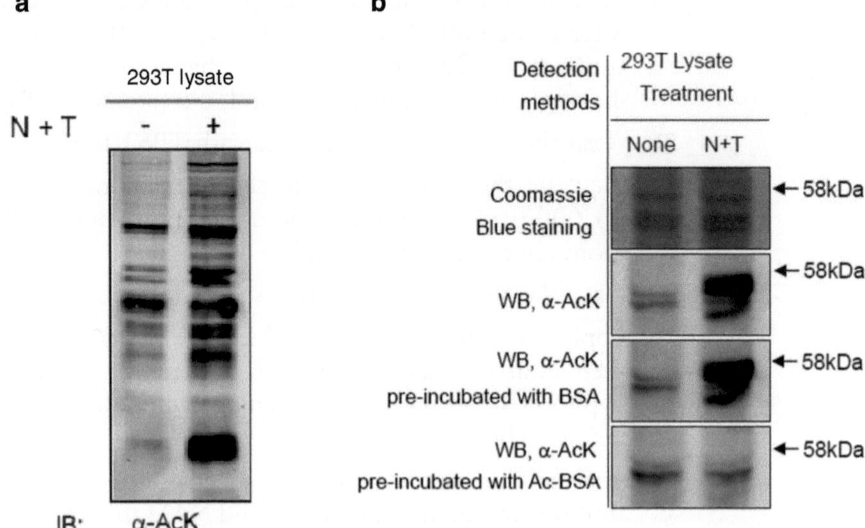

Fig. 2. Characterization of the specificity of pan-anti-acetyllysine antibody (5). (a) HEK293T cells were cultured under either no treatment or nicotinamide plus trichostatin A treatment (N+T) conditions. Equal amount of cell lysates was loaded onto 12% SDS PAGE, and acetylation was probed by homemade pan-anti-acetyllysine antibody. (b) Specificity of the α-AcK antibody. The homemade pan-anti-acetyllysine antibody was co-incubated with bovine serum albumin (BSA) or acetylated BSA during Western blotting process. *Arrows* indicate a prominent tubulin that was detected by the antibody. Co-incubation with acetylated BSA but not the BSA control completely abolished the recognition, demonstrating that our α-AcK antibody specifically recognizes acetyllysine residues in proteins.

37. Add 200 μL antibody-conjugated beads to the peptide solution prepared in step 32; allow 4–8 h for binding at 4°C with rotary shaking (see Note 25).
38. Centrifuge at 1,000 ×*g* briefly (<10 s) and carefully remove the supernatant.
39. Wash beads three times with 1 mL of PBS buffer (see Note 26).
40. Remove PBS buffer completely and add 150 μL of elution buffer (see Note 27).
41. Repeat the elution procedure three times and combine the elution supernatant from each repeat.
42. Vacuum dry the supernatant.
43. Dissolve the eluted peptides in MS sample buffer and subject the sample to liquid chromatography–MS/MS analysis (see Note 28).
44. Look into the spectra of LC–MS/MS and see if there are known acetylated peptides from cells that are enriched. Typically histone peptides should be detected with high confidence (Fig. 3) (see Note 29).

4. Notes

1. The pH of alkaline sodium carbonate solution can be changed after prolonged storage. It is recommended that this solution be prepared freshly to get the best result of modification.
2. The solubility of Tris base in water is about 50 g/100 mL at 25°C, and the solubility is subject to change with temperature. Add enough Tris base when making this solution, and stir for 10 min before use.
3. Both complete and incomplete Freund's adjuvants are commercially available.
4. The age and the body weight usually have impact on the effect of immunization. It is preferred to use rabbits no heavier than 2.5 kg and no elder than 8 weeks old.
5. Acrylamide and bisacrylamide are potent neurotoxins. Use appropriate safety measures, such as protective gloves and safety goggles, and handle under adequate ventilation is strongly recommended. The acrylamide solution can be stored at 4°C to prevent hydrolyzing to acrylic acid and ammonia. The acrylamide mixture, buffer, and water can be prepared in large batches, frozen in aliquots, and used indefinitely.
6. Ammonium persulfate is chemically instable. Make ammonium persulfate solution freshly immediately before use or use stock solution that is stored no longer than 1 week.
7. We formulized peptone-blocking solution to solve the problem of weak signal of anti-acetyllysine antibodies. The high concentration of Tween-20 in the solution can efficiently reduce background of membranes, and the concentration of Tween-20 can be adjusted according to the actual effect during blotting.
8. Given that acetic anhydride can be hydrolyzed to acidic acid and strong acidic condition denatures OVA and causes precipitation thus reduce the efficiency of acetylation due to limited access of acetic anhydride to the lysine resides. Acetic anhydride needs to be added dropwise with constant stirring/shaking to avoid regional high acid conditions and precipitation of OVA. It takes ~10 min to complete this step.
9. High concentration of pyridine denatures and precipitates OVA by creating hydrophobic environment. To avoid this drawback, pyridine is added dropwise with constant shaking to avoid precipitation of OVA. It takes ~30 min to complete this step. Alternatively, lowering the concentration of OVA in the solution will also help. However, lowering the concentration of OVA will eventually cause large volume of final product and will take more effects to separate the acetylated OVA and the rest of the reactants.

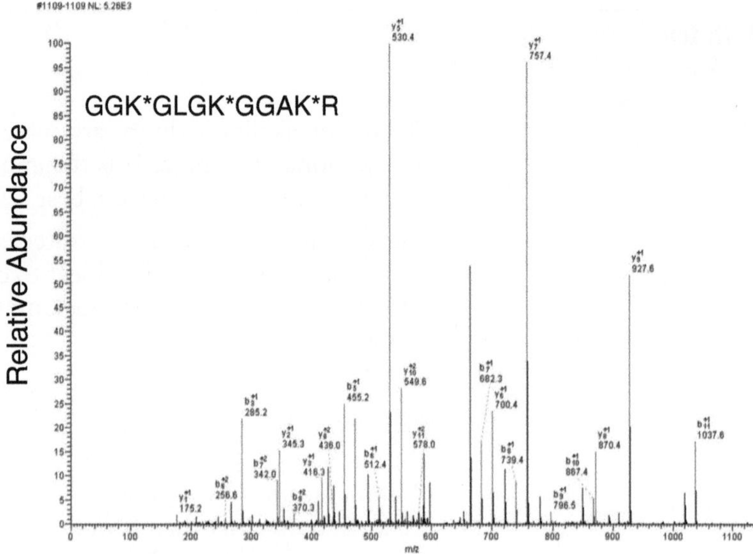

Fig. 3. Pan-specific anti-acetyllysine antibody enriches acetylated histone peptide. An example MS/MS spectrum of pan-specific anti-acetyllysine antibody-enriched histone H4 peptide is shown.

10. Pyridine has absorbance at 280 nm and it runs out of the column right after the protein peak under conditions described here. To get better separation of acetylated OVA and the rest of the reactants, longer column can be used. The negative impact to use longer column, though, is that it will increase the volume of collected OVA containing solution and cause more efforts in the following vacuum-dry procedures. Alternative ways of removing chemicals from acetylated proteins include using centrifugal filters such as Amicon ultra (Millipore).
11. Dried acetylated protein can be stored at 4°C until it is needed for the next step.
12. The titer of the antibody reaches peak in about 2 weeks after final immunization. We found that sacrifice rabbits during days 10–15 result in best antibodies. Do not let final immunization go beyond 20 days before sacrifice rabbits.
13. Store blood at 4°C overnight before continuing to the next step. However, it is recommended that antiserum be separated from blood cell as early as possible to avoid contamination caused by hemolysis.
14. Gentle shaking is recommended for better effect of antigen coating. It is our experience that shaking while coating can facilitate antigen coating.
15. Gentle shaking is recommended for better effect of blocking. It is our experience that shaking boosts blocking effect.

16. Both under- and overdevelopment color will affect the accuracy of the assay significantly; strictly follow the instruction for time of color development.
17. Sulfuric acid is corrosive and exothermic when dissolved in water; sufficient protection such as wearing gloves and face mask while using it is strongly recommended.
18. For best results, the dilution factor should be adjusted with different titer of antiserum. Although many antibodies showed higher specificity and affinity after affinity purification, our experiences showed that affinity purification hurts both specificity and affinity of pan-specific anti-acetylated lysine antibody. We therefore recommend using antiserum directly instead of carrying out affinity purification procedures for this antibody.
19. Acetone precipitates proteins at >80% concentration under low temperature. We recommend to use >90% acetone concentration and prechill acetone at −20°C before use.
20. Higher centrifugation speed will cause tight-packed protein pellet which will cause difficulties in further redissolving steps; do not exceed 1,000 × g when collect the protein pellet at this step.
21. Some proteins may not redissolve even after sonication, and the solution looks turbid. We usually do not do centrifugation to clear the undissolved proteins, as the turbidity can be resolved in the following trypsin digestion steps.
22. Seal the sample containing Eppendorf tubes tightly with Parafilm while heating to avoid burst up of the samples by steam.
23. Excess amount of NH_4HCO_3 will affect the following manipulations; repetitively redissolve and vacuum dry can evaporate NH_4HCO_3 effectively.
24. Before adding NETNA binding buffer, you may or may not see peptide pellet in the tube. A clear film of sample at the bottom of the Eppendorf tube can be seen occasionally. However, if you see sizable white powder in the tube, your sample may still contain NH_4HCO_3 and need to repeat **steps** 30 and 31 before move forward.
25. Theoretically the binding of antibody to protein A beads increases with time; our experience, however, indicates that the conjugation reaches the best result when conjugation time was controlled between 4 and 6 h.
26. Vortex should be avoided. Add PBS to the tube, and shake gently before remove the PBS.
27. Carefully remove residual PBS by a strip of filter paper is a common practice. Care should be taken that naked hands should not touch the filter paper.

28. We centrifuge the sample at 12,000 × *g* for 5 min before injection and only take the upper layer of the sample to be analyzed. This practice can avoid column clot as well as avoid polymer contamination when doing MS analysis.
29. A good affinity pan-specific anti-acetylated lysine antibody should be able to enrich acetylated histone peptides (see Fig. 3).

Acknowledgments

We are grateful for the experimental works by Yan Lin and Wei Yu to develop this protocol. We thank other members of the lab for their insightful discussion. This work is supported by the State Key Development Program for Basic Research of China (2012CB910100), by grant from Chinese National Science Foundation (31030042), by grants from State Key Lab of Genetic Engineering (to S.Z.), and by the Shanghai Key Project for Basic Research (11JC1401100), P. R. China.

References

1. Phillips DM (1963) The presence of acetyl groups of histones. Biochem J 87:258–263
2. Wei G, Roeder RG (1997) Activation of p53 sequence-specific DNA binding by acetylation of the p53 C-terminal domain. Cell 90(4):595–606
3. Kim SC et al (2006) Substrate and functional diversity of lysine acetylation revealed by a proteomics survey. Mol Cell 23(4):607–618
4. Choudhary C et al (2009) Lysine acetylation targets protein complexes and co-regulates major cellular functions. Science 325(5942):834–840
5. Zhao S et al (2010) Regulation of cellular metabolism by protein lysine acetylation. Science 327(5968):1000–1004
6. Wang Q et al (2010) Acetylation of metabolic enzymes coordinates carbon source utilization and metabolic flux. Science 327(5968):1004–1007
7. Zhang J et al (2009) Lysine acetylation is a highly abundant and evolutionarily conserved modification in *Escherichia coli*. Mol Cell Proteomics 8(2):215–225
8. Starai VJ et al (2002) Sir2-dependent activation of acetyl-CoA synthetase by deacetylation of active lysine. Science 298(5602):2390–2392
9. Tao R et al (2010) Sirt3-mediated deacetylation of evolutionarily conserved lysine 122 regulates MnSOD activity in response to stress. Mol Cell 40(6):893–904
10. Li R et al (2010) CobB regulates *Escherichia coli* chemotaxis by deacetylating the response regulator CheY. Mol Microbiol 76(5):1162–1174
11. Norvell A, McMahon SB (2010) Cell biology. Rise of the rival. Science 327(5968):964–965
12. Qiang L et al (2005) Development of a PAN-specific, affinity-purified anti-acetylated lysine antibody for detection, identification, isolation, and intracellular localization of acetylated protein. J Immunoassay Immunochem 26(1):13–23

Chapter 12

Using Functional Proteome Microarrays to Study Protein Lysine Acetylation

Jin-ying Lu, Yu-yi Lin, Jef D. Boeke, and Heng Zhu

Abstract

Emergence of proteome microarray provides a versatile platform to globally explore biological functions of broad significance. In the past decade, researchers have successfully fabricated functional proteome microarrays by printing individually purified proteins at a high-throughput, proteome-wide scale on one single slide. These arrays have been used to profile protein posttranslational modifications, including phosphorylation, ubiquitylation, acetylation, and nitrosylation. In this chapter, we summarize our work of using the yeast proteome microarrays to connect protein lysine acetylation substrates to their upstream modifying enzyme, the nucleosome acetyltransferase of H4 (NuA4), which is the only essential acetyltransferase in yeast. We further prove that the reversible acetylation on critical cell metabolism-related enzymes controls life span in yeast. Our studies represent a paradigm shift for the functional dissection of a crucial acetylation enzyme affecting aging and longevity pathways.

Key words: Protein chip, Posttranslational modification, Lysine acetylation, Functional proteome microarray, Aging, Longevity

1. Introduction

A functional proteome microarray is defined as the construction of high-density protein microarrays composed of the majority (e.g., >60%) of individually purified proteins encoded by an organism (1–3). Unlike the mass spectrometry technologies, in which en masse affinity-purified proteins are subjected to analysis, functional proteome microarrays provide information about direct biochemical and physical interactions between the probe and analytes. For example, proteome microarrays have been used to profile physical interactions between proteins and proteins (4), lipids (4), DNA (5–7), RNA (8), or small molecules (9). In addition to these bind-

Sandra B. Hake and Christian J. Janzen (eds.), *Protein Acetylation: Methods and Protocols*, Methods in Molecular Biology, vol. 981, DOI 10.1007/978-1-62703-305-3_12,

ing assays, proteome microarrays offer a unique platform to perform various types of biochemical reactions that can be used to identify covalently modified substrates by a particular enzyme. Posttranslational modifications (PTMs) of proteins are important regulatory mechanisms in eukaryotes (10–15). While phosphorylation on serine, threonine, and tyrosine residues is well characterized (10–12), modifications on lysine residues, such as ubiquitylation (16); acetylation (17–19); mono-, di-, and trimethylation; SUMOylation (20); and neddylation (21), are now emerging as important PTMs involved in many aspects of cellular functions. Our team is focusing on these mutually exclusive lysine PTMs to better understand the physiological roles and interplay of different signaling pathways.

Here we describe detailed protocols for generating protein samples, printing protein microarrays, and establishing acetylation reactions on these protein chips. Histone acetylation and deacetylation, which are catalyzed by lysine acetyltransferases and deacetylases, respectively, are emerging as critical regulators of chromatin structure and consequent transcription activation. However, it has been hypothesized that many acetyltransferases and deacetylases might also modify nonhistone substrates. For example, the core enzyme, Esa1, of the nucleosome acetyltransferase of H4 (NuA4) complex, is the only essential lysine acetyltransferase in yeast, which strongly suggests that it may target additional nonhistone proteins that are crucial for cells to survive. However, whether such modifications on nonhistone substrates would generate any biological consequences was not fully explored. To address this outstanding question, we first established and performed acetylation reactions on the yeast proteome microarrays using the NuA4 complex in the presence of (^{14}C)-acetyl-CoA as the acetyl group donor (17). Surprisingly, 91 nonhistone proteins were found to be readily acetylated by the NuA4 complex on the array. To further validate these in vitro results, 20 of them were randomly chosen, and 13 of them showed Esa1-dependent acetylation in cells. We further demonstrate that acetylation on two of the substrate proteins is associated with aging pathway and life-span regulation.

2. Materials

2.1. Yeast Culture

1. A high-quality collection of 5,800 yeast open reading frames (ORFs) (93.5% of the total) cloned into a yeast high-copy expression vector using recombination cloning (4). The 5,800 yeast proteins are fused to glutathione S-transferase-polyhistidine (GST-HisX6) at the NH_2 termini and expressed in yeast using an inducible *GAL1* promoter.

2. Yeast synthetic complete media supplement without uracil containing 2% glucose (SC-Ura/glucose) agar plate (Omni, USA).
3. SC-Ura/glucose liquid.
4. 96-Pronger.
5. Yeast synthetic complete media supplement without uracil containing 2% raffinose (SC-Ura/raffinose) liquid.
6. 2 mm diameter glass beads (Fisher Scientific).
7. Automated plate-filling device (QFill2, Genetix, UK).
8. 40% Galactose stock.
9. Ice-cold water (precooled to 4°C).

2.2. Protein Purification in a 96-Well Fashion

1. Zirconia beads (0.5 mm diameter from BSP, Germany).
2. Cap mat used to seal each well.
3. Deep 96-well filter plate (NUNC, USA) sitting on top of a 96-well box.
4. Automated plate-filling device (QFill, Genetix, UK).
5. Cold and clean 96-well box (precooled to 4°C).
6. Glutathione beads (Amersham, USA: 27-4574-01).
7. Coldlysisbuffer(precooledto4°C):50mMTris(hydroxymethyl) aminomethane (stock 1 M), pH 7.5, 100 mM NaCl (stock 5 M), 1 mM ethylene glycol tetraacetic acid (EGTA) (stock 500 mM), 10% glycerol, 0.1% Triton X-100, 0.1% β-mercaptoethanol (BME), 1 mM phenylmethylsulfonyl fluoride (PMSF) (stock 200 mM in isopropanol), 50 μM N-acetyl-l-leucyl-l-leucyl-l-norleucinal (LLnL) (stock 50 mM in DMSO), 1 μM MG-132 (stock 50 mM in DMSO).
8. Roche protease inhibitor tablets (without ethylenediaminetetraacetic acid (EDTA)), two tablets per 50 mL lysis buffer.
9. BME, PMSF, LLnL, MG-132, and inhibitor tablets are added freshly.
10. Roller drum.
11. Wash buffer I: 50 mM Tris-HCl, pH 7.5, 500 mM NaCl, 1 mM EGTA, 10% glycerol, 0.1% Triton X-100, 0.1% BME, 1 mM PMSF. BME and PMSF are added freshly.
12. Wash buffer II: 50 mM HEPES, pH 7.5, 100 mM NaCl, 1 mM EGTA, 10% glycerol, 0.1% BME, 1 mM PMSF. BME and PMSF are added freshly.
13. Elution buffer: 50 mM 4-(2-hydroxyethyl)-1-piperazineethanesulfonic acid (HEPES) (stock 1 M, pH 8.0), 100 mM NaCl, 30% glycerol, 40 mM glutathione (reduced form), 0.03% Triton X-100. Make sure the pH is exactly 8.0. Make freshly before use.
14. 96-Well PCR plate.

2.3. Printing Proteome Microarrays

1. Proteome microarray printer (BioRad: VersArray ChipWriter™ Pro System or ArrayIt Corporation: NanoPrinter LM210).
2. 384-Well plate (Whatman: 7701-5101).
3. FAST slides (Whatman: 10486111A, nitrocellulose-coated slides).
4. Printing pins (ArrayIt Corporation: SMP3).

2.4. Purification of Histone Acetyltransferase (HAT)

1. Extraction buffer + protease inhibitor:

40 mM HEPES, pH 7.5, 350 mM NaCl, 10% glycerol, 0.1% Tween 20, 2 mg/mL leupeptin, 2 mg/mL pepstatin A, 5 mg/mL aprotinin, 1 mM PMSF.

2. 20 mM imidazole buffer: 100 mM NaCl, 20 mM imidazole, pH 7.2, 10% glycerol, 0.1% Tween 20, 2 mg/mL leupeptin, 2 mg/mL pepstatin A, 5 mg/mL aprotinin, 1 mM PMSF.
3. 100 mM imidazole buffer: 100 mM NaCl, 100 mM imidazole, pH 7.2, 10% glycerol, 0.1% Tween 20, 2 mg/mL leupeptin, 2 mg/mL pepstatin A, 5 mg/mL, aprotinin, 1 mM PMSF.
4. Mono Q A buffer: 50 mM Tris–HCl (pH 7.5), 1 mM EDTA, 1 mM BME, 5% glycerol.

2.5. In Vitro Acetylation on a Proteome Microarray

1. Reaction mixture (for 100 μL reaction): 5× histone acetyltransferase (HAT) buffer: 20 μL, (^{14}C)-acetyl-CoA (Amersham 50 μCi/mL: CFA729): 0.5 μL (final 0.25 μCi/mL), 1 M sodium butyrate: 1 μL (final 10 mM), 1 M nicotinamide: 0.5 μL (final 5 mM), 8.3 mM trichostatin A (TSA): 1.2 μL (final 0.1 mM), HAT enzyme: 20 μg, ddH_2O: bring the volume up to 100 μL; in the control reaction, HAT enzyme is left out.
2. 5× HAT buffer: 250 mM Tris–HCl, pH 7.5, 25% glycerol, 5 mM EDTA, 250 mM KCl, 5 mM dithiothreitol (DTT) (stock 1 M), 5 mM PMSF. Add the DTT and PMSF right before use.
3. Phosphate-buffered saline Tween 20 (PBST) (20× stock, Cell Signaling: 9809): 1× PBST contains 3.2 mM Na_2HPO_4, 0.5 mM KH_2PO_4, 1.3 mM KCl, 135 mM NaCl, 0.05% Tween 20, pH 7.4.
4. 50 mM $NaHCO_3$–Na_2CO_3 (pH 9.3): Dissolve 0.572 g of Na_2CO_3 $10H_2O$ and 0.672 g of $NaHCO_3$ in distilled water and bring the total volume to 200 mL.

2.6. In Vitro Acetylation Assay

1. Zirconia beads (0.5 mm diameter from BSP, Germany).
2. Glutathione beads (Amersham, USA: 27-4574-01).
3. Cold lysis buffer (precooled to 4°C): 50 mM Tris(hydroxymethyl) aminomethane (stock 1 M), pH 7.5, 100 mM NaCl (stock 5 M), 1 mM ethylene glycol tetraacetic acid (EGTA) (stock 500 mM),

10% glycerol, 0.1% Triton X-100, 0.1% β-mercaptoethanol (BME), 1 mM phenylmethylsulfonyl fluoride (PMSF) (stock 200 mM in isopropanol), 50 μM *N*-acetyl-l-leucyl-l-leucyl-l-norleucinal (LLnL) (stock 50 mM in DMSO), 1 μM MG-132 (stock 50 mM in DMSO).

4. Roche protease inhibitor tablets (without ethylenediaminetetraacetic acid (EDTA)), two tablets per 50 mL lysis buffer.
5. BME, PMSF, LLnL, MG-132, and inhibitor tablets are added freshly.
6. Wash buffer I: 50 mM Tris, pH 7.5, 500 mM NaCl, 1 mM EGTA, 10% glycerol, 0.1% Triton X-100, 0.1% BME, 1 mM PMSF. BME and PMSF are added freshly.
7. Wash buffer II: 50 mM HEPES, pH 7.5, 100 mM NaCl, 1 mM EGTA, 10% glycerol, 0.1% BME, 1 mM PMSF. BME and PMSF are added freshly.
8. Elution buffer: 50 mM 4-(2-hydroxyethyl)-1-piperazineethanesulfonic acid (HEPES) (stock 1 M, pH 8.0), 100 mM NaCl, 30% glycerol, 40 mM glutathione (reduced form), 0.03% Triton X-100. Make sure the pH is exactly 8.0. Make freshly before use.
9. Vivaspin 500 concentration columns (Sartorius: VS0132).
10. ^{3}H-acetyl-CoA (GE Healthcare 250 μCi/mL: TRK688).
11. Anti-pan-acetyllysine antibody (Cell Signaling: 9681S).
12. nProtein A Sepharose beads (GE Healthcare: 17-5280-01).
13. 4–20% SDS-PAGE (Invitrogen: EC6028BOX).
14. Anti-GST antibody (Chemicon: AB3282).

3. Methods

3.1. Culture of Yeast Cells

1. Inoculate, yeast glycerol stocks harboring inducible *GAL1*-promoter driven, GST-tagged proteins that had been stored in 96-well plates at −80°C, onto a SC-Ura/glucose agar plate using a 96-pronger in the morning.
2. Allow the culture to grow on agar at 30°C for 48 h; visible colonies (2 mm diameter) should then be observed.
3. In the morning, use a 96-pronger to inoculate yeast cells from agar plates to a 96-well 2 mL box in which every well contains 1 mL SC-Ura/glucose liquid media and a 2 mm diameter glass ball, which facilitates the uniform growth.
4. After the culture reaches OD_{600} 6.0 in about 30 h (in the late afternoon) at 30°C with shaking (240 rpm), inoculate 5 μL of

the same strain into 6 mL of SC-Ura/raffinose liquid media in 12-channel boxes. Let the cells grow at 30°C with gentle shaking (90 rpm).

5. After growing overnight, the culture should reach OD_{600} 0.6–1.0. Discard the cultures if the OD_{600} is over 1.0. Using an automated plate-filling device add 300 μL of 40% galactose stock to each well to a final concentration of 2% to induce GST-tagged protein expression of the cells. Induce the cultures at 30°C for 4 h with gentle shaking.
6. Harvest the cells by spinning at 1,500 × *g* for 2 min, and resuspend the cell pellets in 800 μL of cold water (precooled to 4°C). Merge cells from 12-channel boxes into 96-well boxes. Collect cells by spinning, and store immediately in −80°C freezer. The culture can be kept for weeks before protein purification.

3.2. Protein Purification in a 96-Well Fashion

1. Transfer the frozen culture in a 96-well box from −80°C to ice and add 200 μL of zirconia beads (0.5 mm diameter) to each well. While the culture is still frozen, add 400 μL of lysis buffer containing fresh protease inhibitors. Use a cap mat to seal each well. After thawing the culture for 15 min on ice, vortex the cells in the 96-well box 1 min for 12 times with 1 min intervals on ice.
2. Meanwhile, wash the required amount of glutathione beads (20 μL of beads per sample) three times with cold lysis buffer without protease inhibitors, and resuspend 5× of its original volume with lysis buffer containing fresh protease inhibitors. Seal the bottom of deep-well filter plates with melted agar and add 100 μL of prewashed glutathione beads to each well.
3. After 12 times of vortexing, add 400 μL of lysis buffer containing protease inhibitors to each well of the culture plates. After spinning at 1,500 × *g* for 15 min, transfer the supernatant into the filter plates by a robot.
4. Put the filter plates on top of a 96-well box and seal tightly with aluminum foil.
5. Put the plates vertically on a platform shaker at 4°C for 40 min with gentle shaking.
6. Collect the beads and remove the liquid and seal agar by spinning at 750 × *g* for 3 min. Wash beads with 300 μL wash buffer I for three times in the filter plates using Qfill.
7. Wash the beads three times with 300 μL wash buffer II.
8. After complete removal of the buffer by spinning at 750 × *g* for 1 min, tightly seal the bottom of the filter plates with aluminum foil.
9. Seal the plates and store in a −80°C freezer.

10. In the morning before re-array, take the plates out of the −80°C freezer and add 50 μL of elution buffer to each well. Shake the plates vigorously on a platform shaker for 1 h at 4°C.
11. Collect the eluate to a 96-well PCR plate by spinning through the filter plate for 1 min at 1,500 × *g*.
12. Aliquot each purified protein into five 384-well plates using a robot and immediately store in a −80°C freezer.

3.3. Printing Proteome Microarrays

1. Adjust the relative humidity to 30% using a dehumidifier or other means (see Note 1).
2. Meanwhile, set up a printing program. The most important factor in high-density printing is spacing between features. For yeast printing, we use SMP3 pins from ArrayIt with column and row spacing 256 and 240 mm, respectively. After setting the parameters for printing, try a dry run without loading the pins to make sure that the program runs as expected.
3. Load pins into the print head, and make sure that the pins move freely by gravity. As a final check for the printing program and pins, load a 30% glycerol plate, and print a couple of slides. Make sure that each pin leaves glycerol spots on the blotting or preprinted slides.
4. Load slides into each slot in the microarray printer, and wait until humidity drops to 30% before starting printing.
5. After the printing is done, let the slides sit or cure in the printing station for at least 8 h to overnight so that protein binding is maximal. After the curing, the proteome microarray is ready to use. Cured proteome microarrays are stored in a −80°C deep freezer (using freshly made proteome chip, i.e., within 3 months of chip creation, is critical for successful acetylation proteome microarray experiment) (see Note 2).

3.4. Purification of Histone Acetyltransferase (HAT)

Day1

1. Take two tubes of Esa1-FLAG and Gcn5-FLAG cells (7.5 L) from −80°C; thaw on ice.
2. Place in 80 mL bead-beater chamber—make half-full with 0.5 mm beads and fill to mouth with extraction buffer + protease inhibitor (EB + PI).
3. Beat 10× 30 s with 1.5 min rest between pulses at 4°C on ice.
4. Centrifuge at 4°C, 10,270 × *g* for 30 min in GSA.
5. Transfer supernatant to Ti70 tubes.
6. Centrifuge at 4°C, 202,048 × *g* for 1 h and 10 min in SW 40 Ti Rotor.
7. Collect supernatant on ice.

8. Equilibrate 12.5 mL (bead volume = 25 mL with 50% slurry) anti-FLAG M2 resin, then wash 5× 50 mL EB, wash 1× 50 mL EB + PI, spin at 300 × *g* for 1 min at 4°C, and discard supernatant for each wash.
9. Add whole-cell extract to resin and incubate on rotator at 4°C overnight.

Day 2

1. Pour resin into 20 mL column and collect flow through—keep an aliquot.
2. Wash with 5× 50 mL EB + PI's. Make 3× FLAG peptide at a concentration of 0.4 mg/mL in EB + PI (e.g., add 25 mg FLAG peptide in 12.5 mL EB + PI).
3. Elute with 5× 12.5 mL 0.4 mg/mL FLAG peptide.
4. Equilibrate 6.25 mL Ni^{2+}-NTA agarose with 5× 25 mL EB, 1× 25 mL EB + PI; spin at 50 × *g* for 2 min at 4°C; and discard supernatant for each equilibration.
5. Incubate washed Ni^{2+}-NTA agarose with FLAG eluate overnight on rotator at 4°C.

Day 3

1. Pour resin into column and collect flow through, keep an aliquot.
2. Wash with 32 mL EB + PI.
3. Wash with 19 mL of 20 mM imidazole buffer (IB).
4. Elute with 19 mL of 300 mM imidazole buffer (IB).
5. Dialyze against 2 L of Mono Q A buffer for 6 h.
6. Freeze at −80°C (using freshly purified HAT, i.e., within 3 months of purification, is critical for successful acetylation proteome microarray experiment) (see Note 2).

3.5. In Vitro Acetylation on a Proteome Microarray

1. Take out slides from deep freezer, directly plunge into blocking solution as described above, and incubate for 1 h at RT with gentle rotation (see Note 2).
2. Briefly wash the slides with PBST.
3. Equilibrate the slides by washing two times with 1× HAT buffer (without DTT, HDAC inhibitors, or PMSF) at RT for 5 min each.
4. Meanwhile, assemble the reaction mixture.
5. Excess buffer is removed by gently tapping one side of the slide on a paper towel, and apply the reaction mixture evenly across the surface.

6. Using forceps, place lifter slip on the slide. Be careful not to introduce any bubbles. Incubate at 30°C for 1 h.
7. Wash the slide by flooding the slides three times with 50 mM $NaHCO_3$–Na_2CO_3 (pH 9.3).
8. Rinse the slide with PBS.
9. Spin at 712 × *g* for 3 min to dry the slide. Make sure that slides are completely dried.
10. Place the slide in an X-ray cassette, and put X-ray film on the slides with a direct contact. Expose the film as needed. It usually takes two or more weeks (see Note 1).
11. Develop the film. The film is scanned using an office scanner, and the image is processed using Photoshop as follows.
12. Set a HP Scanjet 8300 scanner at the high-sensitivity setting with a resolution of 4,800 dpi, and obtain chip images from the autoradiograph film.
13. Process the captured image of the autoradiograph film, and crop the image size as 6.4 × 4.4 cm and format as jpg, after magnification of image.
14. Open the image in Adobe Photoshop and click the image in order to edit the image.
15. Edit the image with gray scale in 16-bit/channel mode.
16. Invert image (under Adjustments tab).
17. Compress the image to 53.3% with percentage in image size.
18. Save the image in TIFF format for further analysis.
19. Open the image in GenePixPro 6.0 and load the gal file containing each protein's ID.
20. Align the setting with gal file based on each spot from anti-GST profiling and positive control sets.
21. Adjust the size of the grid based on sizes or form of signals on chip.
22. Quantify the signal intensities using GenePixPro 6.0 by calculating the background and foreground signals on each spot.

3.6. In Vitro Acetylation Assays

1. Purify GST-tagged substrates as described in Subheadings 3.1 and 3.2, with slight modification in scale-ups. In brief, grow cells in 0.5 L of SC-Ura/raffinose from OD ~ 0.1 until mid-exponential phase at 30°C. Add a final concentration of 2% galactose to induce protein expression at 30°C for 4 h. Obtain whole-cell extract by breaking cells in cold lysis buffer containing protease inhibitors and HDACs inhibitors. Expand the extract to 100 mL with cold lysis buffer and incubate with 3 mL of glutathione beads (which have been washed three

times with 30 mL of lysis buffer) at 4°C for 1 h with rotation. After binding, wash the glutathione beads four times with 10 mL of wash buffer I, followed by four times of wash with 10 mL of wash buffer II. Incubate the washed beads with 3 mL of elute buffer at 4°C for 1 h with rotation. Centrifuge and collect the eluate, and concentrate the eluted protein to final concentration of 0.1–0.5 μg/μL by ultrafiltration with Vivaspin 500 concentration columns. Determine the final protein concentration by a nanodrop analyzer using the absorbance of the sample at 280 nm.

2. Incubate 25 μL reaction mixture containing ~8 μg of GST-tagged substrates, 1 μL of ^{3}H-acetyl-CoA, 100 μM TSA, 5 mM nicotinamide, 5 mM PMSF and 5 mM DTT in HAT buffer (final 50 mM Tris–HCl, pH 7.5, 5% glycerol, 0.1 mM EDTA, 50 mM KCl), and ~3 μg of purified NuA4 complex for 1 h at 30°C.
3. Analyze the signals of acetylation by scintillation counting.
4. Alternatively, immunoprecipitate the reaction mixture by nProtein A Sepharose beads conjugated with anti-pan-acetyllysine antibody at 4°C for 3 h.
5. Resolve the product by 4–20% SDS-PAGE, and analyze by immunoblotting using a rabbit polyclonal anti-GST antibody (see Notes 3 and 4).

4. Notes

4.1. Make Proteome Microarray

Printing proteome microarrays is quite challenging even though similar equipment has been developed for construction of the DNA microarrays (see Subheading 3.3). This is because fabrication of a functional protein microarray prefers a low ambient temperature (e.g., 4°C), while certain amount of glycerol has to be added to purified proteins as a means to maintain them in aqueous phase during printing. However, glycerol in printed protein spots may cause absorbing of ambient water, which can result in merged features in high-density printing. Thus, formulation of solutions for printing requires that several aspects be considered. To reduce the merged features, printing is done at lower humidity (~30%). We have used both the VersArray (BioRad) and NanoPrinter™ (ArrayIt) microarrayers to print proteome microarrays. The NanoPrinter has a built-in humidity monitor and dehumidifier. Printing pins are also an integral part of the printing process; we use SMP3 or 946MP2 pins from ArrayIt for routine printing.

4.2. Use Acetylation Proteome Microarray to Find NuA4 Substrates

We identified substrates of the NuA4 complex using freshly made yeast proteome chips as described above (see also Subheadings 3.1–3.3) and also freshly purified HAT enzyme (see Subheading 3.4) (4, 16). Acetylated proteins are detected by radiolabeled (^{14}C)-acetyl-CoA as donor reagents. One main advantage of using radioisotopes as the labeling reagent is that it allows detection of de novo modifications by the HAT enzyme (rather than the yet acetylated ones) in the reaction mixture (see Subheading 3.5).

We also performed pilot experiments to check the specificity of HAT enzymes. First, we developed an in vitro acetylation reaction on a glass chip containing histones H3, H4, and BSA. To optimize the reaction conditions and determine the effects of surface chemistry on the sensitivity and specificity of the assay (3), we performed the assay using both the NuA4 and Spt-Ada-Gcn5-acetyltransferase (SAGA) complexes and (^{14}C)-acetyl-CoA as a labeling reagent on three surfaces at various reaction temperatures and time. As illustrated in Fig. 1a, on the FAST (nitrocellulose) slide, the NuA4 and SAGA complexes showed preference for histones H4 and H3, respectively, as expected (22). No detectable signals were observed for BSA. We then carried out acetylation reactions in duplicate with the NuA4 complex on yeast proteome chips containing 5,800 proteins (4). The acetylation signals were visualized by autoradiography after exposure to X-ray film for ~20 days, scanned and saved as TIFF files. The TIFF images were processed by assigning an artificial green color to each brightened spot and analyzed using GenePix software to determine relative acetylation levels. After subtraction of backgrounds, protein spots with median signal intensity greater than 3,000 were arbitrarily chosen as the candidate substrates of NuA4 for further validation studies (see Subheading 3.5).

We found that 91 proteins were readily acetylated by NuA4 (Fig. 1b). Gene ontology (GO) analysis indicated that these candidate substrates were enriched for nucleosome assembly and histone binding activities and depleted for mitochondrial- and membrane-associated proteins, as expected. However, significant proportions ($P < 10^{-4}$) of the candidate substrates are cytoplasmic proteins such as Pck1p. Moreover, no significant enrichment of known protein–protein interactions was found among the NuA4 components and the substrate candidates.

We first tested the ability of the NuA4 complex to acetylate these candidate substrates using standard in vitro acetyltransferase activity assays (23) and confirmed that many substrates, including Pck1p, Cdc34p, Prp19p, Tap42p, Gph1p, and Sip2p, are readily acetylated by NuA4 as detected by immunoprecipitation with a monoclonal anti-pan-acetyllysine antibody (24) (Fig. 1c). We further examined the acetylation status of these proteins in vivo and

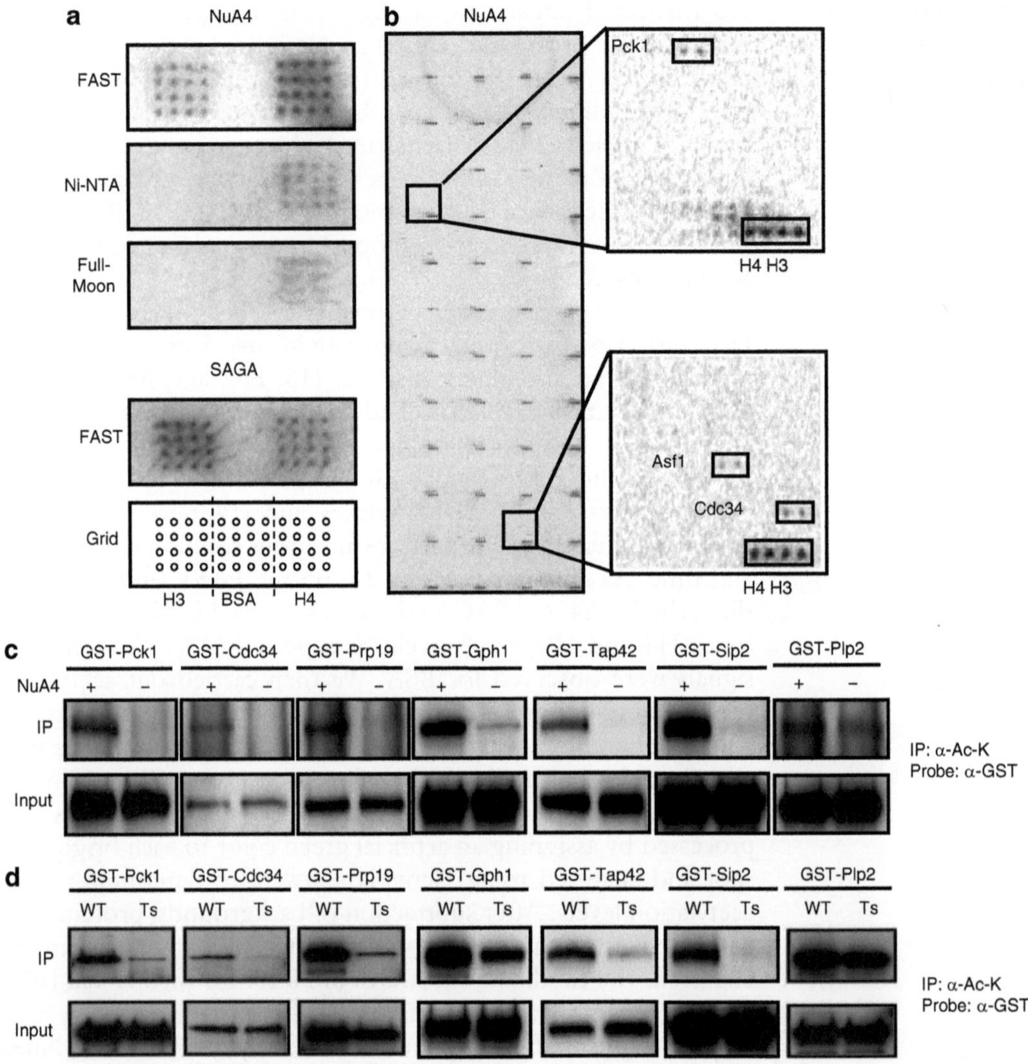

Fig 1. Yeast proteome microarray identifies NuA4 substrates. (**a**) Different slide surface chemistries were compared to determine the best surface for acetylation assays. nitrocellulose surface, nickel-coated surface. (**b**) Acetylation reactions on a yeast proteome microarray on FAST surface. (**c**) In vitro validation of candidate substrates. GST fusion proteins were isolated and reacted with/without (+/−) NuA4 in vitro and immunoprecipitated with anti-pan-acetyllysine (α-Ac-K) followed by immunoblotting with anti-glutathione-S-transferase (α-GST). (**d**) Identification of five in vivo substrates of the NuA4 complex by immunoprecipitation of extracts of yeast cells (previously transformed with the indicated GST fusion construct) with anti-acetyllysine (α-Ac-K) followed by immunoblotting with anti-glutathione-S-transferase (α-GST).

found that all the in vitro substrates were also acetylated in wild-type (WT) cells; their acetylated fractions dramatically diminished in an *ESA1* temperature-sensitive (*esa1-531*) mutant at nonpermissive temperature (Fig. 1d).

4.3. NuA4 Controls Pck1 to Regulate Gluconeogenesis and Chronological Life Span

Among these validated substrates, *PCK1* encodes a well-characterized phosphoenolpyruvate carboxykinase (PEPCK) catalyzing the rate-limiting step in gluconeogenesis by conversion of oxaloacetate and ATP to phosphoenolpyruvate, ADP, and CO_2. The only post-translational modification reported for Pckl is phosphorylation (25). Therefore, we first focused on Pckl to investigate acetylation-mediated effects (see Subheading 3.6).

We found that Pckl acetylation, which is regulated by NuA4 and the counteracting Sir2 deacetylase, enhances its enzyme activity and extends yeast chronological life span, possibly through the augmentation of alcohol handling capacity (17).

4.4. NuA4 Acetylation of Yeast AMP-Activated Protein Kinase Controls Intrinsic Aging Independently of Caloric Restriction

We further showed that NuA4 catalytic mutants have a replicative life-span defect, and AMPK regulatory subunit, *SIP2*, a known replicative life-span regulator and a known NuA4 substrate (17), is involved in NuA4-mediated replicative life-span regulation (see Subheading 3.6).

Sip2 is one of the three regulatory β subunit of yeast AMPK. We found that Sip2 was acetylated and deacetylated by NuA4 and Rpd3, respectively (19). Sip2 acetylation enhances its interaction with the AMPK catalytic subunit, Snfl, and inhibits the kinase activity. Interestingly, Sip2 acetylation declined significantly with aging. Restoration of Sip2 acetylation by deleting the deacetylase Rpd3 or using acetylation mimics slows cellular growth, strengthens the resistance to aging-related oxidative damage, and increases life span. These pro-longevity effects are through suppression of AMPK catalytic activity and its subsequent activation of downstream Sch9 (19).

Our findings suggest that aging, mediated by the NuA4 acetyltransferase and counteracted either by Sir2 or Rpd3 deacetylases, regulates the acetylation of Pckl and Sip2, which in turn changes the gluconeogenesis enzymatic activity and reduces Snfl inhibition by Sip2 and results in life-span shortening.

4.5. Using Functional Proteome Array to Study Aging

Recent years have witnessed a rapid growth in using functional proteome microarrays for basic research. Although the technology is still at a relatively early stage of development, it has become obvious that the proteome microarray platform can and will act as a versatile tool suitable for the large-scale, high-throughput interrogation of specific biologic functions, especially in the areas of profiling PTMs and analyzing signal transduction networks and pathways. As another crucial proteomics technology, recent progress in mass spectrometry has allowed global profiling of PTMs using a shotgun approach. For example, the Zhao, Mann, and Guan groups recently identified numerous acetylated lysine residues in metabolic enzymes in mice and human cells without knowing the upstream acetyltransferases (26–28). In parallel, our team identified many yeast metabolic enzymes as substrates of the NuA4 acetylation complex and bacterial transcription factors as substrates of the Gcn5-like protein acetyltransferase (Pat) by using an *E. coli* K 12 proteome microarray and further checked the actual modification sites using mass spectrometry (18). Therefore, we envision that the combination of the two proteome technologies will have enormous potential both to identify critical regulatory PTMs at the resolution of modified individual amino acids and to identify the enzymes that mediate these effects.

The best example is the application of functional proteome microarrays to study acetylation pathways, which has set a new paradigm for aging research. In the past, most pro-longevity mutations were discovered by systematic gene deletion or RNA interference screens, which mainly revealed abolished or diminished gene functions (29). In our recent publications (17, 19), we used global acetylation proteome microarray screens to study yeast aging and proved that enhanced function of certain genes through PTMs can also lead to longevity. Apparently functional proteome microarray has intensified the research width and depth and helps understand integrative and comprehensive basic biological functions and their relationship to cellular survival and longevity.

Acknowledgments

The work was supported by National Science Council (NSC 98-2314-B-002-031-MY3 to J.-Y.L.), National Taiwan University Hospital (099-001376 to J.-Y.L.), National Taiwan University (99C101-603 to J.-Y.L. and Y.-Y.L.), Liver Disease Prevention & Treatment Research Foundation (J.-Y.L. and Y.-Y.L.), Taiwan, and the NIH Common Fund Grant (U54-RR020839 to H.Z. and J.D.B.), USA.

References

1. Smith MG, Jona G, Ptacek J, Devgan G, Zhu H, Zhu X, Snyder M (2005) Global analysis of protein function using protein microarrays. Mech Ageing Dev 126:171–175
2. Chen C, Zhu H (2006) Protein microarrays. Biotechniques 40:432–439
3. Tao SC, Chen CS, Zhu H (2007) Applications of protein microarray technology. Comb Chem High Throughput Screen 10:706–718
4. Zhu H, Bilgin M, Bangham R, Hall D, Casamayor A, Bertone P, Lan N, Jansen R, Bidlingmaier S, Houfek T, Mitchell T, Miller P, Dean RA, Gerstein M, Snyder M (2001) Global analysis of protein activities using proteome chips. Science 293:2101–2105
5. Hall DA, Zhu H, Zhu X, Royce T, Gerstein M, Snyder M (2004) Regulation of gene expression by a metabolic enzyme. Science 306: 482–484
6. Ho SW, Jona G, Chen CT, Johnston M, Snyder M (2006) Linking DNA-binding proteins to their recognition sequences by using protein microarrays. Pro Natl Acad Sci USA 103(26): 9940–9945
7. Hu S, Xie Z, Onishi A, Yu X, Jiang L, Lin J, Rho HS, Woodard C, Wang H, Jeong JS, Long S, He X, Wade H, Blackshaw S, Qian J, Zhu H (2009) Profiling the human protein–DNA interactome reveals ERK2 as a transcriptional repressor of interferon signalling. Cell 139(3):610–622

8. Zhu J, Gopinath K, Murali A, Yi G, Hayward SD, Zhu H, Kao C (2007) RNA binding proteins that inhibit RNA virus infection. Proc Natl Acad Sci USA 104:3129–3134
9. Huang J, Zhu H, Haggarty SJ, Spring DR, Hwang H, Jin F, Snyder M, Schreiber SL (2004) Finding new components of the target of rapamycin (TOR) signaling network through chemical genetics and proteome chips. Proc Natl Acad Sci USA 101:16594–16599
10. Zhu H, Klemic JF, Chang S, Bertone P, Casamayor A, Klemic KG, Smith D, Gerstein M, Reed MA, Snyder M (2000) Analysis of yeast protein kinases using protein chips. Nat Genet 26:283–289
11. Kafadar KA, Zhu H, Snyder M, Cyert MS (2003) Negative regulation of calcineurin signaling by Hrr25p, a yeast homolog of casein kinase I. Genes Dev 17:2698–2708
12. Ptacek J, Devgan G, Michaud G, Zhu H et al (2005) Global analysis of protein phosphorylation in yeast. Nature 438:679–684
13. Tao SC, Li Y, Zhou J, Qian J, Schnaar RL, Zhang Y, Goldstein IJ, Zhu H, Schneck JP (2008) Lectin microarrays identify cell-specific and functionally significant cell surface glycan markers. Glycobiology 18:761–769
14. Zhu J, Liao G, Shan L, Zhang J, Chen MR, Hayward GS, Hayward SD, Desai P, Zhu H (2009) Protein array identification of substrates of the Epstein-Barr Virus protein kinase BGLF4. J Virol 83:5219–5231
15. Kung L, Tao SC, Qian J, Smith M, Snyder M, Zhu H (2009) Global analysis of the glycoproteome in S. cerevisiae reveals new roles for protein glycosylation. Mol Syst Biol 5:308
16. Lu JY, Lin YY, Tao SC, Zhu J, Pickart CM, Qian J, Zhu H (2008) Functional dissection of a HECT ubiquitin E3 ligase. Mol Cell Proteomics 7:35–45
17. Lin YY, Lu JY, Zhang J, Walter W, Dang W, Wan J, Tao SC, Qian J, Zhao Y, Boeke JD, Berger SL, Zhu H (2009) Protein acetylation microarray reveals NuA4 controls key metabolic target regulating gluconeogenesis. Cell 136:1073–1084
18. Thao S, Chen CS, Zhu H, Escalante-Semerena JC (2010) Nε-lysine acetylation of a bacterial transcription factor inhibits Its DNA-binding activity. PLoS One 5(12):15123
19. Lu JY, Lin YY, Sheu JC, Wu JT, Lee FJ, Chen Y, Lin MI, Chiang FT, Tai TY, Berger SL, Zhao Y, Tsai KS, Zhu H, Chuang LM, Boeke JD (2011) Acetylation of AMPK controls intrinsic aging independently of caloric restriction. Cell 146:969–979
20. Oh YH, Hong MY, Jin Z, Lee T, Han MK, Park S, Kim HS (2007) Chip-based analysis of SUMO (small ubiquitin-like modifier) conjugation to a target protein. Biosens Bioelectron 22(7):1260–1267
21. Del Rincón SV, Rogers J, Widschwendter M, Sun D, Sieburg HB, Spruck C (2010) Development and validation of a method for profiling post-translational modification activities using protein microarrays. PLoS One 5(6): e11332
22. Sterner DE, Berger SL (2000) Acetylation of histones and transcription-related factors. Microbiol Mol Biol Rev 64:435
23. Smith ER, Eisen A, Gu W, Sattah M, Pannuti A, Zhou J, Cook RG, Lucchesi JC, Allis CD (1998) ESA1 is a histone acetyltransferase that is essential for growth in yeast. Proc Natl Acad Sci USA 95:3561–3565
24. Li Y, Yokota T, Gama V, Yoshida T, Gomez JA, Ishikawa K, Sasaguri H, Cohen HY, Sinclair DA, Mizusawa H, Matsuyama S (2007) Bax-inhibiting peptide protects cells from polyglutamine toxicity caused by Ku70 acetylation. Cell Death Differ 14:2058–2067
25. Burlini N, Lamponi S, Radrizzani M, Monti E, Tortora P (1987) Identification of a phosphorylated form of phosphoenolpyruvate carboxykinase from the yeast *Saccharomyces cerevisiae*. Biochim Biophys Acta 930: 220–229
26. Kim SC, Sprung R, Chen Y, Xu Y, Ball H, Pei J, Cheng T, Kho Y, Xiao H, Xiao L, Grishin NV, White M, Yang XJ, Zhao Y (2006) Substrate and functional diversity of lysine acetylation revealed by a proteomics survey. Mol Cell 23:607–618
27. Choudhary C, Kumar C, Gnad F, Nielsen ML, Rehman M, Walther TC, Olsen JV, Mann M (2009) Lysine acetylation targets protein complexes and co-regulates major cellular functions. Science 325(5942):834–840
28. Zhao S, Xu W, Jiang W, Yu W, Lin Y, Zhang T, Yao J, Zhou L, Zeng Y, Li H, Li Y, Shi J, An W, Hancock SM, He F, Qin L, Chin J, Yang P, Chen X, Lei Q, Xiong Y, Guan KL (2010) Regulation of cellular metabolism by protein lysine acetylation. Science 327(5968): 1000–1004
29. Vijg J, Campisi J (2008) Puzzles, promises and a cure for ageing. Nature 454(7208): 1065–1071

Chapter 13

Quantitation of Nucleosome Acetylation and Other Histone Posttranslational Modifications Using Microscale NU-ELISA

Bo Dai, Charles Giardina, and Theodore P. Rasmussen

Abstract

Histone posttranslational modifications (PTMs) are highly important molecular determinants of epigenetic regulatory mechanisms. Histone PTMs associated with nucleosomes are intimately tied to the transcriptional activity or silence of genes. In addition, nucleosomal PTMs participate in the organization of chromatin into higher-order structures and the progression through mitosis. Changes in histone PTMs are also regulated during the course of mammalian development and are altered in pathological states including cancer. Histone acetyl modifications (and also methylation and phosphorylation) are frequently assayed by western blotting (WB), mass spectrometry (MS), and chromatin immunoprecipitation (ChIP). Here we show that an enzyme-linked immunosorbent assay performed on nucleosomes (NU-ELISA) can quickly and effectively yield quantitative detection of global levels of histone acetylation on small samples such as single human embryonic stem cell colonies. The microscale NU-ELISA method presented here can be performed in most laboratories equipped with basic instrumentation for molecular and cellular biology.

Key words: Nucleosome, ELISA, Acetylation, Posttranslational modifications (PTMs), Embryonic stem cells (ESCs), Single-colony assay

1. Introduction

Histone acetylation is a posttranslational modification that can respond dynamically to changes in cellular growth status, development and differentiation, environmental factors, and disease state. Although promoter-specific modifications dictate whether a particular gene is turned on or off, evidence has been obtained that promoter-specific modifications occur on a global acetylation background that may broadly influence which genes can be expressed. Analysis of these global changes can provide important information

Sandra B. Hake and Christian J. Janzen (eds.), *Protein Acetylation: Methods and Protocols*, Methods in Molecular Biology, vol. 981, DOI 10.1007/978-1-62703-305-3_13, © Springer Science+Business Media, LLC 2013

on how a cell is poised to respond to stimuli. However, we presently do not have a good understanding of how shifts in chromatin structure due to large-scale acetylation changes affect gene-specific responses. A better understanding of this level of gene regulation has implications in a wide range of disciplines, ranging from developmental biology to cancer biology. Understanding what global acetylation changes mean is also of increasing interest as pharmacological agents with unique histone deacetylase (HDAC) inhibitory profiles continue to be developed (1–3).

The acetylation status of histones can vary from one cell type to another, and evidence has been obtained that this modification is a key feature of many developmental processes. Embryonic stem cells (ESCs) can be distinguished from some differentiated cell types based on the acetylation of specific lysine residues on histone H3 (4). Moreover, during the process of ESC differentiation, global changes in acetylation occur as the cell makes large-scale changes in its gene expression profile (4). Running the developmental program in reverse (cellular reprogramming) also appears to involve changes in chromatin acetylation "tone," since the efficiency of induced pluripotency stem cell (iPS) production can be increased by HDAC inhibition (5–7). Accurately assessing histone acetylation status could therefore facilitate the development of numerous programming and reprogramming protocols.

One venue in which global histone acetylation may be particularly important is cancer. HDACs are among the most commonly upregulated genes in solid tumors, and their expression has been linked to poor patient outcome (8, 9). HDAC overexpression can induce specific changes in global histone acetylation status, which interfere with cell differentiation and growth control (10). In prostate cancer, the level of acetylation at histone H3 lysine 9 (H3K9ac) and histone H4K12ac (among other modifications) tracks closely with tumor grade, raising the possibility that these modifications could have prognostic value (11, 12). Understanding aberrations in the global acetylation status of neoplastic lesions could potentially provide important information on the types of inhibitors that might be the most effective cancer treatments. Present trends also suggest an increasing interest in histone acetylation in a wide range of fields, well beyond developmental biology and cancer. Nutritional agents, inflammatory responses, and other pathophysiological conditions appear to be accompanied by global shifts in histone acetylation status (13–15). Understanding what these changes mean and how they might be exploited therapeutically is an important research challenge.

Methodologies that can quantify the extent of histone acetylation in a sensitive and straightforward manner are likely to have a wide range of applications, potentially in clinical as well as academic labs. Of particular interest are assays that can distinguish between different modification states; although histone acetylation is thought to generally serve to derepress genes, numerous other reports have highlighted the importance of specific residue acetylations. For this reason, we developed the NU-ELISA approach, a simple quantitative and sensitive approach for monitoring the status of histone acetylation and other PTMs.

2. Materials

2.1. Cell Culture

1. Fibroblast culture medium: 500 mL Dulbecco's modified Eagle's medium (DMEM), 50 mL fetal bovine serum (FBS) (heat inactivated at 56°C), 5 mL nonessential amino acid (NEAA) solution, 5 mL penicillin/streptomycin (P/S) (100×), and 4 μL β-mercaptoethanol.
2. Human ESC culture medium: 200 mL DMEM/F-12, 50 mL knockout serum replacer, 2.5 mL nonessential amino acids (100×), 1.25 mL 200 mM l-glutamine, 1.75 μL β-mercaptoethanol, 500 μL 2 μg/mL basic fibroblast growth factor (bFGF) (stock solution, 10 μg basic fibroblast growth factor (bFGF) in 5 mL 0.1% bovine serum albumin (BSA) in PBS).

2.2. Nucleosome Preparation Materials

1. Micrococcal nuclease (MNase) buffer: 10 mM Tris–Cl pH 7.5, 4 mM $MgCl_2$, and 1 mM $CaCl_2$.
2. MNase: purchased as powder and diluted in MNase buffer to 2 units per 10 μL, aliquots stored at –20°C.
3. Stem cell cutting tools (preferably Vitrolife cat #14601).
4. Dry ice and isopropanol.

2.3. ELISA Components

1. ELISA plates: 96-well plates (preferably NUNC MaxiSorp™).
2. Plate covers: transparent plastic adhesive-coated covers for 96-well plates.
3. Coating buffer: 32 mM Na_2CO_3 and 68 mM $NaHCO_3$.
4. Blocking buffer: 1× phosphate-buffered saline (PBS: 1.06 mM KH_2PO_4, 2.97 mM Na_2HPO_4 $7H_2O$, 155.17 mM NaCl), with 0.05% Tween-20 and 5% BSA.
5. Primary antibody (Ab): acetylation-specific antibodies such as mouse/rabbit anti-acetyl-histone H3 (lysine 9) (H3K9ac) and anti-acetyl-histone H3 (lysine 18) (H3K18ac) and loading control antibodies such as mouse/rabbit anti-histone H2A, H2B, H3, or H4 (these must recognize histones independently of their modification status). Primary antibodies are diluted 1:1,000 in blocking buffer prior to use.
6. Secondary Ab: goat anti-mouse/rabbit Ab (appropriately matched for the species of the primary Ab), diluted 1:5,000

(or even higher dilution as determined empirically) in blocking buffer.

7. Washing buffer: 1× PBS and 0.5% Tween-20, prepared in bulk amount to supply the plate washer.
8. Colorimetric substrate: 3,3′,5,5′-tetramethylbenzidine (TMB). We use 1-Step™ TMB-ELISA (catalog number: 34028; Pierce, Rockford, IL, http://www.piercenet.com) for our assays.
9. Stop reagent: 2 N H_2SO_4.
10. Plate washer: most microtiter plate washers are suitable.
11. Plate reader: a plate reader that can read a 96-well plate at the wave length of 450 nm.

2.4. Nucleosome Quality Control

1. DNA gel analysis: agarose, 100 bp DNA ladder, electrophoresis equipment, phenol, chloroform, ethanol, and gel imaging system.
2. NanoDrop® spectrophotometer.

3. Methods

Note: We developed microscale NU-ELISA for assessment of single hESC colonies. If you are using other cell types, use the normal culture methods appropriate for your cells.

3.1. Human Embryonic Stem Cell Culture

1. Plate mitotically inactivated (4,000 rads) mouse embryonic fibroblast (MEF) feeders.
   Coat a 6-well plate with 1 mL/well of 0.1% gelatin solution overnight at 37°C. Make sure the surface is covered with gelatin. Aspirate the residual gelatin solution immediately before plating feeders. Thaw one vial of mitotically inactivated MEF feeders and transfer the cells to a 15 mL conical tube. Add MEF media up to 10 mL, and centrifuge at 200 × *g* for 5 min. Resuspend the cell pellet in a small volume of MEF media, count the cells, and adjust the cell concentration to 1×10^5/mL. Plate the feeders at 2 mL/well (feeders should be used at a density of 20,000 cells per cm^2 or 1.1×10^6 per 6-well plate). Incubate at 37°C overnight.
2. Grow hESCs on MEF feeders.
   On the second day, thaw one cryovial of hESCs at 37°C carefully, and transfer the contents to a 15 mL conical tube immediately. Gently add 9 mL of hESC media to the tube and mix. Centrifuge at 200 × *g* for 5 min. Aspirate MEF media from the feeder plate, wash the surface with 1× PBS, and aspirate. Add 1 mL hESC media and put the plate back to the incubator.

Add 1 mL hESC media to the pellet and gently mix. While breaking up the pellet, be careful not to disaggregate cell clumps completely, as hESCs will not plate as efficiently as single cells. Pipette the hESCs into washed feeder wells. Tilt plate from front to back and side to side to evenly distribute the cell clumps, and incubate overnight without disturbing, to allow attachment. On the following day after a thaw, it is normal to observe some unattached (floating) cells. On subsequent days, replace the wells with fresh, pre-warmed hESC media daily. After thawing hESCs may take over a week to form colonies large enough to passage. After the second or third split from thaw, cells can be passaged about once a week.

3. Remove nonspecific differentiation from cultures.
Monitor the cell culture daily, and remove any differentiated cells with a pulled Pasteur pipette by gently scraping the differentiated cells so that they detach from the plate and are floating in the media and then aspirate the media, wash with 1× PBS, and replace with fresh hESC media.

3.2. Rapid Nucleosome Chromatin Prep from Small Samples (Microscale or Single Colonies)

1. Grow hESCs for up to 1 week in feeder-free media in a 6-well plate, or grow hESCs on feeders for single colony pickup (see Note 1).
2. Aspirate spent media and wash three times with 1 mL/well 1× PBS.
3. Aspirate the PBS, flash freeze the plate on dry ice soaked in isopropanol, and then put the plate on dry ice for at least 1 h (or overnight at −80°C after the initial dry ice/isopropanol freezing) (see Note 2).
4. Thaw the plate for 1 min at room temperature and then add 50 μL MNase buffer per well and detach the cells with a cell scraper.
5. Collect the scraped cells in a 1.5 mL microcentrifuge tube, add MNase to 5 U/mL, and incubate at 37°C for 12 min, with occasional pipetting.
6. Chromatin can be prepared from single hESC colonies by detaching them with a stem cell cutting tool. Then transfer the colony directly into a 1.5 mL microcentrifuge tube, flash freeze as described above, thaw, and proceed to the MNase digestion as described above.
7. Stop the MNase digestion by adding 1/50 volume of 0.5 mM Na-EDTA, and put the tube on ice.
8. Take out a small aliquot and test DNA concentration at A_{260} with a regular spectrophotometer or NanoDrop® as needed. Save samples for NU-ELISA. Add Tris–HCL, pH 7.5, to 25 mM prior to storage of the samples at −80°C in aliquots. Samples prepared and stored in this way are now ready for ELISA.

3.3. Nucleosome ELISA Assay

1. Coat MaxiSorp™ (Nunc, NY) plates overnight at 4°C or at room temperature for 1 h on a rotator with 50 μL/well nucleosomal chromatin (use fresh samples or frozen samples from −80°C) diluted in coating buffer. Load serial twofold dilutions of nucleosomes from the top row to the bottom row by adding 0.1, 0.05, 0.025, 0.0125, 0.00625, 0.00313, 0.00156, and 0 μg of chromatin (*Note*: Plate covers are needed for this step) (see Note 3).
2. Wash the plates four times for 2 min with 200 μL/well PBS/0.5% Tween-20 at room temperature using an automated plate washer.
3. Block 1 h at 37°C with 100 μL/well PBS/0.05% Tween-20/5% BSA.
4. Remove the blocking buffer, cover the plate, and store at −20°C or go directly to the next step.
5. Add 1° antibodies 50 μL/well diluted 1:1,000 (or as needed) in PBS/0.05% Tween-20/5% BSA and incubate at room temperature for 1 h on an orbital rotator (see Note 4).
6. Wash plates four times for 2 min with 200 μL/well PBS/0.5% Tween-20.
7. Add horseradish peroxidase (HRP)-conjugated 2° antibodies, diluted 1:5,000 (or as appropriate) in PBS/0.05% Tween-20/5% BSA and incubate at room temperature for 1 h on an orbital rotator.
8. Wash plates four times for 2 min with 200 μL/well PBS/0.5% Tween-20.
9. Develop the plates by adding 50 μL/well TMB, incubate for 10 min at RT, and stop the reaction by adding 2 N H_2SO_4. Immediately read the plates with a Bio-Rad microplate reader (or equivalent) at 450 nm (see Note 5). An example of a completed plate is shown in Fig. 1.
10. Export the readings to Excel spreadsheet files. Finished plates may be scanned at 300 dpi using a flatbed scanner.

3.4. Statistical Analysis

1. The raw absorbance values from a plate reader should be in an Excel spreadsheet file. We first subtract every value by the mean of the bottom row, which we defined as background (no nucleosome control) wells.
2. The relative acetylation level (or levels of other PTMs) is determined by the background-corrected value divided by chromatin loading control value (see Note 6) obtained from an identically loaded plate assessed with a general anti-histone Ab that detects one of the four canonical histones independently of its modification status, from the same row and same well. Use values in the linear response range of the assay (see Note 7).

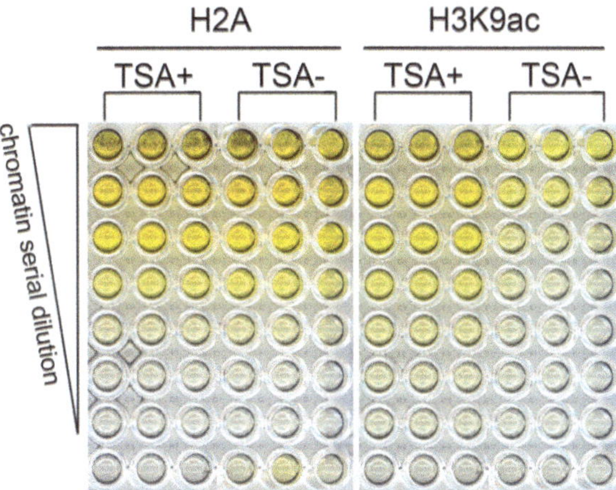

Fig. 1. Representative results of NU-ELISA on TSA-treated mouse ESCs. NU-ELISA assay performed on serially diluted nucleosomes from J1 mouse ESCs that were treated with the histone deacetylase inhibitor trichostatin A (+TSA) at 10 ng/mL or mock-treated (–TSA) for 24 h probed with 1° Ab with general specificity for histone H2A (*left*) as loading controls or specific for acetylation on histone H3 lysine 9 (H3K9ac) content (*right*). Chromatin was prepared in triplicate from independent cultures that were treated or mock treated, and serial twofold dilutions of each sample were coated in vertical columns.

3. A student's *t*-test can be performed to compare two samples. To compare multiple samples, ANOVA is recommended.
4. Graphs can be made based the average of three biological replicates (mean) and standard deviation.

4. Notes

1. The ELISA portion of this method can also be applied to a large-scale cell culture and nucleosome preparations as previously described (4, 16), whereby one can make a series of identically loaded ELISA plates to test a chromatin preparation for a series of PTMs. The microscale method described here is useful when colonies are not homogenous (as is common with hESCs) or when large amount of cells are not available. We have successfully isolated 25 μg of chromatin from a single hESC colony, which can make 125 NU-ELISA assays or coat up to ten plates using twofold dilution series.
2. The microscale NU-ELISA method is also useful for small cell cultures: that is, one well of a 6-well plate, a 24-well plate, or even a 96-well plate. The freeze–thaw method disrupts the cell and nuclear membranes and releases chromatin into the buffer. Addition of 2 U MNase into each sample at 37°C helps to

digest the chromatin into free mono-nucleosomes. After quenching the reaction with 0.5 M Na-EDTA, the crude nucleosome prep can be immediately loaded on an ELISA plate or stored at −80°C for later use.

3. The microscale NU-ELISA method can quantitatively detect histone acetylation on a single hESC colony or even portions of a colony (Fig. 2). However, this method can also be used to interrogate other histone PTMs, such as methylation, phosphorylation, or even histone variants such as macroH2A (4).

4. Since this method is an Ab-based assay, the use of characterized Abs is essential, and it is preferable to choose antibodies that are validated for use in ELISA as specified by the manufacturer. A good loading control Ab is very important for normalization. We screened several histone loading control antibodies, including anti-H2A, H2B, H3, and H4, and found a rabbit polyclonal anti-H2A that works well (IMGENEX IMG-358).

5. After adding the colorimetric substrate TMB, the absorbance reading is time sensitive. Therefore, it is essential to standardize the stop time by adding H_2SO_4. We choose to stop the reaction at 10 min for all assays. After colorimetric development of the plates, air bubbles can produce faulty absorbance readings. A 2 min centrifugation of the plates at 450×*g* will get rid of most of the air bubbles. A needle or pipette tip can also be used to remove air bubbles from a small number of wells.

6. It is important to use assay results from wells that are within the linear range of the assay. In practice, it is very useful to plot the readings of the twofold dilution curve and to choose a well that is midpoint within the linear ELISA range. To quantitatively assess histone acetylation levels for a cell line, we prepare three independent biological replicates (i.e., three independent cell cultures), and for each of these we prepare three technical replicates (three columns of loaded chromatin from the same sample). We routinely find that the technical replicates are highly consistent. Sometimes a spurious signal occurs in one well of an ELISA plate (which we called "black eyes"). If these are noted, then it is best to repeat the assay or move one row up or down, so long as the adjacent well is still in the linear range.

7. To assess global levels of acetylation or PTM levels in a specific sample, we prepare two identically loaded plates and probe one with a loading control Ab (H2A) and the other one with the acetylation-specific Ab (or other PTM Ab). The overall acetylation levels can then be expressed as a ratio of the PTM signal to the overall H2A signal. If one desires to assess 5 PTM levels, for example, then a minimum of six identically loaded plates must be prepared, reserving one for the H2A loading control assay. After normalization, the PTM levels from different

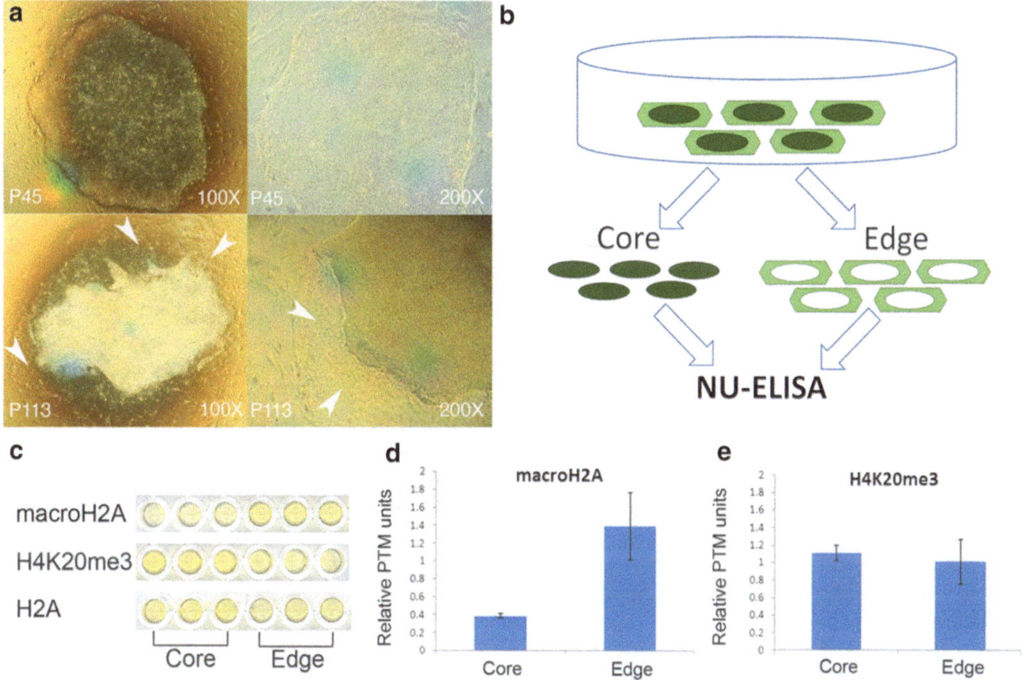

Fig. 2. Use of microscale NU-ELISA to investigate the epigenetic status of outgrowths from very-high-passage hESC line H9. (**a**) Morphology of a single colony of hESC line H9 at passage 45 (normal) and passage 113 (altered morphology). *White arrows* mark flattened outgrowth "edge" cells surrounding the normal "core" of the colony of high-passage H9 (p113). (**b**) Schematic representation for microdissection of the core and edge cells from one well of a six-well plate containing high-passage H9 (P113) ESC colonies. (**c**) Nucleosomes from core and edge cells of H9 ESC P113 were prepared by microscale methods and probed with 1° Abs with specificity for macroH2A, H4K20me3, and general H2A, respectively. (**d** and **e**) Relative macroH2A and H4K20me3 PTM levels in core and edge cells after normalization for chromatin loading based on H2A reactivity. These data are presented as means ± standard deviations from three technical replicates. *Abbreviations: core* pooled core cells from colonies of H9 P113, *edge* pooled edge cells from colonies of H9 P113 of the same wells as core cells, PTM posttranslational modification.

groups of samples are directly comparable; however, since each Ab has a unique avidity for its antigen, it is important not to interpret these ratios in absolute terms. However, when this method is used for two independent biological samples (e.g., a drug-treated cell line versus a control or a genetic knockout cell line versus wild type), the above approach can be used to determine histone residues that are especially sensitive to the treatment.

Acknowledgments

This study was supported by the Connecticut Stem Cell Research Grants 06SCA34 and 09-SCB-UCON-18. This material is based upon work supported by the state of Connecticut under the

Connecticut Stem Cell Research Grants Program. Its contents are solely the responsibility of the authors and do not necessarily represent the official views of the state of Connecticut, the Department of Public Health of the State of Connecticut, or Connecticut Innovations, Inc.

References

1. Balasubramanian S, Verner E, Buggy JJ (2009) Isoform-specific histone deacetylase inhibitors: the next step? Cancer Lett 280:211–221
2. Bieliauskas AV, Pflum MK (2008) Isoform-selective histone deacetylase inhibitors. Chem Soc Rev 37:1402–1413
3. Butler KV, Kozikowski AP (2008) Chemical origins of isoform selectivity in histone deacetylase inhibitors. Curr Pharm Des 14:505–528
4. Dai B, Rasmussen TP (2007) Global epiproteomic signatures distinguish embryonic stem cells from differentiated cells. Stem Cells 25:2567–2574
5. Han J, Sachdev PS, Sidhu KS (2010) A combined epigenetic and non-genetic approach for reprogramming human somatic cells. PLoS One 5:e12297
6. Huangfu D, Osafune K, Maehr R, Guo W, Eijkelenboom A, Chen S, Muhlestein W, Melton DA (2008) Induction of pluripotent stem cells from primary human fibroblasts with only Oct4 and Sox2. Nat Biotechnol 26:1269–1275
7. Huangfu D, Maehr R, Guo W, Eijkelenboom A, Snitow M, Chen AE, Melton DA (2008) Induction of pluripotent stem cells by defined factors is greatly improved by small-molecule compounds. Nat Biotechnol 26:795–797
8. Pilarsky C, Wenzig M, Specht T, Saeger HD, Grutzmann R (2004) Identification and validation of commonly overexpressed genes in solid tumors by comparison of microarray data. Neoplasia 6:744–750
9. Weichert W, Roske A, Niesporek S, Noske A, Buckendahl AC, Dietel M, Gekeler V, Boehm M, Beckers T, Denkert C (2008) Class I histone deacetylase expression has independent prognostic impact in human colorectal cancer: specific role of class I histone deacetylases in vitro and in vivo. Clin Cancer Res 14:1669–1677
10. Godman CA, Joshi R, Tierney BR, Greenspan E, Rasmussen TP, Wang HW, Shin DG, Rosenberg DW, Giardina C (2008) HDAC3 impacts multiple oncogenic pathways in colon cancer cells with effects on Wnt and vitamin D signaling. Cancer Biol Ther 7:1570–1580
11. Seligson DB, Horvath S, McBrian MA, Mah V, Yu H, Tze S, Wang Q, Chia D, Goodglick L, Kurdistani SK (2009) Global levels of histone modifications predict prognosis in different cancers. Am J Pathol 174:1619–1628
12. Seligson DB, Horvath S, Shi T, Yu H, Tze S, Grunstein M, Kurdistani SK (2005) Global histone modification patterns predict risk of prostate cancer recurrence. Nature 435:1262–1266
13. Dashwood RH, Myzak MC, Ho E (2006) Dietary HDAC inhibitors: time to rethink weak ligands in cancer chemoprevention? Carcinogenesis 27:344–349
14. Nian H, Delage B, Ho E, Dashwood RH (2009) Modulation of histone deacetylase activity by dietary isothiocyanates and allyl sulfides: studies with sulforaphane and garlic organosulfur compounds. Environ Mol Mutagen 50:213–221
15. Halili MA, Andrews MR, Sweet MJ, Fairlie DP (2009) Histone deacetylase inhibitors in inflammatory disease. Curr Top Med Chem 9:309–319
16. Dai B, Dahmani F, Cichocki JA, Swanson LC, Rasmussen TP (2011) Detection of post-translational modifications on native intact nucleosomes by ELISA. J Vis Exp. doi:3791/2593

Chapter 14

Preparing Semisynthetic and Fully Synthetic Histones H3 and H4 to Modify the Nucleosome Core

John C. Shimko, Cecil J. Howard, Michael G. Poirier, and Jennifer J. Ottesen

Abstract

The purpose of this chapter is to provide practical chemical ligation procedures to prepare histone proteins suitable for the reconstitution of nucleosomes with specific posttranslational modifications in the nucleosome core. Detailed methods are described for the efficient preparation of semisynthetic histones H3 and H4 with modifications near the C-terminus of the proteins by expressed protein ligation and desulfurization. Additionally, we present optimized protocols for solid phase peptide synthesis combined with sequential native chemical ligation to generate fully synthetic modified histone H3, here in the context of H3 lysine 56 acetylation (H3K56ac).

Key words: Histone, Nucleosome core, Posttranslational modification, Chemical ligation, Protein semisynthesis, Synthetic proteins

1. Introduction

Nucleosomes, the fundamental unit of chromatin, are protein–DNA complexes that package and organize DNA in the nucleus of the cell (1). The core nucleosome particle contains two copies each of four histone proteins: H3, H4, H2A, and H2B. Each of these histone proteins has multiple sites of posttranslational modification (PTM), such that combinations of acetylation, methylation, phosphorylation, ubiquitination, and other modifications regulate the structure and function of chromatin as well as biological processes that depend on interactions with DNA.

While the best characterized histone modifications occur in unstructured N- and C-terminal tail regions that make up

Sandra B. Hake and Christian J. Janzen (eds.), *Protein Acetylation: Methods and Protocols*, Methods in Molecular Biology, vol. 981, DOI 10.1007/978-1-62703-305-3_14,

approximately 30% of the histone sequence, PTMs also occur throughout the structured nucleosome core (2). These modifications are found in distinct regions of the histone–DNA interface and may facilitate varied dynamic events within the nucleosome, such as DNA unwrapping, nucleosome assembly or disassembly, and chromatin remodeling (3, 4).

Chemical ligation is an excellent way to prepare the homogenous samples of precisely modified histone proteins that are necessary to characterize the molecular functions of these modifications within the structured nucleosome core (5). Native chemical ligation (NCL) is the chemoselective condensation of two polypeptides, one with a C-terminal α-thioester and one with a 1,2-aminothiol, typically an N-terminal cysteine, resulting in the formation of a native amide bond at the ligation site. These two segments may be generated using recombinant techniques (expressed protein ligation) or synthetically. Synthetic peptides are not limited to the 20 natural amino acids and can be chemically diverse. For example, peptides can be synthesized with PTMs such as acetylation and phosphorylation. Ligation to generate a full-length protein, such as a histone, site-specifically incorporates these chemical modifications into the histone sequence. These modified histones are then combined with the unmodified, recombinant core histones and reconstituted into nucleosomes and nucleosome arrays.

This chapter describes the techniques used to introduce modifications into the nucleosome core using chemical ligation. We have used these strategies to introduce acetylation and phosphorylation into histones H3 and H4 (6–8), but they are applicable to any modification compatible with peptide synthesis. We describe semisynthetic and fully synthetic approaches to histone preparation and the considerations required for each.

1.1. General Considerations

There are multiple variants of chemical ligation that are applicable to histones (Fig. 1). Expressed protein ligation (EPL) combines synthetic peptide segments with recombinant protein segments. This powerful approach can generate multiple milligrams of material but is limited to the introduction of clusters of modifications close to the N- or C-terminus of a histone protein. In this chapter, we will discuss how this technique has been applied to the preparation of modified histone H3 suitable to study modifications in the nucleosome dyad (3, 6, 7) and acetylated histone H4 to study modifications in the LRS (loss of rDNA silencing) region of the nucleosome (4).

Sequential native chemical ligation is the most flexible technique to introduce PTMs throughout the protein sequence (8). In this approach, each peptide segment is generated synthetically and linked via ligation. Since this total synthesis approach is labor intensive and yields smaller amounts of homogenous material, it is most

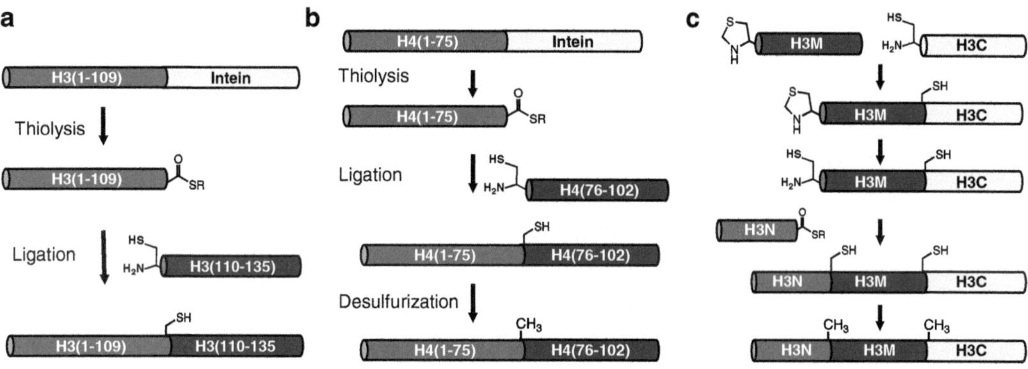

Fig. 1. Ligation strategies for the preparation of acetylated histones. (**a**) Expressed protein ligation (EPL) for generation of modified histone H3. (**b**) EPL–desulfurization strategy for the preparation of modified histone H4. (**c**) Sequential native chemical ligation (Seq-NCL)–desulfurization strategy for the total synthesis of histone H3.

suitable for sensitive biochemical or biophysical characterization of nucleosomes. However, it is the only method by which any combination of PTMs may be introduced across the protein sequence and is limited in scope only by the ability to prepare the appropriate modified peptide segments. This technique has been demonstrated in the preparation of histone H3 acetylated at lysine 56 (H3K56ac).

In each case, consideration must be paid to the selection of appropriate ligation junctions. Native chemical ligation typically requires an N-terminal cysteine for reaction to occur; this leaves a residual Cys at the ligation site. Histone proteins have very few natural cysteines. Most histone H3 variants include a single Cys at position 110, and an additional Cys is present at position 96 in the human histone variants 3.1 and 3.1t (9). Interestingly, both of these Cys are dispensable, and a C110A substitution that is native in yeast H3 sequences is often used for biophysical studies of the nucleosome (10). However, H3-C110 is a convenient ligation site for the semisynthesis of histone H3 with PTMs in the nucleosome dyad and can be used to generate the canonical *X. laevis* histone H3 sequence.

Although a 1,2-aminothiol is required for ligation, ligation junctions are not limited to Cys sites. Desulfurization reactions can be used to convert a ligation-capable 1,2-aminothiol to a cognate amine (11). This is most often used to convert a ligation-site Cys to Ala after ligation is complete, such that in the context of a protein with no essential Cys in the sequence, Ala can be considered a potential ligation site. Histone H4 has no native Cys residues, so H4-His_{75}-Ala_{76} is a convenient junction in a ligation–desulfurization strategy to introduce modifications near the C-terminus of the protein. In the total synthesis of histone H3, the H3-C110A substitution eliminates the only native Cys, and desulfurization after ligation using A47 and A91 as ligation junctions generates a native-like sequence.

2. Materials

2.1. Expression and Purification of Histone Protein Thioesters

1. Plasmid for fusion of desired protein with the *Mxe* GyrA intein, typically inserted in pTXB1 vector (New England Biolabs) which contains a C-terminal chitin-binding domain (CBD) fusion tag, and appropriate reagents for expression in BL21(DE3) cell line of choice.
2. Lysis buffer: 25 mM 4-(2-hydroxyethyl)-1-piperazineethanesulfonic acid (HEPES), pH 7.5, 1.0 mM ethylenediaminetetraacetic acid (EDTA), 1.0 M sodium chloride (NaCl), and 1.0 mM phenylmethylsulfonyl fluoride (PMSF).
3. Triton wash buffer: 25 mM HEPES, pH 7.5, 1.0 mM EDTA, 1.0 M NaCl, and 1% Triton-X.
4. Pellet wash buffer: 25 mM HEPES, pH 7.5, 1.0 mM EDTA, and 1.0 M NaCl.
5. Dimethyl sulfoxide (DMSO).
6. HU-0, HU-100, HU-400, HU-500, and HU-1000 buffer: 6 M urea, 25 mM HEPES, pH 7.5, and 1.0 mM EDTA, with 0, 100, 400, and 500 mM, and 1.0 M NaCl, respectively.
7. 5 mL SP-FF ion exchange column (GE Healthcare).
8. Refolding buffer: 25 mM HEPES, pH 7.5, 1.0 mM EDTA, and 1.0 M NaCl.
9. Spectra/Por dialysis tubing, MWCO 6–8 kDa; gently boil in 1.0 mM EDTA with three buffer changes and store at 4°C.
10. Ultrapure guanidine (Gdn) hydrochloride and urea (MP Biomedical).
11. Sodium 2-mercaptoethanesulfonic acid (MESNA).
12. Vivaspin 5 kDa MWCO protein concentrator (BioExpress).

2.2. Synthetic Peptide Segments

1. For C-terminal segments, use appropriately substituted Boc-AA-PAM-PS or Fmoc-AA-Wang resin (Novabiochem).
2. For H3N and H3M C-terminal thioester peptides, use 3-mercaptopropionamide-MBHA resin prepared according to literature procedures (12).
3. Boc- or Fmoc-protected amino acids, including Fmoc-Lys(Ac)-OH, Boc-Lys(Ac)-OH, Fmoc-Thr(PO(OBzl)OH)-OH (Novabiochem), Boc-L-thiazolidine-4-carboxylic acid (Thz) (Bachem).
4. 2-(1H-benzotriazole-1-yl)-1,1,3,3-tetramethyluronium-hexafluorophosphate (HBTU) or 2-(6-chloro-1H-benzotriazole-1-yl)-1,1,3,3-tetramethylaminium hexafluorophosphate (HCTU).

5. Reagent K: 82.5% trifluoroacetic acid (TFA), 5% H_2O, 5% phenol (w/v), 5% thioanisole, and 2.5% ethanedithiol (EDT).
6. TIS cleavage cocktail: 95% TFA, 2.5% H_2O, and 2.5% triisopropylsilane (TIS); if the peptide sequence contains Met or Thz, 94% TFA, 2.5% H_2O, 2.5% EDT, and 1% TIS may be used.
7. High-performance liquid chromatography (HPLC) apparatus and solvents: buffer A (0.1% TFA in Milli-Q H_2O) and buffer B (90% acetonitrile and 0.1% TFA in Milli-Q H_2O).
8. Supelco Discovery BIO Wide Pore C18 RP-HPLC columns in analytical, semi-preparative, and preparative scales.

2.3. EPL and Sequential NCL

1. Tris(2-carboxyethyl)phosphine (TCEP) stock solution: 1.2 M TCEP, pH 7.5, prepared by dissolving solid TCEP-HCl in 5 N NaOH and adjusting volume and pH as necessary.
2. Sodium 2-mercaptoethanesulfonic acid (MESNA) or 4-mercaptophenylacetic acid (MPAA).
3. Ligation buffer A: 6 M urea, 25 mM HEPES pH 7.5, 1.0 M NaCl, 1.0 mM EDTA, and 50 mM MESNA.
4. Ligation buffer B: 6 M Gdn, 25 mM HEPES pH 7.5, 1.0 M NaCl, 1.0 mM EDTA, and 50 mM MESNA.
5. Ligation buffer C: 6 M Gdn, 100 mM phosphate buffer pH 7.5, 500 mM NaCl, 50 mM MESNA, and 10 mM TCEP.
6. Supelco Discovery BIO Wide Pore C18 RP-HPLC columns in analytical, semi-preparative, and preparative scales.
7. TU buffer: 10 mM tris(hydroxymethyl)aminomethane (Tris), pH 9.0, 7 M urea, 1.0 mM EDTA, and 5 mM β-mercaptoethanol (BME). TU-100 includes 100 mM NaCl, and TU-600 includes 600 mM NaCl.
8. Spectra/Por dialysis tubing, MWCO 6–8 kDa; gently boil in 1.0 mM EDTA with three buffer changes and store at 4°C.
9. TSKgel SP-5PW column (TOSOH Biosciences).

2.4. General Methods

2.4.1. TCA Precipitation

1. Trichloroacetic acid (TCA).
2. 5 N NaOH stock solution.

2.4.2. Thiazolidine Ring Opening

1. Methoxylamine hydrochloride.

2.4.3. Free-Radical Desulfurization to Generate Native Protein Sequences

1. VA-044US (Wako Chemicals).
2. Sodium 2-mercaptoethanesulfonic acid (MESNA).
3. TCEP stock solution: 1.2 M TCEP, pH 7.5.
4. Millipore ZipTip C18 pipette tips, P10 size.

2.4.4. Reduction of Oxidized Methionine

1. Trifluoroacetic acid (TFA).
2. Dimethyl sulfide.
3. Sodium iodide (NaI).

3. Methods

Synthesis or semisynthesis of modified histone proteins is a multi-step process. These methods will describe the preparation of recombinant (Subheading 3.1) and synthetic (Subheading 3.2) polypeptide components for expressed protein ligation (Subheading 3.3) to generate semisynthetic histones H3 and H4 with clusters of modifications as separate procedures. Similarly, the preparation of synthetic peptide segments (Subheading 3.2) for sequential native chemical ligation (Subheading 3.4) to prepare fully synthetic histone H3 will be described. A working knowledge of basic peptide synthesis techniques will be assumed, and only protocols specific to these polypeptides will be presented in detail.

3.1. Preparation of Recombinant Histone Thioesters

Recombinant histone thioesters are prepared by the thiolysis of the histone N-terminal domain fused with a structured intein and a C-terminal chitin-binding domain (CBD) tag. However, these fusion proteins express as misfolded aggregates that accumulate in inclusion bodies. The general steps required to prepare the histone thioester are expression of the fusion protein, purification of inclusion bodies, refolding to generate the functional intein domain, and thiolysis to produce the histone thioester for ligation. Refolding and thiolysis of the histone–intein–CBD is time sensitive. Once the intein is refolded it becomes active and is susceptible to hydrolysis, which diminishes product yield. Also note that buffers should not contain any thiols except where specifically instructed; in particular avoid the use of dithiothreitol or β-mercaptoethanol, which lead to nonproductive intein cleavage.

3.1.1. Expression and Purification of Histone–Intein–CBD

1. Using standard cloning techniques, insert the N-terminal histone segment into pTXB1 plasmid such that the truncated histone domain is directly fused to the N-terminal Cys of the *Mxe* GyrA intein. While expression vectors for several different inteins are commercially available, the *Mxe* GyrA intein is the most suitable due to the ease of refolding.
2. Express the protein in *E. coli* BL21(DE3) cells at 37°C in LB media with 0.1 mg/mL ampicillin, and induce at $OD_{600} = 0.4$ by the addition of IPTG to 0.2 mM final concentration for 4 h. Cells are harvested by centrifugation. The following procedure

was modified from literature conditions (13) to be suitable for intein fusions and assumes 2.5 L starting culture.

3. Resuspend cells in 10–25 mL lysis buffer; lyse cells by French press or sonication.
4. Spin lysate at 20,000 × *g* for 15 min at 4°C. Histone–intein fusion will be found in the pellet.
5. Resuspend pellet in 25 mL Triton wash buffer. Centrifuge at 10,000 × *g* for 10 min at 4°C. Pour off supernatant. Repeat this wash step.
6. Resuspend pellet in 25 mL pellet wash buffer. Centrifuge at 10,000 × *g* for 10 min at 4°C. Pour off supernatant. Repeat the wash step. The pellet may be stored at −80°C.
7. Thaw pellet on ice. Once thawed, add 250 μL DMSO and mince the pellet with a spatula. Allow to soak at 25°C for 30 min.
8. Add 30 mL HU-500 buffer to the pellet and resuspend by mixing for 1 h at room temperature. Centrifuge at 15,000 × *g* for 10 min at 25°C; save the supernatant. Repeat **step** 8.
9. Test each supernatant from **step** 8 for the presence of protein by analysis on a 15% SDS–PAGE gel; if appropriate, combine the supernatants. Dilute with HU-0 to reach 100 mM NaCl or dialyze overnight against HU-100 buffer.
10. Split in half, and purify each batch by ion exchange over 5 mL SP-FF column according to manufacturer's instructions. Load sample, wash with HU-100 buffer, and elute protein with HU-500 buffer. Analyze eluent by SDS–PAGE; combine fractions containing histone–intein–CBD.

3.1.2. Refolding and Thiolysis of Histone–Intein–CBD

1. Dilute purified protein to 30–50 mL using HU-1000 and dialyze fractions overnight against 4 L of refolding buffer at 4°C. Some protein will precipitate, but the majority should be properly refolded protein in the supernatant. If sufficient protein is lost, the pellet may be resuspended in HU-1000 and refolded by dialysis a second time.
2. In a 50 mL conical tube, add MESNA to the refolded histone–intein–CBD to a final concentration of 100 mM. Nutate at 4°C for 18–24 h.
3. For H3 thioester, add ultrapure urea, adjust buffer components to compensate for the change in volume, and modify pH to reach a final ligation buffer A composition.
4. For H4 thioester, add ultrapure guanidine, adjust buffer components to compensate for the change in volume, and modify pH to reach final ligation buffer B composition.

5. Assess cleavage by SDS–PAGE analysis on a 15% SDS–PAGE gel. This requires TCA precipitation for histone H4 to prepare a gel sample (Procedure 3.5.1).
6. Concentrate protein using a Vivaspin centrifugal concentrator, MWCO 5 kDa, at 4°C until histone thioester reaches a final concentration >1 mg/mL.
7. When calculating concentration by UV absorbance, the majority of Abs_{280} is attributable to the aromatic residues of the intein and chitin-binding domain. Thioester amounts are estimated by SDS–PAGE assessment of thioester cleavage, calculated against total histone–int–CBD + int–CBD with extinction coefficients according to Table 1.
8. Flash freeze the concentrated thioester in appropriately sized aliquots (typically 100 μg, 250 μg, 500 μg, and 1 mg) and store at −80°C for use in ligation reactions.

3.2. Synthetic Peptides for EPL or NCL to Generate Modified Histone H3 or H4

Peptides designed to introduce modifications into the C-terminus of histone protein H3 or H4 require an N-terminal Cys and may be synthesized with a free carboxyl terminus. Throughout these methods, we assume familiarity with standard solid phase peptide synthesis (SPPS) techniques. Acetylated peptides may be prepared by standard Fmoc- or Boc-methods; phosphorylated peptides should be synthesized by Fmoc-SPPS. These peptides may also be ordered from a commercial service if facilities for peptide synthesis are not available locally.

Three component polypeptides are required for sequential native chemical ligation to produce fully synthetic histone H3 (Table 2). These peptides are each ~45 amino acids in length and can be synthetically challenging. The requirement for an α-thioester in peptides H3N and H3M provides an additional challenge. This section will describe the syntheses of the peptides used to generate H3K56ac by Boc-SPPS chemistry, as described in Shimko et al. (8). Appropriate peptides for the preparation of other modified histones may be synthesized similarly.

3.2.1. Peptide H3(110–135) for EPL of Modified Histone H3

1. Peptide H3(110–135) CAIHAKRVTIMPKDIQLARRIRGERA should be synthesized with any desired modifications (typically a combination of K115ac, K122ac, pT118) on Fmoc-Ala-Wang resin using standard Fmoc-SPPS protocols with HBTU or HCTU activation (see Note 1).
2. Acetyllysine is introduced as Fmoc-Lys(Ac)-OH using standard coupling conditions.
3. Phosphorylated Thr is introduced as Fmoc-Thr(PO(OBzl)OH)-OH.
4. Phosphorylated H3(110–135)pT118 should be cleaved from resin by treatment with Reagent K for 3 h. Other H3(110–135) peptides are cleaved from resin using the TIS cleavage

Table 1
Physical properties of H3 and H4 intein fusion, thioester, and full-length proteins[a]

Protein	Mass (M + H)+	$\varepsilon_{280\,nm}$ (cm^{-1}/M)	Protein	Mass (M + H)+
H3				
H3(1–109)–intein–CBD	40,143 *m/z*	38,600	H3-K115ac,K122ac	15,356 *m/z*
H3(1–109) thioester	12,394 *m/z*	3,840	H3-K56ac(C110A)	15,282 *m/z*
H3(1–109)-OH	12,286 *m/z*	3,840	H3-pT118	15,352 *m/z*
H3(1–135)	15,273 *m/z*	3,960		
H4				
H4(1–75)–intein–CBD	36,106 *m/z*	37,320	H4-A76C,K77ac,K79ac	11,352 *m/z*
H4(1–75) thioester	8,376 *m/z*	2,560	H4-K77ac,K79ac	11,320 *m/z*
H4(1–75)-OH	8,268 *m/z*	2,560	H4-A76C,pT80	11,348 *m/z*
H4(1–102)	11,237 *m/z*	5,120	H4-pT80	11,316 *m/z*
Intein–CBD	27,857 *m/z*	34,760		

[a]$\varepsilon_{280\,nm}$ is calculated for samples in 6 M Gdn

Table 2
Sequence of peptide segments utilized in sequential ligation

Peptide	Description	Sequence
H3N	H3(1–46) thioester	ARTKQTARKSTGGKAPRKQLATKAARKSAPAT GGVKKPHRYRPGTV-**SR**
H3M	H3(47–90)-A47Thz,K56ac thioester	(Thz)LREIRRYQ(**Kac**)STELLIRKLPFQRLVREIA QDFKTDLRFQSSAVM-**SR**
H3C	H3(91–135)-A91C,C110A	CLQEASEAYLVALFEDTNLAAIHAKRVTIMPK DIQLARRIRGERA-**OH**

cocktail for 2 h. Reduce the TFA volume to 10% with a stream of nitrogen gas, then precipitate with cold diethyl ether, and wash three times with additional cold ether.

5. Resuspend peptides in 25% HPLC buffer B and lyophilize to dryness as a white powder.
6. H3(110–135) peptides are purified by RP-HPLC with a gradient of 18–33% buffer B.
7. After purification, peptide should be dissolved in Milli-Q H_2O, divided into appropriate aliquots for use in expressed protein ligation at the desired scale (typically 100 μg, 500 μg, 1 mg, and 5 mg aliquots), and lyophilized to dryness. Peptide may be stored at −80°C until used.

3.2.2. Peptide H4(76–102) for EPL of Modified Histone H4

1. PeptideH4(76–102)CKRKTVTAMDVVYALKRQGRTLYGFGG is synthesized with any desired modifications (typically a combination of K77ac, K79ac, K91ac, pT80) on Fmoc-Gly-Wang resin. Synthesis considerations as per Subheading 3.2.1 apply (see also Note 1).
2. Purify modified H4(76–102) peptides by RP-HPLC over a C18 column using a 25–40% buffer B gradient.

3.2.3. Synthesis of Peptide H3C

1. The total synthesis of histone H3 requires a single non-thioester peptide, peptide H3C: H3(91–135) A91C,C110A (see Notes 1 and 2).
2. Peptide H3C is prepared with a carboxylate terminus on Boc-Ala-PAM-PS resin.
3. H3C may be prepared by Boc-SPPS using in situ neutralization protocols (14). Single coupling cycles can be used for the addition of the first 20 amino acids; double coupling cycles should be used for each subsequent residue. Acetylation/capping cycles should be used throughout to minimize deletion sequences and simplify purification.
4. H3C may be prepared by Fmoc-SPPS using HCTU activation. We recommend double coupling cycles after the first 20 amino acids and the use of acetylation/capping cycles to simplify purification.
5. Purify H3C by RP-HPLC using a gradient of 25–45% buffer B. Among the columns we have tested, Supelco Discovery BIO Wide Pore C18 columns provide the best resolution and yield for these histone peptides.

3.2.4. Synthesis of Thioester Peptides H3N and H3M

1. Peptide H3N corresponds to H3(1–46) thioester. H3M is the central segment of fully synthetic H3, with both a masked N-terminal Cys and a C-terminal thioester, and corresponds to H3(47–90)-A47Thz,K56ac (see Notes 1 and 2).
2. Thioester peptides H3N and H3M may be prepared by Boc-SPPS on mercaptopropionamide substituted resin (12) using in situ neutralization protocols with single coupling cycles for the first 20 amino acids, double coupling cycles through the remainder of the sequence, and capping cycles to simplify purification. The acetylated lysine is introduced as Boc-Lys(Ac)-OH. For peptide H3M, the N-terminal Cys is introduced as Boc-thiazolidine-OH to avoid intramolecular ligation.
3. Purify H3N and H3M by RP-HPLC with gradients of 10–25% buffer B and 35–55% buffer B, respectively.
4. Synthesis of thioester peptides H3N and H3M by Fmoc-SPPS is possible but is beyond the scope of this protocol. Briefly, modified Dawson Dbz resin (15, 16) may be used to prepare thioester peptides H3N and H3M.

3.3. Expressed Protein Ligation Approaches to Histones H3 and H4

3.3.1. EPL to Generate Histone H3

1. 10 molar equivalents of H3(110–135) peptide with appropriate modification is dissolved in a minimal-volume ligation buffer A and added to H3(1–109) thioester mixture from Protocol 3.1.2. Add TCEP from a 1.2 M stock solution to a final concentration of 20 mM.
2. The pH of the final ligation mixture should be assessed by spotting 0.2 μL of the reaction mixture onto pH paper at 0 and 2 h. If necessary, pH should be adjusted to 7.5 by the addition of 1.0 M HCl or NaOH.
3. Reaction proceeds overnight or until no further ligation is assessed by analysis on a 15% SDS–PAGE gel (see Note 3). Care should be taken to transfer the gel to stain or a fixing solution immediately in order to minimize diffusion of peptide from the gel prior to analysis.
4. Dialyze products into Milli-Q H_2O overnight.
5. Transfer contents of the dialysis bag to a 50 mL conical tube, taking care to recover any precipitate, and lyophilize to dryness.
6. Resuspend products in TU-100 and purify by ion exchange HPLC over a TSKgel SP-5PW column with a gradient of TU-100 to TU-600 (600 mM NaCl).
7. Assess purification fractions by 15% SDS–PAGE. The full-length H3 product will not separate fully from residual truncated H3 (see Note 4). If the semisynthetic histone is to be used to generate histone octamer for nucleosome reconstitutions, each fraction containing more than 50:50 ratio of full-length protein to truncated histone should be combined. The histone will be further purified during histone octamer refolding since the peptide comprises an essential interface in the folded octamer.

3.3.2. EPL–Desulfurization to Generate Histone H4

1. Dissolve 5–10 molar equivalents of H4(76–102) peptide with appropriate modifications in minimal-volume ligation buffer B and added to H4(1–75) thioester mixture from Protocol 3.1.2. Add TCEP from a 1.2 M stock solution to a final concentration of 20 mM. Reaction pH should be assessed as per Protocol 3.3.1.
2. Allow the reaction to proceed until complete as measured by SDS–PAGE, typically 6–12 h (see Note 3).
3. Desulfurization is carried out with the addition of TCEP and VA-044US as per Protocol 3.5.3 (see Note 5).
4. Histone H4 may be purified either by ion exchange over a TSKgel SP-5PW column as per Subheading 3.3.1, steps 4–6, or by RP-HPLC using a gradient of 35–65% buffer B. Typically, acetylated H4 is compatible with RP-HPLC purification, but

H4-pT80 co-elutes with int–CBD by RP-HPLC so must be purified by ion exchange (see Note 5).

5. If the modified histones will be used to refold histone octamers, H4(1–75) may be co-purified with full-length protein to maximize yield (see Note 4).

3.4. Sequential Ligation to Generate Histone H3

3.4.1. Ligation of H3M to H3C

1. Resuspend H3M peptide and 2.2 molar equivalents of H3C peptide independently in ligation buffer C, monitoring solubility carefully.
2. Combine resuspended peptides and allow ligation to proceed with mixing by nutation at room temperature (see Note 3).
3. Reaction progress may be monitored by RP-HPLC. However, peptide H3M and product H3MC are only mildly soluble in acidic RP-HPLC conditions, so chromatograms may be deceptive. We typically monitor reaction both by SDS–PAGE following TCA precipitation (Subheading 3.5.1) as well as RP-HPLC. Reaction is allowed to proceed until no further product formation is observed (typically 2 days).
4. After reaction is complete, add methoxylamine to the crude ligation mixture (400 mM) according to Protocol 3.5.2 to convert Thz to Cys. Deprotection typically takes 6 h for this ligation product.
5. Purify the Cys-H3MC product by RP-HPLC over a 35–65% buffer B gradient (see Note 5). Analyze fractions by MALDI-TOF MS. Make a small-volume test pool by proportional combination of fractions to analyze by SDS–PAGE to confirm the composition of the combined product. If test pool is satisfactory, combine fractions and lyophilize to dryness.

3.4.2. Ligation of H3N to H3MC

Peptide H3N has a β-branched Val as the C-terminal amino acid. This ligation terminus has poor kinetics (17), and the reaction can take up to 5 days to reach maximal yield, depending on the choice of co-thiol for this reaction. Ligation is slow with the MESNA thioester and typically does not go to completion. Ligation with the MPAA thioester typically proceeds faster and generates more of the desired product, but MPAA quenches desulfurization and must be removed prior to the conversion step. This protocol is written to reflect use of MESNA:

1. Independently resuspend H3MC and 5 molar equivalents of H3N in the minimal volume of ligation buffer C required to fully dissolve the pellets.
2. Combine the peptide aliquots to initiate ligation, and allow to react while mixing by nutation at room temperature (see Note 3).
3. Monitor reaction progress by SDS–PAGE, using TCA precipitation (Subheading 3.5.1) to remove guanidine. Ligation is

continued until no further product formation is observed (typically 2 days with MPAA, 5 days with MESNA) (see Notes 6 and 7).

4. Once no further progress is observed, the crude ligation mixture may be desulfurized as per Protocol 3.5.3.
5. Following desulfurization, purify using RP-HPLC over a gradient of 35–65% buffer B. The crude desulfurization mixture should be brought to approximately 0.1 mg/mL at 35% buffer B loading conditions by addition of HPLC buffers. Solubility of full-length H3 can be problematic, and the sample should be centrifuged thoroughly to eliminate any particulates. The sample should not be filtered, as nonspecific interactions result in unacceptable yield losses.
6. Any pellet remaining after centrifugation should be analyzed to assess product content; precipitate should be resuspended in 35% buffer B for subsequent purification as required.
7. After RP-HPLC, fractions are analyzed by MALDI-TOF MS. A small-volume proportional test pool should be assessed by analytical RP-HPLC and SDS–PAGE as per Protocol 3.4.1 to guide selection of fractions to pool and lyophilize to dryness as the final product.
8. Product yield should be calculated by UV–Vis spectroscopy using extinction coefficients from Table 1.
9. Oxidized methionine may form during ligation or desulfurization. If methionine oxidation (Met(O)) is detected, Met(O)-containing protein may be co-purified with the non-oxidized sample and reduced following Protocol 3.5.4 (see Note 2).

3.5. General Methods

3.5.1. TCA Precipitation of Samples for SDS–PAGE

Several buffer systems used in these protocols require guanidine, which is not compatible with SDS–PAGE analysis of thiolysis and ligation progress. TCA precipitation is used to isolate total protein content from buffer components in a sample:

1. Dilute 1–5 μL sample with Milli-Q H_2O to 100 μL.
2. Add 25 μL 100% (w/v) TCA, mix, and incubate at 4°C for 10 min.
3. Centrifuge at 16,000×g and carefully remove supernatant without disturbing pellet.
4. Protein pellet may be resuspended directly without washing in SDS–PAGE loading buffer. Residual TCA is neutralized with the addition of NaOH as needed (typically 1 μL 5 N NaOH).

3.5.2. Conversion of Thz to Cys by Treatment with Methoxylamine

The condensation of an N-terminal Cys with formaldehyde generates thiazolidine, which serves as a masked N-terminal Cys during sequential ligation. The Cys can be regenerated by addition of an excess of methoxylamine under acidic conditions.

1. Methoxylamine HCl is added to a final concentration of 0.2–0.4 M. and pH is adjusted to ~4. In a typical buffered ligation reaction, addition of the acidic reagent should depress pH sufficiently, but HCl can be added dropwise as required.
2. Reaction proceeds until complete as assessed by RP-HPLC and MALDI-TOF MS; the loss of a methylene is observed as −12 Da.

3.5.3. Desulfurization of Cys to Native Ala Residue

1. To the crude ligation mixture, add 1.2 M TCEP stock solution, pH 7.5, to a final concentration of 300 mM TCEP. Add MESNA to compensate for the increase of volume, to a minimum concentration of 50 mM (see Note 7).
2. Sparge with argon gas for 30 min. If the target protein contains Met, insufficient sparging will result in partial oxidation to generate the +16 Da Met(O) species which may be reduced after final protein purification (see Protocol 3.5.4).
3. Add VA-044US to a final concentration of 10 mM and incubate at 42°C until reaction is complete as assayed by MALDI-TOF MS (typically 4 h).
4. Millipore ZipTip C18 pipette tips should be used to remove buffer components from 1 μL of the desulfurization reaction prior to analysis by MALDI-TOF MS. Manufacturer's protocols should be altered slightly: after sample loading, wash three times with 30% buffer B, and elute with 5 μL 70% buffer B. Typically 1 μL of this eluant is required for MALDI-TOF analysis.
5. Each Cys to Ala conversion results in a loss of 32 Da from the protein mass.

3.5.4. Reduction of Oxidized Methionine

Met(O) is generated by the exposure of Met-containing peptides or proteins to atmospheric oxygen. Met can be regenerated from Met(O) by the following protocol, which was modified from the literature (18) for optimal use with histone proteins (8):

1. Resuspend lyophilized protein at 0.5 mg/mL in TFA, on ice.
2. Add 10% (v/v) dimethyl sulfide.
3. Initiate reduction by adding NaI. Depending on the reaction scale, NaI may be added directly as a solid or from a 2 M stock solution in H_2O to a final concentration of 45 mM.
4. Incubate on ice for 1 h with occasional mixing.
5. Precipitate the protein with cold diethyl ether and lyophilize. No wash steps are required prior to histone octamer refolding, if reduction is carried out on a full-length histone.

4. Notes

1. Care should be taken to avoid racemization of N-terminal Cys (19). We introduce this residue as the Boc-thiazolidine-OH derivative (Bachem) to maintain stereochemistry and to allow for the removal of N-terminal protection during cleavage. Procedure 3.5.2 carried out on the crude peptide mixture unmasks the N-terminal Cys prior to purification.
2. Peptides may also be synthesized with Norleucine (Nle) in place of Met. The Met to Nle substitution eliminates oxidation of Met as a possible side product and removes the need for protocol 3.5.4.
3. The native chemical ligation reaction is sensitive to pH and should be continuously monitored over the course of a ligation to verify that a pH of ~7.5 is continuously maintained.
4. Semisynthetic histone with synthetic αC helix, such as H3 and H4 when prepared by these protocols, can be with the truncated recombinant histone sequence. After purification, each fraction containing greater than 50% full-length protein should be combined. Since the synthetic segment comprises the αC helix, an essential interface in the folded octamer, only full-length histone will generate the full refolded histone octamer. Size exclusion purification of the octamer then removes any residual truncated protein. To set up refolding, histone content should be calculated such that recombinant histones are equimolar to the sum of full-length semisynthetic and truncated histone domain.
5. Histone proteins may be marginally soluble under desulfurization conditions or with addition of acetonitrile. Any precipitate which forms should be reclaimed, analyzed, and treated as required to maximize yield.
6. Trace aldehydes from glass or from buffer components can react with N-terminal Cys over the course of a lengthy ligation to form a Thz side product, typically observed as a shift in RP-HPLC retention and +12 Da by MALDI-TOF MS. If Thz is detected, add 200 mM methoxylamine and incubate until reversion to Cys is complete (typically 4 h). Ligation may be reinitiated by raising the pH to 7.5.
7. The common MPAA co-thiol quenches free-radical desulfurization and must be completely removed by dialysis prior to this protocol.

Acknowledgment

This work was supported by NSF grant MCB-0845695 (JJO), seed funding from ACS IRG-6700344 (JJO) and NIH grant GM083055 (MGP and JJO). We also acknowledge Michelle B. Ferdinand and Christopher E. Smith for contributions to protocol development.

References

1. Kornberg RD, Thomas JO (1974) Chromatin structure; oligomers of the histones. Science 184:865–868
2. Mersfelder EL, Parthun MR (2006) The tale beyond the tail: histone core domain modifications and the regulation of chromatin structure. Nucleic Acids Res 34:2653–2662
3. Javaid S, Manohar M, Punja N, Mooney A, Ottesen JJ, Poirier MG, Fishel R (2009) Nucleosome remodeling by hMSH2-hMSH6. Mol Cell 36:1086–1094
4. Simon M, North JA, Shimko JC, Forties RA, Ferdinand MB, Manohar M, Zhang M, Fishel R, Ottesen JJ, Poirier MG (2011) Histone fold modifications control nucleosome unwrapping and disassembly. Proc Natl Acad Sci USA 108:12711–12716
5. Chatterjee C, Muir TW (2010) Chemical approaches for studying histone modifications. J Biol Chem 285:11045–11050
6. Manohar M, Mooney AM, North JA, Nakkula RJ, Picking JW, Edon A, Fishel R, Poirier MG, Ottesen JJ (2009) Acetylation of histone H3 at the nucleosome dyad alters DNA–histone binding. J Biol Chem 284:23312–23321
7. North JA, Javaid S, Ferdinand MB, Chatterjee N, Picking JW, Shoffner M, Nakkula RJ, Bartholomew B, Ottesen JJ, Fishel R, Poirier MG (2011) Phosphorylation of histone H3(T118) alters nucleosome dynamics and remodeling. Nucleic Acids Res 39:6465–6474
8. Shimko JC, North JA, Bruns AN, Poirier MG, Ottesen JJ (2011) Preparation of fully synthetic histone H3 reveals that acetyl-lysine 56 facilitates protein binding within nucleosomes. J Mol Biol 408:187–204
9. Hake SB, Allis CD (2006) Histone H3 variants and their potential role in indexing mammalian genomes: The "H3 barcode hypothesis". Proc Natl Acad Sci U S A 103:6428–6435
10. Luger K, Mäder AW, Richmond RK, Sargent DF, Richmond TJ (1997) Crystal structure of the nucleosome core particle at 2.8 A resolution. Nature 389:251–260
11. Wan Q, Danishefsky SJ (2007) Free-radical-based, specific desulfurization of cysteine: a powerful advance in the synthesis of polypeptides and glycopolypeptides. Angew Chem Int Ed 46:9248–9252
12. Camarero JA, Muir TW (2001) Native chemical ligation of polypeptides. Curr Protoc Protein Sci 18.14.1–18.14.21
13. Luger K, Rechsteiner TJ, Richmond TJ (1999) Expression and purification of recombinant histones and nucleosome reconstitution. Methods Mol Biol 119:1–16
14. Schnolzer M, Alewood P, Jones A, Alewood D, Kent SB (1992) In situ neutralization in Boc-chemistry solid phase peptide synthesis. Int J Pept Protein Res 40:180–193
15. Blanco-Canosa JB, Dawson PE (2008) An efficient Fmoc-SPPS approach for the generation of thioester peptide precursors for use in native chemical ligation. Angew Chem Int Ed 47:6851–6855
16. Mahto SK, Howard CJ, Shimko JC, Ottesen JJ (2011) A reversible protection strategy to improve Fmoc-SPPS of peptide thioesters by the N-Acylurea approach. ChemBioChem 12:2488–2494
17. Hackeng TM, Griffin JH, Dawson PE (1999) Protein synthesis by native chemical ligation: expanded scope by using straightforward methodology. Proc Natl Acad Sci USA 96:10068–10073
18. Hackenberger CPR (2006) The reduction of oxidized methionine residues in peptide thioesters with NH4I-Me2S. Org Biomol Chem 4:2291–2295
19. Han YX, Albericio F, Barany G (1997) Occurrence and minimization of cysteine racemization during stepwise solid-phase peptide synthesis. J Org Chem 62:4307–4312

Chapter 15

Production of Amino-Terminally Acetylated Recombinant Proteins in *E. coli*

Matthew Johnson, Michael A. Geeves, and Daniel P. Mulvihill

Abstract

The majority of proteins in eukaryote cells are subjected to amino-terminal acetylation. This co-translational modification can affect the stability of a protein and also regulate its biological function. Amino-terminally acetylated recombinant proteins cannot be produced using prokaryote expression systems, such as *E. coli*, as these cells lack the appropriate N-α-terminal acetyltransferase complexes. Here we describe a simple protocol that allows the recombinant expression and purification of NatB-dependent amino-terminally acetylated proteins from *E. coli*.

Key words: NatB, Amino-terminal acetylation, N-α-terminal acetyltransferases, *E. coli*, Fission yeast, Recombinant protein

1. Introduction

Prokaryotes, such as *E. coli*, are commonly the cell type of choice for recombinant protein production during research and biomedical applications, as they provide a cheap and simple system for producing large quantities of protein. However, one drawback is that these cells lack the necessary machinery to undertake posttranslational modifications of proteins, such as amino-terminal acetylation, that normally occur within a eukaryote cell.

N-terminal acetylation is one of the most common modifications of proteins in eukaryotes, which involves the transfer of an acetyl group from acetyl coenzyme A to the amino-terminal amino acid of a protein (1). The addition of the acetyl group to the N-terminal residue neutralizes positive charges, which can affect the function and stability of the protein as well as regulate either interactions with other proteins or further posttranslational modifications such as phosphorylation (e.g., troponin T) or ubiquitination.

Sandra B. Hake and Christian J. Janzen (eds.), *Protein Acetylation: Methods and Protocols*, Methods in Molecular Biology, vol. 981, DOI 10.1007/978-1-62703-305-3_15,

The acetylation of amino-terminal amino acids is catalyzed by a group of protein complexes called N-α-terminal acetyltransferases (NATs). The NATs have been extensively studied in the budding yeast *Saccharomyces cerevisiae*, in which three major classes of NAT have been identified: NatA, NatB, and NatC. These complexes have since been shown to be conserved and acetylate the majority of the human proteome (2). The NatB complex consists of two subunits: a 23 kDa catalytic subunit, Naa20, and a 92 kDa auxiliary subunit, Naa25 (3). NatB target proteins always have an N-terminal methionine residue, which is followed by an acidic or arginine residue, and previous studies have shown that all proteins with these sequences are acetylated (1).

Here we describe a simple method developed in this lab for generating amino-terminally acetylated recombinant proteins in *E. coli* cells (4). By co-expressing genes encoding for proteins that are normally substrates of the NatB N-α-acetyltransferase complex (i.e., protein sequence starts with M-D-, M-E-, M-N-) together with genes encoding for the fission yeast NatB complex, it is possible to produce large quantities of amino-terminally acetylated proteins in *E. coli* with significant cost time befits over eukaryote expression and chemical acetylation alternatives.

2. Materials

1. LB medium: 10 g tryptone, 10 g sodium chloride (NaCl), 5 g yeast extract, and 20 g agar (if solid medium required). Make up to 1 L with distilled water and autoclave for 15 min. Antibiotics added to medium after autoclaving and cooling.
2. NZY medium: 10 g NZ amine casein hydrolysate, 5 g NaCl, and 5 g yeast extract. Make up to 1 L with distilled water and autoclave for 15 min. The following were added after autoclaving and cooling and just prior to use: 12.5 mL of sterile 1 M magnesium chloride ($MgCl_2$), 12.5 mL of sterile 1 M magnesium sulfate ($MgSO_4$), 20 mL of sterile 20% (w/v) glucose, and appropriate antibiotics.
3. Antibiotics: for every mL media volume, add 1 μL from the appropriate antibiotic stock solutions which are made up at the following concentrations and can be stored in 1 mL aliquots at -20°C: 50 mg/mL ampicillin (dissolved in H_2O), 38 mg/mL chloramphenicol (dissolved in ethanol), and 50 mg/mL of kanamycin (dissolved in H_2O).
4. SOC medium (1 L): 20 g tryptone, 5 g yeast extract, 2 mL of 5 M NaCl, 2.5 mL of 1 M potassium chloride (KCl), 10 mL of 1 M $MgCl_2$, 10 mL of 1 M $MgSO_4$, and 30 mL of 1 M glucose.
5. *E. coli* expression strain, e.g., BL21 DE3.
6. pNatB plasmid: pACYCduet-*naa20+*-*naa25+* (see Note 1).

3. Methods

3.1. Preparation of BL21-DE3-pNatB Competent Cells

1. Use an aliquot of the pNatB plasmid to transform a lab stock of BL21-DE3 *E. coli* competent cells to generate the BL21-DE3-pNatB strain.
2. Set up a 5 mL overnight shaking (220 rpm) culture from a single-colony inoculum of BL21-DE3 in LB media supplemented with chloramphenicol.
3. Add 1 mL of this culture to 0.5 mL of sterile 60% (v/v) glycerol. Then mix and transfer to a 2 mL cryotube before storing in a −80°C freezer.
4. Streak out BL21-DE3-pNatB *E. coli* cells from the −80°C stocks onto a LB + chloramphenicol plate and incubate overnight at 37°C.
5. Grow a 5 mL overnight shaking (220 rpm) culture from a single-colony inoculum of the BL21-DE3-pNatB cells in LB media supplemented with chloramphenicol.
6. In the next morning use 0.5 mL of this pre-culture to inoculate 50 mL of fresh LB medium (+ chloramphenicol) within a sterile 250 mL Erlenmeyer flask.
7. Grow cells in a shaking (180 rpm) incubator at 37°C until the culture reaches a 578 nm optical density of 0.6–0.8.
8. Cool the cells on ice for 10 min, before centrifuging at 2,700 × *g* at 4°C for 10 min.
9. Place 0.5 mL microfuge tubes on dry ice to chill for aliquots in step 11.
10. Gently resuspend the pelleted cells in 10 mL of sterile solution containing 0.1 M $CaCl_2$ and 10% glycerol. Incubate on ice for 15 min.
11. Centrifuge cells at 2,700 × *g* at 4°C for 10 min and resuspend the pelleted cells in 1 mL sterile of cold 0.1 M $CaCl_2$ and 10% glycerol solution. Dispense competent cells as 50 μL aliquots into microfuge tubes (which have been resting on dry ice) and then store at −80°C.

3.2. Transformation of BL21-DE3-pNatB Cells

1. Defrost an aliquot of BL21-DE3-pNatB competent cells on ice.
2. Add up to 5 μL of the plasmid from which the protein to be acetylated can be expressed, and mix by gently stirring with a pipette tip (see Note 2).
3. Incubate cells on ice for 20 min before subjecting to a 90 s heat shock at 42°C, and then return immediately onto ice for 2 min.

4. Add 200 μL of SOC medium to the cells and incubate at 37°C with shaking (220 rpm) for 1 h.
5. Plate cells onto LB agar plate supplemented with chloramphenicol and appropriate antibiotic for plasmid containing gene encoding for amino-terminal acetylation target protein.

3.3. Expression of Acetylated Recombinant Proteins in E. coli

1. Set up a culture for protein expression as soon as colonies have formed on the BL21-DE3-pNatB transformation plate (see Note 3).

   If protein is not toxic to *E. coli* cells, go through steps 2–4. However, if protein is toxic or you find its expression is repressed in the cell, go straight to step 5.
2. Set up a 50 mL starter culture in NZY media (see Note 4) supplemented with appropriate antibiotics in a 250 mL Erlenmeyer flask using a single-colony inoculum from the BL21-DE3-pNatB transformation plate.
3. Grow starter culture overnight at 37°C with shaking (see Note 5).
4. Inoculate a larger volume (i.e., 1 L) culture using 1 mL inoculum from the starter culture for every 200 mL culture volume (i.e., for a 1 L culture, use a 5 mL inoculum). Culture cells at 37°C with vigorous shaking (220 rpm). Go straight to step 6.
5. Inoculate 200 mL of NZY supplemented with appropriate antibiotics in a 500 mL Erlenmeyer flask using a single-colony inoculum from the BL21-DE3-pNatB transformation plate from step 1.
6. Culture at 37°C (see Note 6) until the 595 nm optical density reaches value between 0.4 and 0.5.
7. Induce expression of NatB complex by adding 1 mL of 100 mg/mL IPTG/L culture volume. Take a 1 mL sample of cells, which you should pellet and store at −20°C for running on a gel to compare recombinant protein expression.
8. If gene expressing target protein is not under control of T7 promoter, induce expression by addition of appropriate molecule (see Note 6).
9. Take samples at hourly time points to check production (see Note 7).
10. Grow cells for 2–4 h from induction and then harvest by centrifugation at 4,000×g at 4°C for 30 min. Store pellets at −20°C until protein purification is undertaken.
11. Run an SDS-PAGE gel of samples taken at induction and post-induction. Coomassie stain the gel and check NatB complex and target protein that have expressed (see, e.g., Fig. 1).

3.4. Confirmation of Amino-Terminal Acetylation

1. Purify target protein using the same method as you normally would use for the unacetylated form.

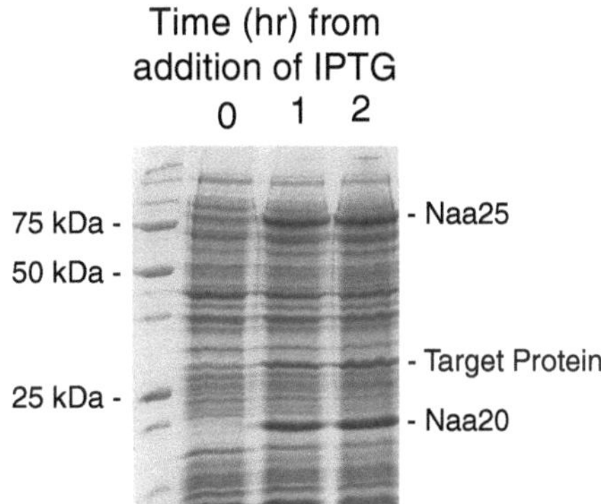

Fig. 1. Coomassie-stained SDS-PAGE gel showing induction of the NatB complex and a target protein. Once a fresh culture of BL21-DE3-pNatB pTarget cells had reached an $OD_{595\,nm}$ of 0.4, IPTG was added to a final concentration 100 μg/mL. Samples of cells were taken at 0, 1, and 2 h from addition of IPTG, boiled in SDS buffer, and run on an SDS-PAGE gel. Intense protein bands can be observed in the each of the post-induction samples which have migrated with masses corresponding to the NatB catalytic subunit, Naa20 (20.5 kDa); the NatB regulatory subunit, Naa25 (92.4 kDa); and the target protein.

2. It is worth running a gel to compare the migration of equivalent amounts of unacetylated and acetylated proteins, as differences in migration can give an indication of acetylation efficiency. Acetylation can affect stability of amino-terminal secondary structure, which can be reflected in the ability of a protein to migrate through an SDS-PAGE matrix (5).
3. Determine size of potentially acetylated protein using electron-spray mass spectroscopy, and compare size with unacetylated protein. Acetylation will increase protein mass by 42 Da (see, e.g., Fig. 2).
4. If appropriate you may wish to test acetylation using a functional assay for the target protein, comparing with unacetylated and purified endogenous acetylated protein (Notes 8–11).

4. Notes

1. The pNatB (pACYCduet-*naa20*$^+$-*naa25*$^+$) plasmid (4), allowing co-expression of the catalytic (Naa20) and regulatory (Naa25) subunits of the fission yeast NatB complex, is freely available upon request to researchers working in nonprofit organizations. Send requests to d.p.mulvihill@kent.ac.uk.
2. Ensure plasmid from which target protein is expressed confers resistance to an antibiotic, which is distinct from that of pNatB (i.e., an antibiotic other than chloramphenicol).
3. In our experience, expression and acetylation is most efficient when freshly transformed cells are used.

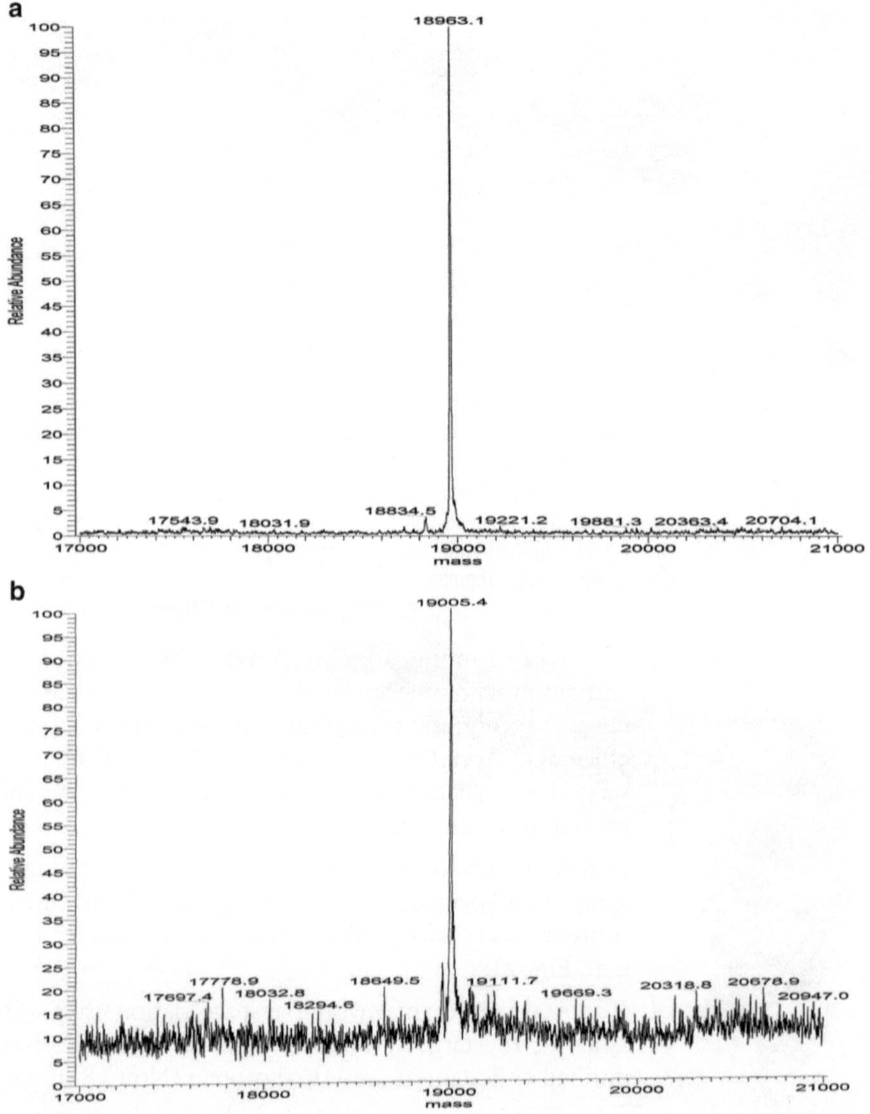

Fig. 2. Deconvolved mass spectra of fission yeast tropomyosin (Cdc8) purified from *E. coli* either lacking (**a**) or containing (**b**) the pNatB plasmid. Using established methods (5), untagged Cdc8 was purified from either (**a**) BL21-DE3 pJC20cdc8 or (**b**) BL21-DE3-pNatB pJC20cdc8 cells. Subsequent electron-spray mass spectroscopy of the purified proteins confirmed that all of the Cdc8 purified from the BL21-DE3-pNatB cells had an additional 42 Da mass (**b**), the expected mass of the acetyl group, when compared to Cdc8 from BL21-DE3 cells (**a**). Amino-terminal acetylation was also confirmed by functional assays, western blot analysis, and the inability to sequence the peptide from the amino terminus.

4. Expression and acetylation efficiency can be affected by choice of media. For many proteins yield is improved by growing cells in NZY when compared to LB medium. However we have received reports that in some cases minimal media improve the solubility and acetylation efficiency for some proteins.

5. We have found that even expression of minimal quantities of some acetylated proteins can be toxic to the *E. coli* cell (i.e., from leaky expression from many un-induced promoters) and prolonged growth/culture leads to dramatic reduction in protein yields. Therefore, for many proteins acetylation and expression yield can be dramatically improved by leaving out the starter culture step and instead inoculating a large volume culture with a single colony from a plate of freshly transformed BL21-DE3-pNatB cells.
6. Solubility and expression of acetylated proteins are sometimes improved by growing bacterial cells at lower temperatures (i.e., 20°C or 25°C). Undertake small-scale test inductions at different temperatures if you need to improved yield of acetylated protein.
7. Using T7 promoters to control expression of target protein and NatB complex allows simultaneous and efficient induction. However sequential induction may be desirable if acetylation yield is low. Therefore you may wish to consider putting the target gene under the control of a different promoter (e.g., arabinose) and induce its expression ~30 min after the NatB complex has been expressed.
8. Confirm that the expression of the NatB complex does not abolish the expression of the target protein. You may find that expression is reduced; however this is to be expected.
9. This method has been used successfully to produce recombinant amino-terminally acetylated proteins from *E. coli* with each of the three NatB recognition sequences (i.e., M-E-, M-D-, and M-N-) ((4) + unpublished results).
10. If you find that you normally obtain higher protein yields when expressing your target protein of interest in a specific *E. coli* strain (e.g., BL21 Star or Rosetta DE3), it is likely that you will obtain higher yield of the acetylated form in the same strain co-transformed with the pNatB plasmid.
11. Remember, if you are using affinity tags during the purification of your protein, ensure you tag the protein at the carboxyl terminus and not at the amino terminus you are attempting to acetylate.

Acknowledgment

Work in these labs is supported by the BBSRC, Royal Society, and Wellcome Trust.

References

1. Polevoda B, Sherman F (2003) N-terminal acetyltransferases and sequence requirements for N-terminal acetylation of eukaryotic proteins. J Mol Biol 325:595–622
2. Arnesen T, Van Damme P, Polevoda B, Helsens K, Evjenth R, Colaert N, Varhaug JE, Vandekerckhove J, Lillehaug JR, Sherman F, Gevaert K (2009) Proteomics analyses reveal the evolutionary conservation and divergence of N-terminal acetyltransferases from yeast and humans. Proc Natl Acad Sci U S A 106:8157–8162
3. Polevoda B, Arnesen T, Sherman F (2009) A synopsis of eukaryotic Nalpha-terminal acetyltransferases: nomenclature, subunits and substrates. BMC Proc 3(Suppl 6):S2
4. Johnson M, Coulton AT, Geeves MA, Mulvihill DP (2010) Targeted amino-terminal acetylation of recombinant proteins in *E. coli*. PLoS One 5:e15801
5. Skoumpla K, Coulton AT, Lehman W, Geeves MA, Mulvihill DP (2007) Acetylation regulates tropomyosin function in the fission yeast *Schizosaccharomyces pombe*. J Cell Sci 120:1635–1645

Chapter 16

Identification of Lysine Acetyltransferase Substrates Using Bioorthogonal Chemical Proteomics

Markus Grammel and Howard C. Hang

Abstract

Bioorthogonal chemical proteomics is a valuable method to identify enzyme-specific substrates, a challenging task by traditional biochemical standards. The addition of recombinant enzyme and alkynyl chemical reporter to complex protein mixtures, such as cell lysates, allows the detection and identification of modified substrates. Proteins that have been modified with the chemical reporter can be selectively labeled with fluorescent dyes for detection or affinity tags for biochemical enrichment and subsequent identification by mass spectrometry. Here, we describe the detection and identification of substrates of the lysine acetyltransferase p300 in nuclear extracts using the chemical reporter 4-pentynoyl-CoA.

Key words: Acetylation, Bioorthogonal chemical proteomics, Click chemistry, 4-Pentynoyl-CoA

1. Introduction

The study of posttranslational modifications (PTMs) still poses a great analytical challenge for biological research. In strong contrast to many nucleic acid-directed analytical methods, which allow analyte amplification by template-based cloning and PCR approaches, PTMs are particularly difficult to analyze from a chemical standpoint, because they represent the proverbial needle in the haystack. Three main factors complicate the analysis: in essence all proteins are chemically identical (possess the same functional groups), the dynamic range of protein copy numbers spans multiple orders of magnitude, and many PTMs are present only on a small fraction of all copies of an individual gene product. Protein lysine acetylation, an important PTM with various roles in epigenetics, cancer biology, and other fields of biology, is further characterized by the lack of an intrinsic chemical property for affinity enrichment (e.g., phosphorylated peptides bind to titanium dioxide) (1–3). These facts require analytical reagents with exquisite

Sandra B. Hake and Christian J. Janzen (eds.), *Protein Acetylation: Methods and Protocols*, Methods in Molecular Biology, vol. 981, DOI 10.1007/978-1-62703-305-3_16, © Springer Science+Business Media, LLC 2013

chemical selectivity for protein lysine acetylation. Traditionally, acetylation has been detected by radioactive metabolic labeling or immunoblotting with PTM-specific antibodies (see also Chapter 11). However, both methods suffer from various drawbacks (e.g., hazardous materials, long incubation times, lack of specificity). While tandem mass spectrometry has facilitated the identification of thousands of new acetylation substrates, it requires the preceding enrichment of proteins carrying the PTM, which in many cases again requires acetylation-specific antibodies (1). Furthermore it is still exceedingly difficult to identify the corresponding acetyltransferase for individual acetylation substrates.

Bioorthogonal alkynyl chemical reporters provide a way to circumvent many of these problems and allow rapid detection of PTMs as well as protein substrate identification by mass spectrometry (4). These reporters mimic natural metabolites or cofactors and are enzymatically installed on protein substrates instead of the natural PTM. Once they are covalently bound to their protein substrate, they provide a chemical handle (alkyne), which allows further chemical functionalization of the substrate. The alkyne can be highly selectively reacted with azide-containing reagents by the Cu(I)-catalyzed azide-alkyne cycloaddition (CuAAC) to form a triazole, which covalently connects the modified protein with the azide-containing reagent (5) (Fig. 2).

We have recently developed alkynyl-acetate and alkynyl-acetyl-CoA-based chemical reporters for the analysis of protein lysine acetylation (6) (Fig. 1). These reporters are incorporated instead of the native acetyl group and allow the detection of acetylated proteins by reacting the labeled substrates with the azide-fluorescent dye azido-rhodamine (az-rho, Fig. 1) by CuAAC. In addition, acetylation substrates that have been labeled with the chemical reporter can be ligated to azido-azo-biotin (Fig. 1), a cleavable biotin affinity enrichment tag, in the same manner. Azido-azo-biotin-modified proteins can be selectively enriched with streptavidin agarose beads and selectively eluted through reduction of the azo-bond with the reducing agent sodium dithionite (7). The eluted proteins and their sites of modification with the chemical reporter can be subsequently identified using tandem mass spectrometry. While the alkynyl-acetate-based salts (e.g., sodium pent-4-ynoate) are readily incorporated in proteomes by metabolic labeling, the alkynyl-acetyl-CoA-based cofactor analogs serve as versatile reporters for in vitro protein lysine acetylation.

These CoA-based reporters allow in vitro labeling of individual substrates by recombinant enzymes such as the lysine acetyl transferase (KAT) p300. In particular, 4-pentynoyl-CoA serves as an efficient chemical reporter for the KAT p300 (6). We could show that p300 uses 4-pentynoyl-CoA to label histone proteins, a bona fide p300 substrate, as well as histone peptides in vitro. Furthermore, we could demonstrate the in vitro labeling of extracted core histones

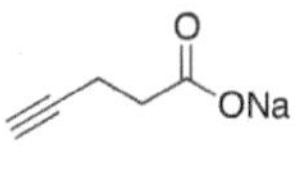

sodium pent-4-ynoate

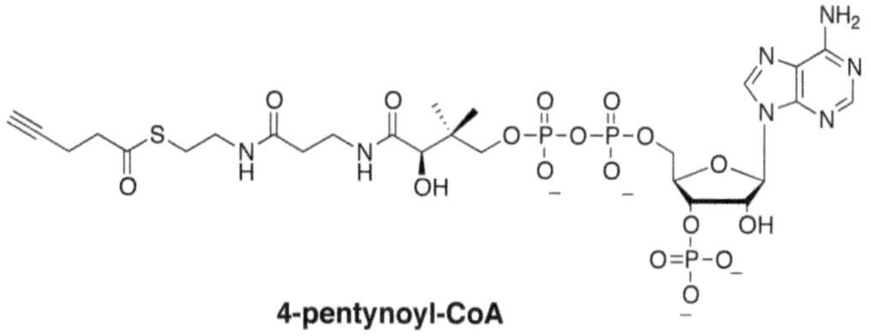

4-pentynoyl-CoA

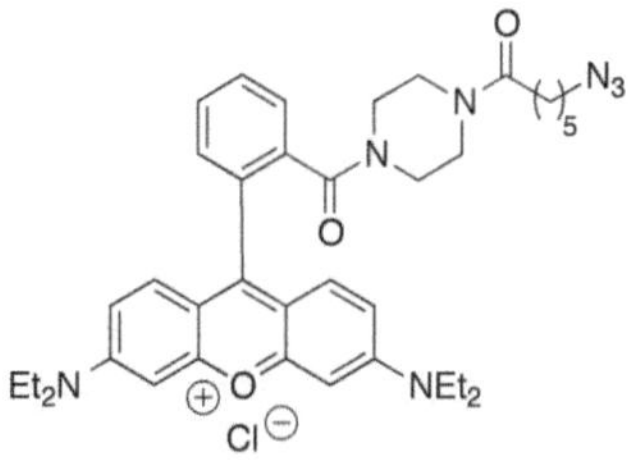

azido-rhodamine

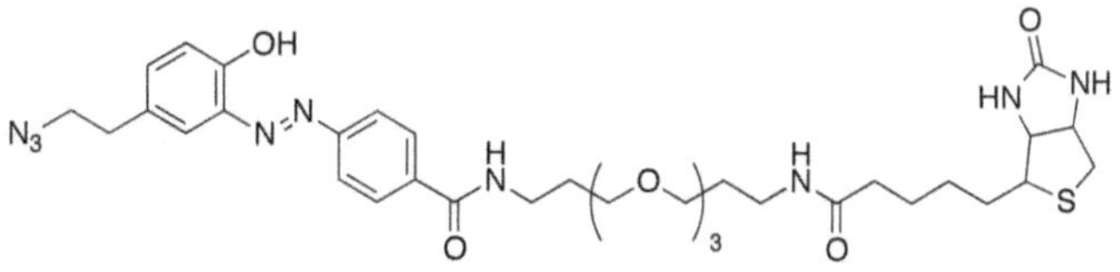

azido-azo-biotin

Fig. 1. Important reagents. Sodium pent-4-ynoate—chemical reporter for protein acetylation (used for metabolic labeling), 4-pentynoyl-CoA—chemical reporter for protein acetylation (used for in vitro labeling in cell lysates), azido-rhodamine—azide-bearing fluorescence dye for visualization of acetylation proteins by in-gel fluorescence scanning, and azido-azo-biotin—azide-bearing biotin affinity tag for the enrichment of acetylated proteins.

by p300 (8) (Fig. 3a). Extending this approach, p300 in combination with 4-pentynoyl-CoA labels distinct protein populations when added to nuclear extracts of mammalian cells (8) (Fig. 3b). These proteins can be identified by mass spectrometry upon ligation to azido-azo-biotin, affinity enrichment, and mass spectrometry-based protein identification. Here, we present a detailed protocol for p300 substrate labeling in cell lysates, visualization of 4-pentynoyl-CoA-modified proteins by in-gel fluorescence gel scanning, and sample preparation for identification of p300-specific protein substrates through affinity enrichment with streptavidin beads (Fig. 2).

2. Materials

2.1. In-Gel Fluorescence Analysis of In Vitro Labeled p300 Substrates

1. 2 mM 4-pentynoyl-CoA in water (Yang et al. (6) or Cayman Chemical #10547).
2. 50 mM Tris–HCl, pH 7.9, 10% glycerol.
3. Methanol, chloroform, acetone.
4. Vortex bench mixer.
5. Table centrifuge (Eppendorf Centrifuge 5417R).

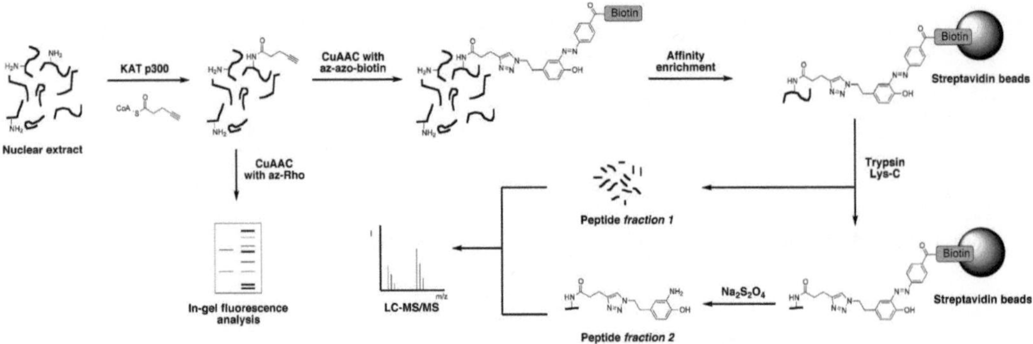

Fig. 2. Labeling of p300 substrates with 4-pentynoyl-CoA chemical reporter in nuclear extracts. The lysine acetyltransferase (KAT) p300 is added to one sample of nuclear extract and omitted in the negative control sample. KAT p300 transfers the 4-pentynoyl reporter group from 4-pentynoyl-CoA onto lysine target residues on its cognate protein substrates. Cu(I)-catalyzed alkyne-azide cycloaddition (CuAAC) is subsequently used to tag all labeled proteins with either azido-rhodamine (az-Rho) for in-gel fluorescence analysis or azido-azo-biotin (az-azo-biotin) for mass spectrometry-based protein identification. The biotin moiety allows affinity enrichment of 4-pentynoyl-CoA-labeled proteins on streptavidin agarose beads. Captured proteins are digested with trypsin and Lys-C on beads, and the resulting peptide fraction 1 is collected for LC–MS/MS analysis. The remaining peptides, still bound to the affinity beads, contain the sites of modification. These peptides (fraction 2) are then eluted with sodium dithionite ($Na_2S_2O_4$), yielding the characteristic product of the azo-bond reduction, which can be identified by LC–MS/MS analysis (Reprint with permission by Bioorg Med Chem Lett).

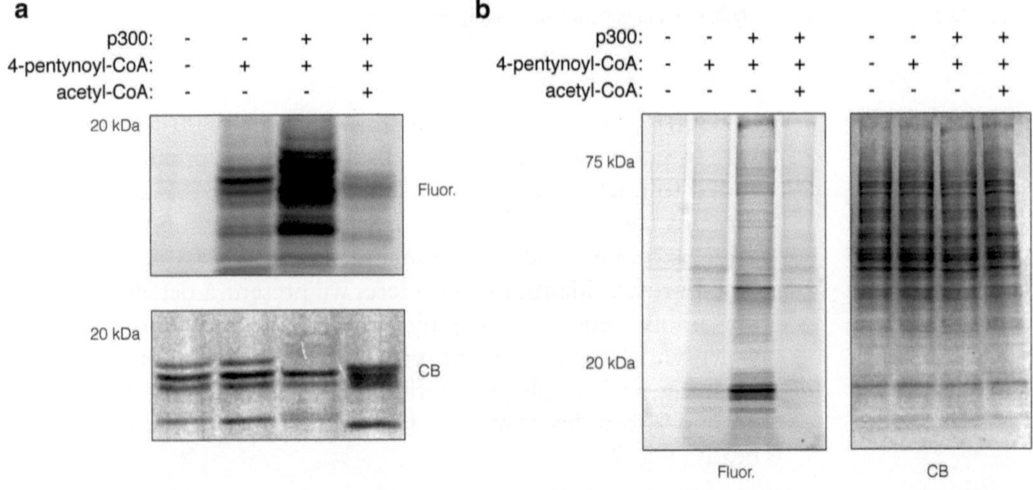

Fig. 3. In-gel fluorescence analysis of p300-specific labeling of extracted core histones and nuclear extracts with 4-pentynoyl-CoA. (**a**) In-gel fluorescence analysis of p300-catalyzed 4-pentynoylated core histone proteins. p300-catalyzed acylation reactions were carried out with 5 μg extracted core histones, 50 μM 4-pentynoyl-CoA (lanes 2–4), and 125 ng p300 (lanes 3–4) in 50 mM, pH 7.9 Tris buffer (+10% glycerol) for 2 h incubation time at 30°C. In lane 4, 500 μM acetyl-CoA was preincubated with core histones in the presence of p300 for 10 min followed by the addition of 50 μM 4-pentynoyl-CoA and co-incubation for additional 2 h. The crude reaction products were subjected to CuAAC with az-Rho for 1 h followed by separation on SDS-PAGE and in-gel fluorescence scanning. (**b**) In-gel fluorescence analysis of the p300-catalyzed 4-pentynoylated proteins in HeLa cell nuclear extract. p300-catalyzed acylation reactions were carried out with 10 μg HeLa cell nuclear extract, 50 μM 4-pentynoyl-CoA (lanes 2–4), and 250 ng p300 (lanes 3–4) in 50 mM, pH 7.9 Tris buffer (+10% glycerol) for 2 h incubation time at 30°C. In lane 4, 500 μM acetyl-CoA was preincubated with nuclear extract in the presence of p300 for 10 min followed by the addition of 50 μM 4-pentynoyl-CoA and co-incubation for additional 2 h. The crude reaction products were subjected to CuAAC with az-Rho for 1 h followed by separation on SDS-PAGE and in-gel fluorescence scanning. Lanes are counted consecutively from to . Coomassie blue (CB), fluorescence (Fluor.) (Reprint with permission by Bioorg Med Chem Lett).

6. 4% SDS in phosphate-buffered saline (PBS).
7. Water bath tabletop sonicator.
8. 10 mM azido-rhodamine (9) or alternatively tetramethylrhodamine-5-carbonyl azide ((10) or Life Technologies) in dimethyl sulfoxide (DMSO).
9. 50 mM tris(2-carboxyethyl)phosphine (TCEP) in water.
10. 2 mM ris((1-benzyl-1H-1,2,3-triazol-4-yl)methyl)amine (TBTA) in 4:1 (v/v) *n*-butanol/DMSO.
11. 50 mM $CuSO_4$ in water.
12. 4× Laemmli sample buffer.
13. 2-Mercaptoethanol.
14. Fluorescent protein ladder (e.g., Precision Plus Dual Color, BioRad).
15. Criterion Precast Gels 4–20% Tris–HCl 18-well (BioRad).
16. Destaining solution: 50% (v/v) water, 40% (v/v) methanol, 10% (v/v) acetic acid.
17. Fluorescence gel scanner with a 532 nm excitation and 580 nm filter and 30 nm band-pass (e.g., Typhoon 9400 scanner, Amersham Biosciences).

2.2. Proteomic Identification of 4-Pentynoylated Proteins as well as Their 4-Pentynoylation Sites

See Subheading 2.1 and in addition:

1. 10 mM azido-azo-biotin (6) or alternatively azide-biotin ((10) or Life Technologies; does not allow selective elution with sodium dithionite but on bead capture and digest).
2. 6 M urea, 2 M thiourea in 10 mM HEPES, pH 8.0.
3. 100 mM DTT in water.
4. 550 mM iodoacetamide in water.
5. High capacity streptavidin agarose resin (Thermo).
6. 0.2% SDS in PBS.
7. 4.5 M urea, 1.5 M thiourea in 10 mM HEPES, pH 8.0.
8. Lysyl endopeptidase (Lys-C) mass spectrometry grade (Wako), trypsin endoproteinase mass spectrometry grade (Thermo).
9. 25 mM sodium dithionite in PBS.
10. SPE cartridges (Sep-Pak Vac 1 cm^3 (100 mg) C8 cartridge, Waters).
11. Trifluoroacetic acid (TFA).
12. 70% Acetonitrile (ACN), 0.1% TFA in water.
13. 0.1% TFA in water.
14. 70% ACN in water.

3. Methods

3.1. In-Gel Fluorescence Analysis of In Vitro Labeled p300 Substrates

1. Incubate 125 ng KAT p300 (125 ng/10 μg nuclear extract) with 10 μg nuclear extract (11) and 100 μM 4-pentynoyl-CoA (see Note 1) in 50 mM Tris–HCl, pH 7.9, 10% glycerol for 2 h at 30°C in a total volume of 15 μL in a 1.5 mL microcentrifuge tube (see Note 2).
2. Stop acylation reaction by methanol chloroform precipitation. Bring reaction mix to a total volume of 100 μL by addition of H_2O and add 400 μL methanol and 100 μL chloroform (see Note 3). Vortex vigorously. Add 300 μL of H_2O and vortex vigorously. Spin tubes for 5 min at 20,000 × *g* at room temperature and carefully discard the upper aqueous phase (see Note 4). Add 1 mL of ice-cold acetone and invert a few times. Spin tubes for 10 min at 20,000 × *g* at 4°C and carefully discard the supernatant (see Note 5) without dislodging the formed protein pellet. Dry the protein pellet for 5 min at room temperature.
3. Resolubilize the protein pellets in 15 μL 4% SDS in PBS by brief sonication (see Note 6).
4. Prepare a click chemistry master mix by adding the following reagents (use a multiple of the listed volumes, depending on the number of samples; see Note 7):
   (a) 7.5 μL of 4% SDS in PBS.
   (b) 0.25 μL of 10 mM azido-rhodamine.
   (c) 0.5 μL of 50 mM TCEP.
   (d) 1.25 μL of 2 mM TBTA.
   (e) 0.5 μL of 50 mM $CuSO_4$.
5. Add 10 μL of click chemistry master mix to each sample and vortex vigorously. Incubate for 1 h at room temperature on benchtop.
6. Stop click chemistry reaction by methanol chloroform precipitation (see step 2).
7. Resolubilize the protein pellets in 15 μL 4% SDS in PBS by brief sonication.
8. Add 5 μL 4× Laemmli buffer and 1 μL 2-mercaptoethanol; vortex and incubate the samples for 5 min at 95°C.
9. Spin samples for 30 s at 20,000 × *g* in a tabletop centrifuge at room temperature.
10. Load samples and fluorescent protein ladder on 4–20% Tris–HCl SDS-PAGE gel and run at 150 V until blue dye front has left the gel (see Note 8).

11. Destain the gel for 1 h in destaining solution by rocking on shaker to remove traces of unreacted rhodamine and loading buffer.
12. Soak gel in H_2O for 30 min.
13. Scan the gel using a fluorescence gel scanner.
14. Stain the gel with Coomassie blue to visualize the total protein load.

3.2. Proteomic Identification of 4-Pentynoylated Proteins as well as Their 4-Pentynoylation Sites

3.2.1. Acylation Reaction

1. For large-scale proteomic analysis, incubate 12.5 μg KAT p300 with 1.0 mg nuclear extract and 100 μM 4-pentynoyl-CoA in 50 mM Tris–HCl, pH 7.9, 10% glycerol for 2 h at 30°C in a total volume of 1 mL in a 15 mL centrifuge tube (see Note 9).
2. Stop the acylation reaction by methanol chloroform precipitation. Add 4 mL methanol and 1 mL chloroform. Vortex vigorously. Add 3 mL of H_2O and vortex vigorously. Spin tubes for 30 min at 5,200×*g* at 4°C and carefully discard the upper aqueous phase. Add 10 mL of ice-cold acetone and invert a few times. Spin tubes for 30 min at 5,200×*g* at 4°C and carefully discard the supernatant without dislodging the formed protein pellet. Dry the protein pellet for 5 min at room temperature.
3. Resolubilize the protein pellets in 400 μL 4% SDS in PBS by brief sonication.

3.2.2. Click Chemistry

1. Prepare a click chemistry master mix by adding the following reagents (use a multiple of the listed volumes, depending on the number of samples; see Note 10):
   (a) 500 μL of 4% SDS in PBS.
   (b) 10 μL of 10 mM azido-azo-biotin.
   (c) 20 μL of 50 mM TCEP.
   (d) 50 μL of 2 mM TBTA.
   (e) 20 μL of 50 mM $CuSO_4$.
2. Add 600 μL of click chemistry master mix to each sample and vortex vigorously. Incubate for 2 h at room temperature on benchtop.
3. Stop click chemistry reaction by methanol chloroform precipitation (see step 2).

3.2.3. Protein Reduction and Alkylation

1. Resolubilize protein pellets in 1 mL 6 M/2 M urea/thiourea in 10 mM HEPES, pH 8.0.
2. Add 10 μL 100 mM DTT stock solution and incubate proteins for 40 min at room temperature on end-over-end rotator (see Note 11).
3. Add 10 μL 550 mM iodoacetamide stock solution and incubate for 30 min at room temperature on end-over-end rotator in the dark.

3.2.4. Protein Enrichment, On-Bead Digest, and Elution

1. Transfer protein solution to 100 μL prewashed streptavidin beads in a 1.5 mL microcentrifuge tube (see Note 12).
2. Incubate beads with protein solution on end-over-end rotator for 1.5 h at room temperature in the dark.
3. Wash beads three times each with 1 mL 6 M/2 M urea/thiourea in 10 mM HEPES, pH 8.0, 0.2% SDS in PBS and PBS. Wash beads once more with 4.5 M/1.5 M urea/thiourea in 10 mM HEPES, pH 8.0.
4. Resuspend beads in 200 μL 4.5 M/1.5 M urea/thiourea in 10 mM HEPES, pH 8.0 containing 10 μg Lys-C and incubate at room temperature for 4 h in shaker at 1,000 rpm on Eppendorf thermomixer (see Note 13).
5. Dilute the urea concentration by adding 600 μL 10 mM HEPES, pH 8.0 and incubate with 10 μg trypsin at room temperature in shaker (1,000 rpm) for 18 h (see Note 14).
6. Spin tubes for 1 min at 2,000 × *g* at room temperature and collect supernatant ("fraction 1" (Fig. 2); see Note 15).
7. Add 1 mL PBS, invert tubes to wash beads, and spin tubes for 1 min at 2,000 × *g* at room temperature. Discard supernatant.
8. Add 300 μL 25 mM sodium dithionite in PBS to the beads and incubate for 30 min at room temperature on benchtop.
9. Spin tubes for 1 min at 2,000 × *g* at room temperature and collect supernatant ("fraction 2" (Fig. 2); see Note 16).
10. Repeat steps 9 and 10 and pool elutes.

3.2.5. Desalting

1. Acidify peptide samples with 10% trifluoroacetic acid (TFA) to pH 2 (see Note 17).
2. Place SPE cartridge in 15 mL tube.
3. Add 1 mL methanol and spin at 1,500 × *g* for 1 min.
4. Add 0.5 mL 70% acetonitrile (ACN), 0.1% TFA and spin at 1,500 × *g* for 1 min.
5. Add 0.5 mL 0.1% TFA and spin at 1,500 × *g* for 1 min.
6. Repeat step 4.
7. Load peptide samples and spin at 1,500 × *g* for 1 min.
8. Recover flow-through and load again.
9. Repeat step 7 twice.
10. Add 0.5 mL 0.1% TFA and spin at 150 × *g* for 3 min.
11. Repeat step 10 twice (see Note 18).
12. Transfer the cartridge to a new tube and add 0.5 mL 70% ACN and spin at 150 × *g* for 3 min. The eluate contains the desalted peptides.

4. Notes

1. 4-Pentynoyl-CoA can be stored as a 2 mM stock solution in H_2O at −80°C.
2. Depending on the KAT, the buffer conditions might have to be changed. Typically the same buffer conditions as for regular in vitro acetylation reactions for a particular KAT can be used. The total amount of 4-pentynoyl-CoA as well as the ratio of KAT to cell lysate needs to be titrated for each individual enzyme cell lysate pair. It is crucial to dialyze the generated cell lysate before use for in vitro acylation to remove any free endogenous cofactor.
3. Chloroform needs to be handled in a chemical fume hood.
4. The precipitated protein should be visible as a very small white flake at the interphase between aqueous and organic phase.
5. The collected supernatant, which contains acetone and chloroform, needs to be disposed in waste for halogenated organic solvents.
6. Brief sonication (15–30 s should be sufficient) in a water bath tabletop sonicator.
7. Prepare one or two more volumes of master mix than required to have enough volume for all samples. A 10 mM solution of azido-rhodamine in DMSO can be stored at 4°C. A 2 mM solution of TBTA in n-butanol:DMSO (4:1) can be stored at 4°C. The aqueous solutions of TCEP and $CuSO_4$ have to be prepared freshly every time. Vortex vigorously after addition of each reagent.
8. Empty wells should be filled with an equal volume of blank loading buffer (4× Laemmli buffer diluted 1:4 in 4% SDS in PBS) to avoid protein spreading during the run. The fluorescent protein ladder may require prior dilution to reduce the fluorescence intensity. We dilute Precision Plus Dual Color (BioRad) protein ladder 1:1,000 in 4× Laemmli buffer diluted 1:4 in 4% SDS in PBS and load 20 μL.
9. It is essential to include a negative control, which lacks the added recombinant enzyme, but is otherwise treated absolutely identically.
10. A 10 mM stock solution of azido-azo-biotin can be stored at −20°C.
11. Reduction and alkylation of cysteines ensure linear tryptic peptides without disulfide bonds.
12. Bead equilibration, binding, and washing should be conducted in dolphin microcentrifuge tubes to reduce loss during processing. Wash and equilibrate streptavidin beads with three

1 mL washes of 6 M/2 M urea/thiourea in 10 mM HEPES, pH 8.0. Drain beads completely before adding alkylated protein solutions.

13. In in-gel and in-solution proteolytic digests for shotgun proteomics, a wide range of ratios of protease to total protein are commonly used (w/w of 1:25 to 1:100). Since the total amount of proteins retained on the beads cannot be easily determined, we base the amount of protease on the total protein input. We have successfully applied a protease to total input ratio (w/w) of 1:100 for Lys-C and trypsin.
14. To retain trypsin activity, the urea concentration has to be diluted fourfold.
15. This fraction (peptide fraction 1, Fig. 2) contains the majority of peptides. These peptides should not contain any sides of 4-pentynoylation.
16. This fraction (peptide fraction 2, Fig. 2) contains peptides that carry the 4-pentynoylation site and were directly bound to the streptavidin beads.
17. Peptide solutions have to be acidified for effective retention by SPE cartridges.
18. Peptides can be stored at this stage for many weeks at 4°C.

References

1. Witze ES, Old WM, Resing KA, Ahn NG (2007) Mapping protein post-translational modifications with mass spectrometry. Nat Methods 4:798–806
2. Choudhary C, Kumar C, Gnad F, Nielsen ML, Rehman M, Walther TC, Olsen JV, Mann M (2009) Lysine acetylation targets protein complexes and co-regulates major cellular functions. Science 325:834–840
3. Yang Y-Y, Hang HC (2011) Chemical approaches for the detection and synthesis of acetylated proteins. Chembiochem 12: 314–322
4. Prescher JA, Bertozzi CR (2005) Chemistry in living systems. Nat Chem Biol 1:13–21
5. Sletten EM, Bertozzi CR (2009) Bioorthogonal chemistry: fishing for selectivity in a sea of functionality. Angew Chem Int Ed Engl 48:6974–6998
6. Yang Y-Y, Ascano JM, Hang HC (2010) Bioorthogonal chemical reporters for monitoring protein acetylation. J Am Chem Soc 132: 3640–3641
7. Yang Y-Y, Grammel M, Raghavan AS, Charron G, Hang HC (2010) Comparative analysis of cleavable azobenzene-based affinity tags for bioorthogonal chemical proteomics. Chem Biol 17:1212–1222
8. Yu-Ying Y, Markus G, Howard HC (2011) Identification of lysine acetyltransferase p300 substrates using 4-pentynoyl-coenzyme A and bioorthogonal proteomics. Bioorg Med Chem Lett 21:4976–4979
9. Charron G, Zhang MM, Yount JS, Wilson J, Raghavan AS, Shamir E, Hang HC (2009) Robust fluorescent detection of protein fatty-acylation with chemical reporters. J Am Chem Soc 131:4967–4975
10. Martin BR, Cravatt BF (2009) Large-scale profiling of protein palmitoylation in mammalian cells. Nat Methods 6:135–138
11. Dignam JD, Lebovitz RM, Roeder RG (1983) Accurate transcription initiation by RNA polymerase II in a soluble extract from isolated mammalian nuclei. Nucleic Acids Res 11: 1475–1489

Chapter 17

Nonradioactive In Vitro Assays for Histone Deacetylases

Alexander-Thomas Hauser, Julia M. Gajer (née Wagner), and Manfred Jung

Abstract

The effect of histone deacetylases (HDACs) on normal and aberrant gene expression has been studied widely, making these enzymes interesting targets for the treatment of cancer and other diseases. In this chapter, we present in vitro assays that are commonly used to detect HDAC activity that do not rely on radioactive substrates and are amenable for high-throughput testing in microtiter plates. The major focus is on in vitro screening, but we also provide protocols to monitor HDAC activity from cancer cells and peripheral white blood cells. We will discuss the advantages and drawbacks of the respective protocols and give general hints and suggestions that are valuable to obtain reliable and reproducible results.

Key words: Histone deacetylases, In vitro assays, Cellular assays, Fluorescence-based assays

1. Introduction

Histones are basic proteins that together with DNA and nonhistone proteins form the nucleosomes as repetitive entities of the chromatin (1, 2). Histone tails that are protruding out of the nucleosome core are rich in basic amino acids like lysine and arginine and also contain hydroxyl groups of tyrosines, serines, and threonines. They are therefore accessible for different posttranslational modifications including acetylation, methylation, or phosphorylation (3).

The reversible acetylation of histone lysines is one of the best-studied posttranslational histone modifications. The steady state of lysine acetylation is maintained by histone acetyltransferases (HATs) on the one hand and histone deacetylases (HDACs) on the other hand. The HDACs can be separated in four different classes (I–IV)

Sandra B. Hake and Christian J. Janzen (eds.), *Protein Acetylation: Methods and Protocols*, Methods in Molecular Biology, vol. 981, DOI 10.1007/978-1-62703-305-3_17,

based on their homology to yeast transcriptional repressors (4). The 11 histone deacetylases from the classes I, II, and IV, which are zinc-dependent amidohydrolases (5, 6), are referred to as "classical" HDACs, whereas class III HDACs are called sirtuins. The sirtuins convert their lysine substrates via a mechanism that requires NAD^+, and the human enzyme family comprises seven members (7).

Aberrant histone deacetylation is accepted as a key driving force in the development of several diseases such as cancer, and an inhibition of histone deacetylases has been shown to effect apoptosis, cell growth arrest, cellular differentiation, and anti-angiogenesis (8, 9). Therefore, HDAC inhibitors have emerged as promising anticancer drugs over the last years, and so far, already two inhibitors of the "classical" zinc-dependent HDACs have been approved for the treatment of cutaneous T-cell lymphoma. A variety of assay setups have been developed that have accompanied the characterization of HDAC inhibitors and the enzymes themselves.

The first assays that were performed to detect the activity of histone deacetylases have been based on the measurement of released tritiated acetic acid from radiolabeled histones or oligopeptide substrates. These protocols comprise incubation of enzyme samples with proteins or oligopeptides that have previously been labeled with (^{3}H) acetate. After an extraction with ethyl acetate and a centrifugation step, an aliquot of the overlaying organic phase is added to a liquid scintillation cocktail and counted for radioactivity in a scintillation spectrophotometer (10). With, e.g., the scintillation proximity assay (SPA), such radioactive HDAC assays also were developed in a homogeneous manner (11), but all these assay setups suffered from the same disadvantages affiliated with radioactivity, such as decontamination of laboratory equipment, exposure of personnel to radiation, and the problematical waste disposal.

Thus, efforts have been made to develop nonisotopic deacetylase substrates that allow for a nonradioactive detection of enzyme activity. In this chapter, we want to introduce and specify in vitro assay protocols that include the use of fluorescence-labeled substrates enabling convenient and reproducible measurement of enzyme activity.

Fluorescence-based in vitro assays have become valuable tools for monitoring histone deacetylase activity respectively its inhibition. This is mostly used for screening potential inhibitors in random screening (12) but also for measuring deacetylase activity in tissue or blood samples from animal testing (13) or in clinical studies (14). We focused here on the most common assay principles that we regularly use in our group. Further variants are reviewed elsewhere (15, 16). Assay kits are commercially available for many of these purposes, but the protocols here provide a much more cost effective (even if, e.g., MAL is bought from commercial sources) approach, and the protocols can be easier refined and optimized to a specific study if the nature of all reagents is known. Therefore, these assays will continue to serve as valuable tools in drug discovery and development.

2. Materials

For 2.1 and 2.2, lysine derivatives fluorescently labeled with 7-amino-4-methyl coumarin (AMC) are used that have been developed in our group. Specifically for rat liver HDACs or recombinant HDACs, we are using either Boc-Lys(Ac)-AMC (MAL) (17) or Z-Lys(Ac)-AMC (ZMAL) (18). Often ZMAL gives higher conversion rates and also has a broader acceptance by different HDACs but MAL usually works as well. MAL is commercially available, e.g., from Calbiochem/EMD Biosciences (Nr. 382155, Histone Deacetylase Substrate, Fluorogenic).

In sirtuin assays, ZMAL works well but MAL does not. For HDAC8, we and others (19) have found that a trifluoroacetylated substrate is necessary, and we use the requisite analog Z-Lys(F_3Ac)-AMC (ZMTFAL) (20) below.

For the fluorescence polarization (FP) assays we use a substrate from the literature that is derived from a p53 peptide sequence (21). The exact sequence is Ac-Glu-Glu-Lys(Biotin)-Gly-Gln-Ser-Thr-Ser-Ser-His-Ser-Lys(Ac)-Nle-Ser-Thr-Glu-Gly-Lys(6-TAMRA)-Glu-Glu-NH_2, in the following called Ac-p53-TAMRA. It was custom synthesized (PSL, Heidelberg, Germany).

2.1. Histone Deacetylase Assay: Homogeneous (HDASH Assay)

1. Histone deacetylase preparation (cellular lysate or purified enzyme; e.g., in phosphate buffer, *must be without* primary amines like, e.g., Tris–HCl).
2. Incubation buffer: 1.4 mM NaH_2PO_4, 18.6 mM Na_2HPO_4, 0.25 mM EDTA, 10 mM NaCl, 10% (v/v) glycerol, 10 mM mercaptoethanol, pH 7.9.
3. Dimethyl sulfoxide (DMSO, Fluka, assay grade).
4. Boc-Lys(Ac)-AMC (MAL).
5. Inhibitor stock solutions in DMSO.
6. Stock solution of Trichostatin A (TSA): 3.3 mM TSA in DMSO.
7. Borate buffer: H_3BO_3 (6.18 g/L), adjusted to pH 9.5 with 1 M NaOH.
8. Naphthalene dicarboxaldehyde (NDA).
9. Stop solution: 190 μL borate buffer, 5 μL TSA (3.3 μM), and 5 μL of a naphthalene dicarboxaldehyde solution (3 mg/mL ethanol) for each well.

10. Microplate reader with an excitation filter of 330 nm and an emission filter of 390 nm.
11. Black 96-well microtiter plates (e.g., Optiplate 96F, PerkinElmer).

2.2. Trypsin Assays

For the trypsin assays, the buffer does not have to be phosphate-based; it could also be, e.g., Tris-based.

2.2.1. Preparation of HDACs

1. Histone deacetylase preparation (e.g., in phosphate buffer).
2. Incubation buffer: 1.4 mM NaH_2PO_4, 18.6 mM Na_2HPO_4, 0.25 mM EDTA, 10 mM NaCl, 10% (v/v) glycerol, 10 mM mercaptoethanol, pH 7.9.
3. DMSO (Fluka, assay grade).
4. Inhibitor stock solutions in DMSO.
5. Boc-Lys(Ac)-AMC (MAL).
6. 7-Amino-4-methyl coumarin (AMC) stock solution: 12.6 mM AMC in DMSO.
7. Stock solution of TSA: 3.3 mM TSA in DMSO.
8. Trypsin buffer: 50 mM Tris–HCl, pH 8.0, 100 mM NaCl.
9. Trypsin, solved in trypsin buffer to a concentration of 6 mg/mL.
10. Stop solution containing 10 μL trypsin solution (6 mg/mL), 5 μL TSA dilution (33 μM), and 45 μL trypsin buffer for each well.
11. Microplate reader with an excitation filter of 390 nm and an emission filter of 460 nm.
12. Black 96-well microtiter plates (e.g., Optiplate 96F, PerkinElmer).

2.2.2. HDAC8

1. HDAC8 preparation (e.g., in phosphate buffer).
2. Incubation buffer: 50 mM KH_2PO_4, 15 mM Tris–HCl, 3 mM $MgSO_4 \cdot 7H_2O$, 10 mM KCl, pH 7.5.
3. DMSO (Fluka, assay grade).
4. Inhibitor stock solutions in DMSO.
5. Z-Lys(F_3Ac)-AMC (ZMTFAL).
6. Cremophor in ethanol (25/75%).
7. AMC stock solution: 12.6 mM AMC in DMSO.
8. TSA stock solution: 3.3 mM TSA in DMSO.
9. Trypsin buffer: 50 mM Tris–HCl, pH 8.0, 100 mM NaCl.
10. Trypsin, solved in trypsin buffer to a concentration of 6 mg/mL.
11. Stop solution containing 5 μL trypsin solution (6 mg/mL), 2.5 μL TSA dilution (33 μM), and 22.5 μL trypsin buffer per well.

12. Microplate reader with an excitation filter of 390 nm and an emission filter of 460 nm.
13. Black 96-well half-area microtiter plates (e.g., ½ Area Plate 96F, PerkinElmer).

2.2.3. Sirtuins

1. Sirtuin preparation (e.g., in phosphate buffer).
2. Incubation buffer: 25 mM Tris–HCl, 137 mM NaCl, 2.7 mM KCl, 1 mM $MgCl_2$, pH 8.0.
3. DMSO (Fluka, assay grade).
4. Inhibitor stock solutions in DMSO.
5. Z-Lys(Ac)-AMC (ZMAL).
6. AMC stock solution: 12.6 mM AMC in DMSO.
7. Nicotinamide dinucleotide (NAD^+, Sigma-Aldrich).
8. Nicotinamide (Sigma).
9. Trypsin (Sigma), solved in trypsin buffer to a concentration of 6 mg/mL.
10. Trypsin buffer (pH 8.0): 50 mM Tris–HCl, 100 mM NaCl.
11. Stop solution containing 10 μL trypsin solution (6 mg/mL), 4 μL nicotinamide solution (120 mM in DMSO), and 46 μL trypsin buffer to each well.
12. Microplate reader with an excitation filter of 390 nm and an emission filter of 460 nm.
13. Black 96-well microtiter plates (e.g., Optiplate 96F, PerkinElmer).

2.2.4. Detection of HDAC Activity in Cancer Cell Lines

1. Cells from an adherently growing cancer cell line.
2. Cell culture media.
3. Dimethyl sulfoxide (DMSO, Fluka, assay grade).
4. Inhibitor stock solutions in DMSO.
5. Boc-Lys(Ac)-AMC (MAL).
6. Igepal-CA630 stock solution; 10% Igepal-CA630 in H_2O deionized.
7. HDAC assay buffer: 50 mM Tris–HCl pH 8.0, 137 mM NaCl, 1 mM $MgCl_2$, 2.7 mM KCl.
8. TSA stock solution: 100 μM TSA in DMSO.
9. Trypsin solution: 10 mg/mL trypsin in H_2O deionized.
10. Stop solution containing 5 μL trypsin solution, 0.5 μL TSA stock solution, and 44.5 μL HDAC assay buffer per well.
11. Microplate reader with an excitation filter of 390 nm and an emission filter of 460 nm.
12. Black 96-well tissue culture-suited microtiter plates (e.g., PAA 96-well microplates black, PS, surface treated, sterile).

2.2.5. Detection of HDAC Activity in Peripheral White Blood Cells

1. Fresh whole blood stored in heparin-coated tubes.
2. Hypotonic lysis buffer: 20 mM HEPES, pH 7.5, 20 mM NaCl, 5 mM $MgCl_2$.
3. Cell culture medium RPMI 1640.
4. Centrifuge suitable for falcon tubes.
5. Boc-Lys(Ac)-AMC (MAL).
6. Igepal-CA630 stock solution: 10% Igepal-CA630 in H_2O deionized.
7. HDAC assay buffer: 50 mM Tris–HCl, pH 8.0, 137 mM NaCl, 1 mM $MgCl_2$, 2.7 mM KCl.
8. TSA stock solution: 100 μM TSA in DMSO.
9. Trypsin solution: 10 mg/mL trypsin in H_2O deionized.
10. Stop solution containing 5 μL trypsin (10 mg/mL), 0.5 μL TSA (100 μM), and 44.5 μL HDAC assay buffer per well.
11. Microplate reader with an excitation filter of 390 nm and an emission filter of 460 nm.
12. Black 96-well tissue culture-suited microtiter plates (e.g., PAA 96-well microplates black, PS, surface treated, sterile).

2.3. Fluorescence Polarization (FP) Assay

1. Sirtuin preparation (e.g., in phosphate buffer).
2. Incubation buffer: 25 mM Tris–HCl, 137 mM NaCl, 2.7 mM KCl, 1 mM $MgCl_2$, pH 8.0.
3. DMSO (Fluka, assay grade).
4. Inhibitor stock solutions in DMSO.
5. Substrate Ac-p53-TAMRA (sequence see above).
6. Nicotinamide dinucleotide (NAD^+).
7. NAD^+ stock solution: 140 mM NAD^+ in DMSO.
8. Trypsin, solved in trypsin buffer to a concentration of 0.1 mg/mL.
9. Streptavidin stock solution: 70 μM streptavidin in trypsin buffer, diluted to 2.1 μM in incubation buffer.
10. Stop solution I: 7 μL trypsin solution, 2 μL nicotinamide solution, 2.5 μL streptavidin solution, and 8.5 μL trypsin buffer.
11. Stop solution II: 7 μL trypsin solution, 2 μL nicotinamide solution, and 11 μL trypsin buffer.
12. Microplate reader, appropriate for the measurement of fluorescence polarization with an excitation filter of 531 nm and emission filters (s-pol, p-pol) of 595 nm.
13. Black 96-well half-area microtiter plates (e.g., ½ Area Plate 96F, PerkinElmer).

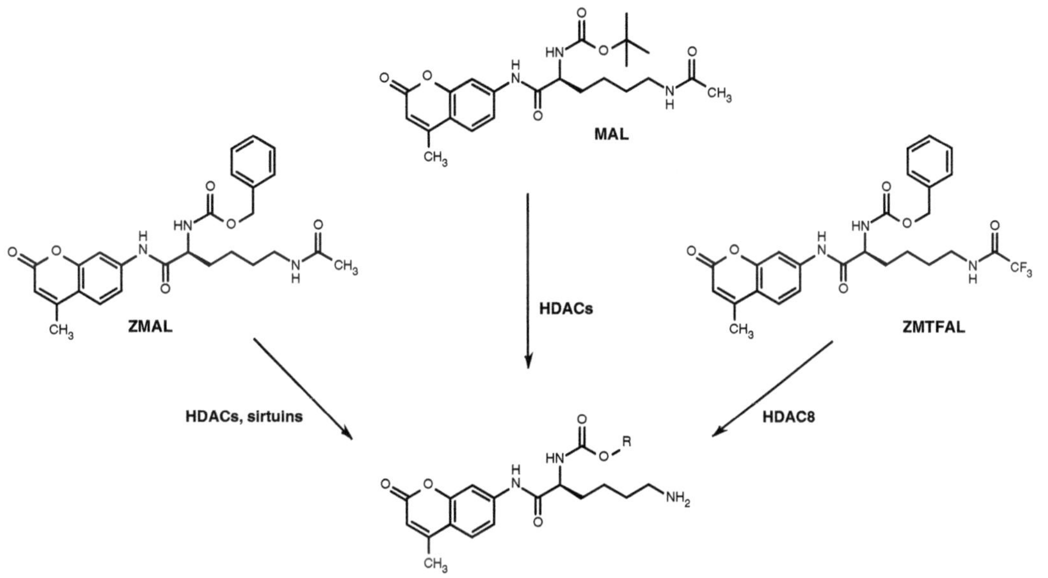

Fig. 1. Structures of the small molecule histone deacetylase substrates MAL, ZMAL, and ZMTFAL and their conversion by the respective histone deacetylases.

3. Methods

3.1. Histone Deacetylase Assay: Homogeneous (HDASH Assay)

In the search for inexpensive nonisotopic HDAC substrates, Boc(Ac)Lys-AMC (also called MAL, see Fig. 1) was one of the first lysine-derived substrates that was developed for the use of fluorescence-based HDAC activity assays (17).

The deacetylated metabolite ML that is generated in a time-dependent manner after incubation with a histone deacetylase preparation shows the same fluorescence properties as the parent compound MAL. Therefore, the conversion originally could not be monitored in a homogenous setup. But by including a derivatization step into the assay protocol, the development of a homogenous HDAC assay system, the so-called HDASH assay, was achieved (22). In this derivatization step, the addition of the amine detection reagent naphthalene dicarboxaldehyde (NDA) leads to quenching of the fluorescence of the metabolite ML (see Fig. 2). Thus, the remaining amount of not converted substrate MAL as a function of HDAC activity can be quantified without the need of any separation steps. The assays work also with other similar substrates, such as ZMAL.

1. Prepare and aliquot a stock solution of MAL in DMSO (12.6 mM) (see Note 1). Dilute 5 μL of this MAL stock solution in freshly prepared incubation buffer (see Note 2) to a total volume of 500 μL. Keep this solution on ice until use (see Note 3).

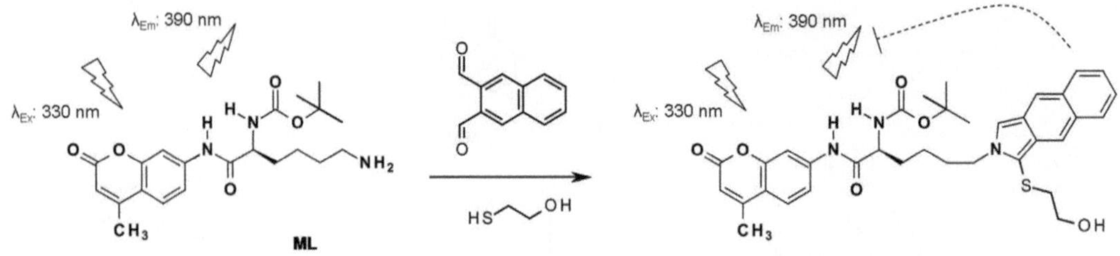

Fig. 2. Detection principle of the HDASH assay. Quenching of HDAC metabolite ML by chemical derivatization with NDA.

2. Freshly prepare a solution of the reference inhibitor TSA (3.3 mM) or dilute a previously prepared and aliquoted stock solution of the reference inhibitor TSA in DMSO to the concentration of 3.3 mM (see Note 1).
3. Dilute the DMSO inhibitor stock solutions with DMSO to the needed concentration (see Note 5).
4. Transfer 40 mL of the aliquoted enzyme preparation (see Note 4) into each well of the microtiter plate (see Note 6). Deep-freeze unused enzyme preparation (see Note 7).
5. Add 5 mL of the diluted inhibitors. As a negative control, add 5 mL of the TSA dilution, and as positive control, use 5 mL DMSO instead (see Note 8).
6. Start the enzyme reaction by transferring 5 mL of the MAL dilution to each well of the microtiter plate (see Note 9).
7. Incubate and gently shake (140 rpm) the microplate for 90 min at a temperature of 37°C with closed lid.
8. Freshly prepare (see Note 1) and add the stop solution.
9. Immediately measure fluorescence intensity ($l_{Ex} = 330$ nm, $l_{Ex} = 390$ nm) in a plate reader.

3.2. Trypsin Assay

The turnover of MAL that is catalyzed by a preparation showing HDAC activity can alternatively be detected with an assay setup that includes a second enzymatic incubation step. In this step, a detection reagent that contains the endopeptidase trypsin is added to the reaction batch (23). Remarkably, trypsin accepts the deacetylated metabolite ML as a substrate but not the acetylated parent substrate. When cleaved by the endopeptidase, ML releases its fluorophore aminomethyl coumarin (AMC), which shows different fluorescence properties enabling quantification without any further extraction or separation steps (*see* Fig. 3). Alternatively, other similar substrates with a lysine or oligopeptide structure (24, 25) can be used in a similar fashion.

In attempt to discover optimized substrates in terms of conversion time and rate and in terms of HDAC subtype selectivity, several analogs of MAL had been synthesized. For example, the

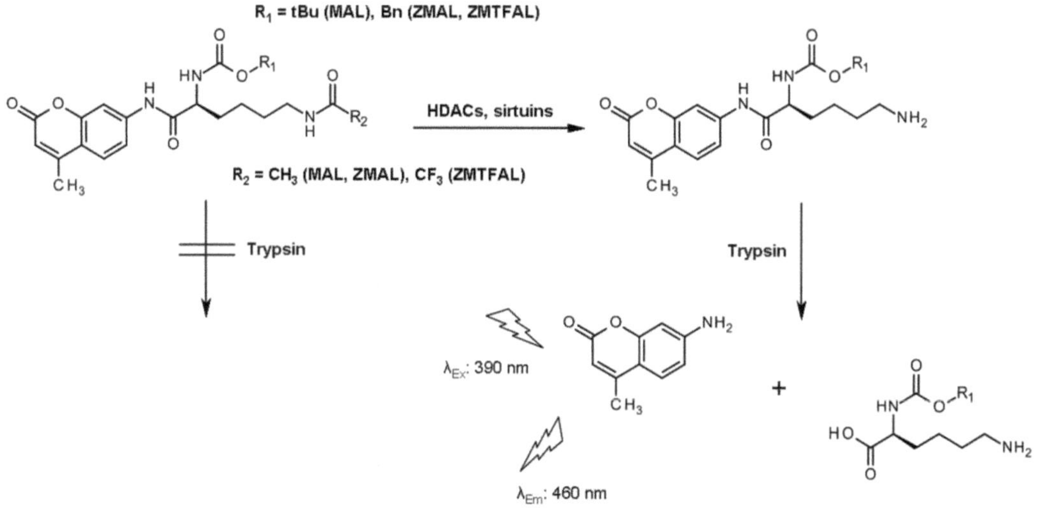

Fig. 3. Detection principle of the trypsin assay. Cleavage of the deacetylated metabolites by trypsin leads to release of AMC.

deacetylation of Z(Ac)Lys-AMC (also called ZMAL; see Fig. 1) is taking place faster and to a greater extent as compared to MAL, and it is accepted as a substrate by a wider range of histone deacetylases. So, ZMAL is—unlike MAL—also deacetylated by class III HDACs (sirtuins). The HDASH assay (see Subheading 3.1) cannot be applied to characterize sirtuins, but as ZML, the deacetylated metabolite of ZMAL, is well accepted as substrate of trypsin, the assay protocol with the endopeptidase incubation step offers a convenient way to measure sirtuin activity in a homogenous setup.

MAL and ZMAL show differing selectivity towards different HDAC subtypes; this is apparent in the fact that ZMAL is accepted equally well as a substrate by HDAC1 (class I) and HDAC6 (class IIa), while MAL shows moderate subtype selectivity towards HDAC6. To dissect the subtype selectivity of HDAC inhibitors, further lysine-derived substrates with an even greater HDAC subtype selectivity have been developed (20). HDAC8 is a special case, as it could be shown that acetyl-lysine substrates like MAL or ZMAL are very poorly converted by this class I HDAC subtype. Replacement of the ε-acetyl group with a trifluoroacetyl moiety like in ZMTFAL (see Fig. 1), however, leads to considerably better conversions by HDAC8 (26). But there are also reports on an acetylated oligopeptide as a HDAC8 substrate in the literature (27).

These fluorescent substrates can also be used in cell-based assays. With a modified protocol, they work well with cancer cell lines and can thus serve for the cellular characterization of HDAC inhibitors. In this setting, together with the substrate, a nonionic tenside for the cell lysis is added to the cells. In the original protocol, the lysis was performed after an incubation period (28), but work in our group shows that the results are very similar.

Another application is the measurement of HDAC activity from peripheral white blood cells. In a first step these cells are isolated from whole blood by a hypotonic lysis of the red blood cells. In a second step the isolated cells are incubated with the same substrate solution that is used for the cancer cell line assay. This assay format is a valuable tool for the determination of the pharmacodynamic effects of HDAC inhibitors in clinical trials. The isolation of the peripheral white blood cells can alternatively be performed by centrifugation on Ficoll-Paque Plus (Amersham, Little Chalfont, UK), which may possibly lead to more homogeneous cell suspensions.

3.2.1. HDACs

1. Prepare and aliquot a 12.6 mM stock solution of MAL in DMSO (see Note 1). Dilute 5 μL of this MAL stock solution in 495 μL freshly prepared incubation buffer (see Note 2).
2. Dilute 1 μL of the AMC stock solution to a final volume of 100 μL in incubation buffer. Keep these solutions on ice until use (see Note 3).
3. Dilute the DMSO inhibitor stock solutions in DMSO to the needed concentrations (see Note 5).
4. Fill the wells of the microtiter plate with 32 μL incubation buffer and 55 μL buffer in case of the positive control (see Note 6).
5. Add 5 μL of the AMC dilution to wells of the positive control.
6. Transfer 20 μL enzyme preparation to the wells of the microplate (except for the positive control). Deep-freeze unused enzyme preparation (see Note 7).
7. Add 3 μL of the inhibitor dilutions to the wells; in case of the enzyme control, use 3 μL DMSO instead (see Note 8).
8. Start the enzyme reaction by adding 5 μL of the MAL dilution to each well, except for the positive control (see Note 9).
9. Incubate the plate for 90 min at 37°C under stirring (120 rpm).
10. Add 60 μL of the freshly prepared stop solution (see Note 1).
11. Incubate and stir (120 rpm) the plate for another 20 min at 37°C.
12. After this second incubation step, measure fluorescence intensity (λ_{Ex} = 390 nm, λ_{Em} = 460 nm) in a plate reader.

3.2.2. HDAC8

1. Prepare and aliquot a stock solution of ZMTFAL in DMSO (12.6 mM) (see Note 1). Dilute 25 μL of this ZMTFAL stock solution in a mixture of 12.5 μL freshly prepared incubation buffer (see Note 2) and 12.5 μL Cremophor in EtOH (25/75%). Take 20 μL of this predilution and add 380 μL

incubation buffer. Dilute this second predilution with incubation buffer at the ratio of 1:1. Keep this final solution on ice until use (see Note 3).

2. Dilute 5 μL of the AMC stock solution to a final volume of 1,000 μL in incubation buffer. Then take 10 μL of this dilution and add 20 μL incubation buffer. Keep this solution on ice until use (see Note 3).
3. Dilute the DMSO inhibitor stock solutions in DMSO to the needed concentrations (see Note 5).
4. Add 12.5 μL incubation buffer to the wells of the plate; use 25 μL buffer for the wells of the positive control (see Note 6).
5. For the positive control, add 5 μL of the AMC dilution to the respective well.
6. Transfer 10 μL of the HDAC8 preparation to the wells of the microtiter plate (apart from the positive control). Deep-freeze unused enzyme preparation (see Note 7).
7. Add 2.5 μL of the inhibitor dilutions to the wells, except for the positive control and the enzyme control. For the enzyme control use 2.5 μL DMSO instead (see Note 8).
8. Add 5 μL ZMTFAL dilution to each well (apart from the positive control) to start the enzyme reaction (see Note 9).
9. Incubate the microtiter plate at a temperature of 37°C for 90 min under stirring (120 rpm).
10. Freshly prepare (see Note 1) and add 30 μL of the stop solution to each well.
11. Incubate the plate for another 20 min at 37°C under shaking at 120 rpm with a closed lid.
12. Finally, measure fluorescence intensity at an emission wavelength of 460 nm after excitation at 390 nm in a plate reader.

3.2.3. Sirtuins

1. Prepare and aliquot a 12.6 mM stock solution of ZMAL in DMSO (see Note 1). Dilute this ZMAL stock solution to a concentration of 126 μM in freshly prepared (see Note 2) incubation buffer. Keep this dilution on ice until use (see Note 3).
2. Do the same with the 12.6 mM AMC stock solution. Also keep this solution on ice until it is used (see Note 3).
3. Freshly prepare (see Note 1) a solution of 6 mM NAD^+ in incubation buffer and store this dilution on ice until use (see Note 3) .
4. Dilute the DMSO inhibitor stock solutions in DMSO to the wanted concentrations (see Note 5).
5. Transfer 17 μL incubation buffer to the wells; take 50 μL buffer in case of the positive control.

6. Add AMC dilution (5 μL) to the wells of the positive control.
7. Transfer 30 μL sirtuin preparation to each well (apart from the wells of the positive control). Quick-freeze unused enzyme preparation (see Note 7).
8. Add 3 μL of the inhibitor dilutions to the wells; in case of the enzyme control, use 3 μL DMSO instead (see Note 8).
9. Add 5 μL of the ZMAL dilution to each well (except for the positive control).
10. By adding 5 μL of the NAD^+ solution to each well, the enzyme reaction is started (see Note 9).
11. Incubate the plate at 37°C for 4 h with a tightly closed lid under moderate shaking (140 rpm).
12. Add 60 μL of the freshly prepared stop solution (see Note 1).
13. Incubate and gently stir (140 rpm) the plate for 20 more min at a temperature of 37°C.
14. After that time, measure fluorescence intensity ($\lambda_{Ex} = 390$ nm, $\lambda_{Em} = 460$ nm) in a plate reader.

3.2.4. Detection of HDAC Activity from Cancer Cell Lines

1. Seed cells at an appropriate density in 96-well tissue culture-suited microtiter plates (e.g., for MCF-7 breast adenocarcinoma cells 5×10^4 cells/well) and incubate for 24 h at 37°C and 5% CO_2.
2. Dilute inhibitor stock solutions in cell culture media to the desired concentrations. Make sure that the final concentration of DMSO does not exceed 1% (see Note 10).
3. Aspirate the supernatant tissue culture media from the wells and add 50 μL of the inhibitor solution.
4. Incubate for 2 h at 37°C and 5% CO_2.
5. Prepare and aliquot a 30 nM stock solution of MAL in DMSO (see Note 1).
6. Freshly prepare (see Note 1) the substrate solution containing 0.5 μL MAL stock solution and 0.5 μL Igepal-CA630 stock solution in 49 μL cell culture media per well.
7. Aspirate the inhibitor solution from the cells and add the substrate solution (see Note 9).
8. Incubate and stir (100 rpm) for another 2 h at 37°C.
9. Finally add 50 μL of the freshly prepared stop solution (see Note 1).
10. Incubate and stir (100 rpm) the plate for 20 min at 37°C.
11. Measure fluorescence intensity ($\lambda_{Ex} = 390$ nm, $\lambda_{Em} = 460$ nm) in a plate reader.

3.2.5. Detection of HDAC Activity from Peripheral White Blood Cells

Isolation of Peripheral White Blood Cells

1. Take an aliquot of the blood sample and add four times the volume of hypotonic lysis buffer in a falcon tube (e.g., 1 mL blood + 4 mL hypotonic lysis buffer).
2. Incubate on ice for 15 min.
3. Centrifuge for 10 min at 400 × *g*.
4. Discard the supernatant. If the pellet is still red, add two times the volume of hypotonic lysis buffer (based on the original volume of your blood sample) and incubate again on ice for 5–10 min. After that, centrifuge for 10 min at 400 × *g*.
5. If you have a clear or very pale red pellet, wash twice with RPMI 1640 and centrifuge for 5 min at 200 × *g*.
6. Resuspend the pellet in the same volume of RPMI 1640 as your original blood sample.

3.2.6. HDAC Activity Assay

1. Add 49 μL of your cell suspension containing 1.6×10^6 cells/mL to a microtiter plate (see Note 6).
2. Start the enzyme reaction by adding 1 μL substrate solution containing 0.5 μL freshly prepared MAL stock solution (30 nM) and 0.5 μL Igepal-CA630 stock solution to each well (see Note 9).
3. Incubate and stir (100 rpm) at 37°C for 2 h.
4. Freshly prepare the stop solution (see Note 1).
5. Add 50 μL of the stop solution to each well and incubate and stir (100 rpm) for another 20 min at 37°C.
6. Measure fluorescence intensity (λ_{Ex} = 390 nm, λ_{Em} = 460 nm) in a plate reader.

3.3. Fluorescence Polarization (FP) Assay

Assays with a readout based on the measurement of fluorescence intensity always are prone to false positive or negative results due to screening compounds with intrinsic fluorescence or quenching properties, respectively. One way to reduce these problems is the use of assay setups that are based on fluorescence polarization (FP) (29). In FP-based readouts, linear polarized excitation light is applied and the polarization of the emitted light is measured. The degree of polarization increases with decreasing molecular rotation, correlating with molecular size (30). This means that fluorescence polarization of small fluorescent compounds increases with the size of the fraction of molecules that is tightly bound to a macromolecule, e.g., a target protein. FP assays are homogenous setups that are less susceptible to quenching and intrinsic fluorescence effects.

In the FP assay protocol that is commonly used to detect sirtuin activity (21), a peptide substrate that is structurally based on the deacetylase substrate acetyl-p53 is used. The peptide substrate is biotinylated on the N-terminus and carries a fluorescent tag on the C-terminal end. Fig. 4 schematically shows the procedure of this coupled enzyme assay. Although the substrate is derived from the Sirtl substrate Ac-p53, it works equally well with Sirt2. Surprisingly, in our hands, it did not work with HDACs, even though Ac-p53 is also reported to be an HDAC substrate.

In a first step, the substrate is deacetylated when incubated with a preparation that comprises a suitable deacetylase activity. After that, the endopeptidase trypsin is able to cleave the deacetylated substrate at the newly exposed lysine residue. The original acetylated and uncleaved fluorescent peptide substrate is still biotinylated, while the smaller fluorescent cleavage product is not carrying the biotin tag any more. The addition of streptavidin now leads to an enhancement of size difference between the original acetylated substrate and the cleavage product due to the strong biotin–streptavidin interactions only with the acetylated substrate. Finally, the inhibition can be quantified by measuring the fluorescence polarization ($\lambda_{Ex}=650$ nm; $\lambda_{Em}=680$ nm).

1. Freshly prepare a 0.3 mM stock solution of the substrate Ac-p53-TAMRA in incubation buffer (see Note 1). Dilute this stock solution at the ratio of 1:300, before diluting this predilution again at the ratio of 1:1. Store this solution on ice until it is used (see Note 3).
2. Prepare a solution of 1.5 mM NAD^+ in incubation buffer and keep this dilution on ice until use (see Note 3).
3. Dilute the DMSO inhibitor stock solutions in DMSO to the desired concentrations (see Note 5).
4. Transfer 17.5 mL incubation buffer to the wells; take 37.5 mL buffer in case of the controls without streptavidin.
5. Transfer 20 mL sirtuin preparation to each well (apart from the wells of the control). Quick-freeze unused enzyme preparation (see Note 7).
6. Add 2.5 mL inhibitor dilution to the wells; use 2.5 mL DMSO for the wells of the enzyme control instead (see Note 8).
7. Add 5 mL of the Ac-p53-TAMRA dilution to each well.
8. The enzyme reaction is started by adding 5 mL of the NAD^+ solution to each well (see Note 9).
9. The plate is incubated under stirring (120 rpm) for 3 h at a temperature of 37°C with a tightly closed lid.
10. Freshly prepare stop solutions I and II (see Note 1).
11. After the incubation time, the wells for the minimal signal control are filled with 20 mL of the stop solution II. All other wells (inhibitor testing, enzyme control, and maximal signal control) are filled with 20 mL of the stop solution I.

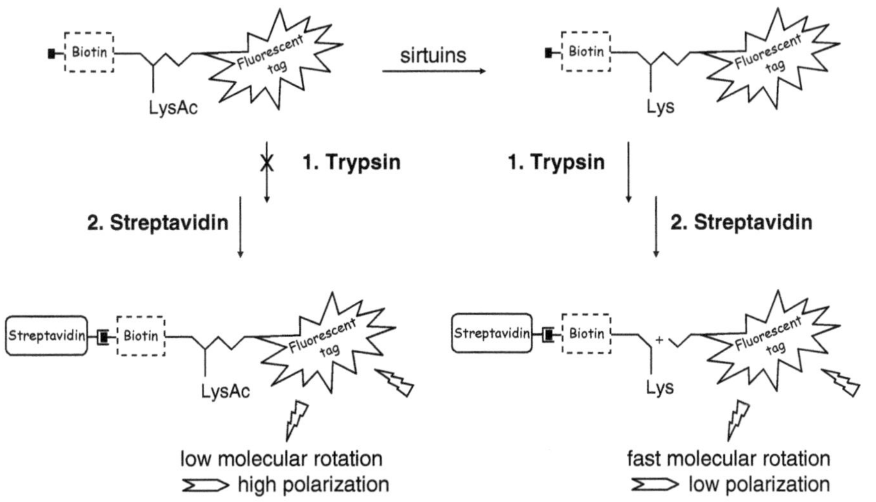

Fig. 4. Detection principle of the FP assay. Addition of streptavidin increases the difference in size between the original peptide substrate and the deacetylated and cleaved metabolite.

12. Incubate the plate for 20 min under gentle shaking (120 rpm) at a temperature of 37°C.
13. Finally, measure fluorescence polarization in a plate reader.

4. Notes

The presented assays are convenient methods to study the characteristics of histone deacetylases and their modulators. As they are homogenous systems that are carried out in microtiter plates, they also are amenable for high-throughput testing of large inhibitor libraries. In order to obtain reproducible and reliable results, one should pay regard to the following suggestions.

1. All solutions (substrates, reagents, stop solutions, etc.) should be prepared just before the assay is performed. Solutions that already have been prepared to an earlier point of time (e.g., inhibitor stock solutions) should be stored at a temperature of -20 °C or better -80 °C.
2. Also all buffer solutions should always be prepared not too long before performing the assay and stored in the refrigerator until they are used. In some cases, it may be reasonable to percolate them through a sterile filter to avoid the occurrence of suspended particles.
3. Keep all solutions on ice while performing the assay.
4. Enzymes usually are sensitive to freeze and thaw cycles, so all enzyme preparations should be portioned into small aliquots.

5. The same is true for stock solutions of inhibitors or substrates.
6. When filling the microtiter plate, it is helpful to start with the solutions with the highest volume in order to enhance the accuracy of pipetting.
7. Remaining enzyme preparations or inhibitor stock solutions that have been thawed but were not used in the assay should be quick-frozen with the help of liquid nitrogen.
8. The enzyme inhibitors usually are solved in organic solvents (in most cases, dimethyl sulfoxide, DMSO). Therefore, it is important to always add the same volume of solvent to the enzyme control as used for the inhibitor solutions.
9. As the enzyme reaction starts with the addition of the substrate (or a co-substrate), this is supposed to be the final step of each assay, which should be performed rapidly.
10. In cellular assays, the final concentration of DMSO should not exceed 1%, as cells get influenced by higher concentrations of DMSO.

Acknowledgment

Work on HDAC assays in blood cells is funded by the Deutsche Forschungsgemeinschaft (DFG, SPP1463, Ju 295/9-1).

References

1. Luger K, Mader AW, Richmond RK, Sargent DF, Richmond TJ (1997) Crystal structure of the nucleosome core particle at 2.8 A resolution. Nature 389:251–260
2. Kornberg RD, Lorch Y (1999) Twenty-five years of the nucleosome, fundamental particle of the eukaryote chromosome. Cell 98:285–294
3. Berger SL (2007) The complex language of chromatin regulation during transcription. Nature 447:407–412
4. Yang XJ, Seto E (2008) The Rpd3/Hda1 family of lysine deacetylases: from bacteria and yeast to mice and men. Nat Rev Mol Cell Biol 9:206–218
5. Finnin MS, Donigian JR, Cohen A, Richon VM, Rifkind RA, Marks PA et al (1999) Structures of a histone deacetylase homologue bound to the TSA and SAHA inhibitors. Nature 401:188–193
6. Somoza JR, Skene RJ, Katz BA, Mol C, Ho JD, Jennings AJ et al (2004) Structural snapshots of human HDAC8 provide insights into the class I histone deacetylases. Structure 12:1325–1334
7. Smith JS, Avalos J, Celic I, Muhammad S, Wolberger C, Boeke JD (2002) SIR2 family of NAD(+)-dependent protein deacetylases. Methods Enzymol 353:282–300
8. Marks P, Rifkind RA, Richon VM, Breslow R, Miller T, Kelly WK (2001) Histone deacetylases and cancer: causes and therapies. Nat Rev Cancer 1:194–202
9. Marks PA (2007) Discovery and development of SAHA as an anticancer agent. Oncogene 26:1351–1356
10. Kolle D, Brosch G, Lechner T, Lusser A, Loidl P (1998) Biochemical methods for analysis of histone deacetylases. Methods 15:323–331
11. Nare B, Allocco JJ, Kuningas R, Galuska S, Myers RW, Bednarek MA et al (1999) Development of a scintillation proximity assay for histone deacetylase using a biotinylated peptide derived from histone-H4. Anal Biochem 267:390–396
12. Wegener D, Hildmann C, Riester D, Schober A, Meyer-Almes FJ, Deubzer H et al (2008) Identification of novel small-molecule histone

deacetylase inhibitors by medium-throughput screening using a fluorogenic assay. Biochem J 413:143–150

13. Fenic I, Hossain HM, Sonnack V, Tchatalbachev S, Thierer F, Trapp J et al (2008) In vivo application of histone deacetylase inhibitor trichostatin-A impairs murine male meiosis. J Androl 29:172–185
14. Garcia-Manero G, Assouline S, Cortes J, Estrov Z, Kantarjian H, Yang H et al (2008) Phase 1 study of the oral isotype specific histone deacetylase inhibitor MGCD0103 in leukemia. Blood 112:981–989
15. Sippl W, Jung M (eds) (2009) Epigenetic targets in drug discovery, 1st edn. Wiley-VCH, Weinheim
16. Hauser AT, Jung M, Jung M (2009) Assays for histone deacetylases. Curr Top Med Chem 9:227–234
17. Hoffmann K, Brosch G, Loidl P, Jung M (1999) A non-isotopic assay for histone deacetylase activity. Nucleic Acids Res 27:2057–2058
18. Heltweg B, Dequiedt F, Verdin E, Jung M (2003) A non isotopic substrate f or assaying both human zinc and NAD+-dependent histone deacetylases. Anal Biochem 319:42–48
19. Hildmann C, Wegener D, Riester D, Hempel R, Schober A, Merana J et al (2006) Substrate and inhibitor specificity of class 1 and class 2 histone deacetylases. J Biotechnol 124:258–270
20. Heltweg B, Dequiedt F, Marshall BL, Brauch C, Yoshida M, Nishino N et al (2004) Subtype selective substrates for histone deacetylases. J Med Chem 47:5235–5243
21. Milne JC, Lambert PD, Schenk S, Carney DP, Smith JJ, Gagne DJ et al (2007) Small molecule activators of SIRT1 as therapeutics for the treatment of type 2 diabetes. Nature 450:712–716
22. Heltweg B, Jung M (2003) A homogeneous nonisotopic histone deacetylase activity assay. J Biomol Screen 8:89–95
23. Heltweg B, Trapp J, Jung M (2005) In vitro assays for the determination of histone deacetylase activity. Methods 36:332–337
24. Wegener D, Hildmann C, Riester D, Schwienhorst A (2003) Improved fluorogenic histone deacetylase assay for high-throughput screening applications. Anal Biochem 321:202–208
25. Wegener D, Wirsching F, Riester D, Schwienhorst A (2003) A fluorogenic histone deacetylase assay well suited for high-throughput activity screening. Chem Biol 10:61–68
26. Riester D, Wegener D, Hildmann C, Schwienhorst A (2004) Members of the histone deacetylase superfamily differ in substrate specificity towards small synthetic substrates. Biochem Biophys Res Commun 324:1116–1123
27. Vannini A, Volpari C, Gallinari P, Jones P, Mattu M, Carfi A et al (2007) Substrate binding to histone deacetylases as shown by the crystal structure of the HDAC8-substrate complex. EMBO Rep 8:879–884
28. Bonfils C, Kalita A, Dubay M, Siu LL, Carducci MA, Reid G et al (2008) Evaluation of the pharmacodynamic effects of MGCD0103 from preclinical models to human using a novel HDAC enzyme assay. Clin Cancer Res 14:3441–3449
29. Burke TJ, Loniello KR, Beebe JA, Ervin KM (2003) Development and application of fluorescence polarization assays in drug discovery. Comb Chem High Throughput Screen 6:183–194
30. Jameson DM, Croney JC (2003) Fluorescence polarization: past, present and future. Comb Chem High Throughput Screen 6:167–173

Chapter 18

The Fluorescence-Based Acetylation Assay Using Thiol-Sensitive Probes

Tielong Gao, Chao Yang, and Yujun George Zheng

Abstract

Lysine acetyltransferases (KATs) catalyze the acetylation of specific lysine residues in histone and nonhistone proteins. The enzymatic activities of KATs are involved in a broad spectrum of cellular processes. Thus far, the reaction of KAT catalysis has been studied by various bioanalytical methods such as radioisotopic labeling, spectrophotometric and fluorometric measurements, and antibody-dependent immunosorbent assays. In particular, the fluorescent method has the advantage of simplicity for implementation, fast assay speed, fine signal to noise ratio, and superior sensitivity. We describe here the technical protocols of using thiol-sensitive fluorogenic probes for the fluorescent analysis of enzymatic activities of KATs, with males on the first (MOF) as an exemplary KAT enzyme. 7-Diethylamino-3-(4′-maleimidylphenyl)-4-methylcoumarin (CPM) is selected as the KAT probe owing to its fast reaction kinetics with coenzyme A (CoA) and excellent fluorogenicity upon thiol conjugation. The fluorescence-based acetylation assay is well suited for both kinetic characterization of KAT catalysis and KAT inhibitor investigation.

Key words: Lysine acetyltransferase, KAT, HAT, Acetylation, Fluorescence, MYST, Histone

1. Introduction

Protein lysine acetylation is a key posttranslational modification that regulates diverse biological cascades such as gene expression, signal transduction, and metabolism. The reaction is catalyzed by lysine acetyltransferases (KATs), also known as histone acetyltransferases (HATs), which transfer the acetyl group from acetyl-coenzyme A (Ac-CoA) to the epsilon amino group of specific lysine residues, with CoA produced as a by-product. Based on sequence and functional similarity, KAT proteins are divided into several major families: the GCN5-related N-acetyltransferase (GNAT), the p300/CBP family, the MYST family, and RTT109 (1, 2). On the chromatin template, lysine acetylation can loosen nucleosome structure and promote the accessibility of transcription factors to

Sandra B. Hake and Christian J. Janzen (eds.), *Protein Acetylation: Methods and Protocols*, Methods in Molecular Biology, vol. 981, DOI 10.1007/978-1-62703-305-3_18,

local and global genetic loci, leading to active gene expression (3). Furthermore, recent study showed that protein acetylation affects a variety of cellular processes beyond the scope of chromatin biology (4, 5). Moreover, accumulated evidence reveals that KAT activities are deregulated in many disease states (6, 7). Effective biochemical assays are essential experimental techniques for detecting and analyzing the activity of KATs from various resources. An ideal assay should combine high sensitivity, fast speed, simplicity, and cost-effectiveness. The classic radiometric assay relies on (^{3}H)- or (^{14}C)-labeled Ac-CoA to study KAT activities and is well suited for kinetic and functional characterization of KATs. However, radioactive assays suffer from several drawbacks such as expensive radioactive materials, environmental safety of radioactivity handling, and experimental inflexibility (8). Fluorescence-based assay is one attractive alternative strategy to analyze acetyltransferase activity without involving hazardous radioactive chemicals while providing great flexibility. In general, fluorescence measurement is rapid, sensitive, and potentially continuous.

The most common format of fluorescence-based acetylation assay is to analyze and quantify the side-product CoA produced in the enzymatic turnover. The fluorogenic dye, 7-diethylamino-3-(4′-maleimidylphenyl)-4-methylcoumarin (CPM), has very weak fluorescence but strongly fluoresces upon reacting with thiol motif-containing compounds such as CoA. We describe here the procedure of using CPM for fluorescent analysis of KAT catalysis, using MOF as an illustrative example. The fluorescent method is sensitive for kinetic characterization of KAT activity and is also amenable to investigation and screening of KAT inhibitors.

2. Materials

Ultrapure water was prepared by passing deionized water through a MILLI-Q System (quality controlled at a resistivity of 18 MΩ cm at 25°C) and was used for all the experiments. All reagents were of the highest grade commercially available. The thiol-sensitive probe CPM was purchased from Invitrogen. Ac-CoA and CoA were purchased from Sigma. Lys-CoA was synthesized as previously reported (9).

2.1. Ac-CoA

An enrichment procedure was conducted to eliminate any residual amounts of CoA in the commercial Ac-CoA. 3 mg Ac-CoA was treated with 10 μL acetic anhydride for 10 min at room temperature and then quenched with 1 mL of 100 mM N-2-hydroxyethylpiperazine-N′-2-ethanesulfonic acid (HEPES) buffer, pH 8.0. The enriched Ac-CoA was divided into small aliquots and frozen at −80°C.

2.2. Preparation of Buffers

1. 1 M HEPES buffer, pH 8.0: Weigh and transfer 238.3 g HEPES to a 1-L graduated beaker. Add water to a volume of 900 mL. Stir to dissolve the solid, and then adjust pH to 8.0 with 6 N NaOH. Make up to 1 L with water and store the buffer at 4°C.
2. 0.5 M ethylenediaminetetraacetic acid (EDTA), pH 8.0: Weigh 186.1 g EDTA disodium salt dihydrate and transfer to 1 L graduated beaker. Add water to a volume of 900 mL. Mix and adjust pH to 8.0 with 6 N HCl. Make up to 1 L with water and store the buffer at 25°C.
3. 2× acetylation reaction buffer (2× RB): Mix 100 mL HEPES buffer, 2 mL EDTA in a 1 L graduated beaker. Add water to a volume of 800 mL. Make up to 1 L with water and store the 2× RB buffer at 4°C. The prepared 2× RB contained 100 mM HEPES and 1 mM EDTA.

2.3. Stock Solutions of the Reagents

1. 2 mM CoA stock solution: 15.4 mg of CoA hydrate (C4282, Sigma) was dissolved in 10 mL H_2O in a plastic tube. The solution was divided into small aliquots and frozen at −80°C.
2. 1 mM CPM stock solution: 4.02 mg of CPM (Invitrogen) was dissolved in 10 mL dimethyl sulfoxide (DMSO) in a glass tube. The solution was divided into small aliquots and frozen at −20°C.
3. The 20-amino-acid peptide derived from the N-terminal region of histone H4, namely ac-SGRGKGGKGLGKGGAKRHRK (H4-20), was synthesized using the Fmoc-based solid phase peptide synthesis (SPPS) protocol and purified with C-18 reversed phase HPLC, using a gradient of H_2O:CH_3CN containing 0.05% trifluoroacetic acid. The purity was confirmed with MALDI-MS as previously described (10). A stock solution of 10 mM H4-20 was prepared in water (6.1 mg dissolved in 300 μL) and stored at −20°C.

2.4. Expression of Recombinant MOF

The His10×-tagged MOF protein (125–458) was expressed from the pET19b vector in *E. coli* BL21 (DE3). The protein was purified on Ni-NTA beads. Protein concentrations were determined using the Bradford assay. Proteins were flash frozen in a storage buffer containing 25 mM HEPES (pH 7.0), 500 mM NaCl, 1 mM EDTA, and 10% glycerol and stored at −80°C.

3. Methods

3.1. Fluorescent Response of the Probe Reacting with CoA

1. 40 μL 2× RB were mixed with 24 μL H_2O. Then, 16 μL CoA at a series of concentrations (0, 1, 10, or 20 μM in H_2O) were added and mixed.

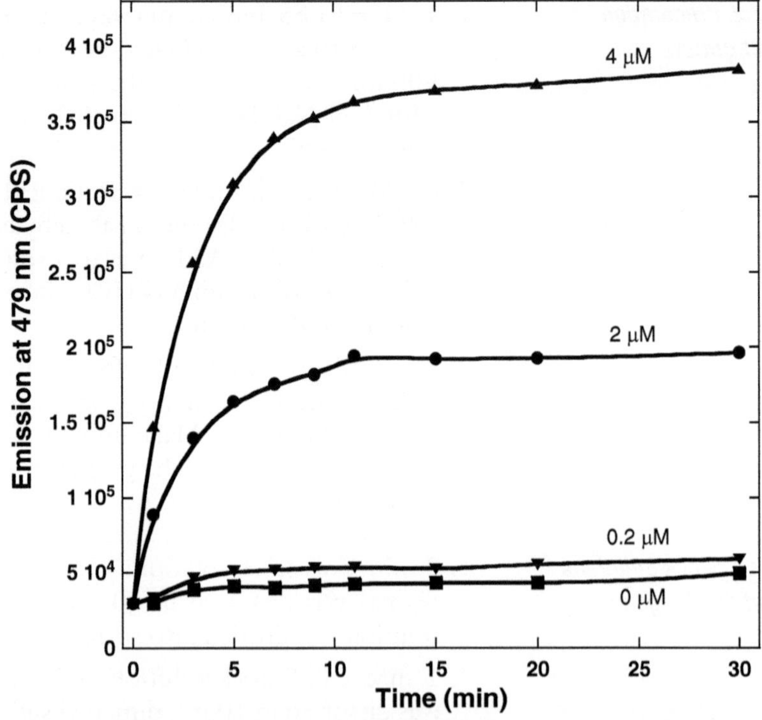

Fig. 1. The change of the fluorescence intensity of CPM as a function of time reacting with CoA. The reaction was carried out at room temperature in 1× RB containing 50% (v/v) DMSO. CPM concentration was 5 μM and CoA concentrations were selected as 0, 0.2, 2, and 4 μM.

2. 8 μL of 100 μM CPM was mixed with 72 μL DMSO. The mixture was added to the CoA solution and rapidly mixed.
3. Fluorescence emission at 479 nm was recorded at different time points with excitation at 390 nm (see Note 1). As shown in Fig. 1, the fluorescence emission increased as CPM reacted with CoA and reached a plateau after 10 min, indicating that the reaction was complete.

3.2. Fluorescence of CPM at Varied Concentrations of CoA

1. In separate aliquots, 40 μL 2× RB were mixed with 24 μL H_2O. Then, 16 μL CoA with different concentrations (0, 0.15625, 0.3125, 0.625, 1.25, 2.5, 5, 10, 20 μM) were added to each aliquot.
2. 8 μL of 100 μM CPM were mixed with 72 μL DMSO. The mixture was added to and mixed with the above solution rapidly.
3. The mixture was incubated at room temperature in darkness for 12 min, and then fluorescence intensity at 479 nm was measured with excitation at 390 nm. As seen in Fig. 2, the intensity at 479 nm increased as CoA concentration went higher and also good linearity was observed (note: the background was

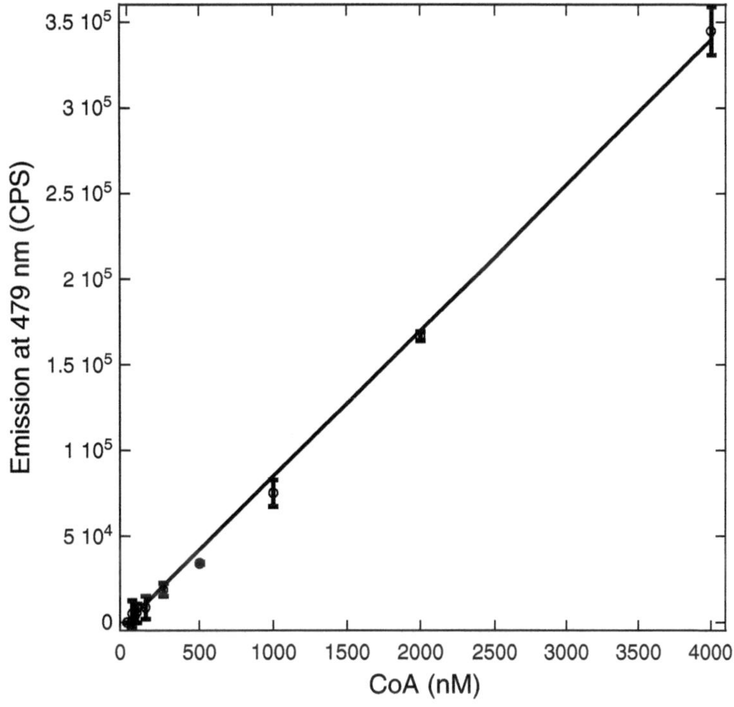

Fig. 2. The linear relationship of the fluorescence intensity of CPM at 479 nm with the concentration of CoA. CPM concentration was kept at 5 μM. The reaction was carried out in 1× RB with 50% DMSO and the reaction time was 12 min.

subtracted before fitting). The linear equation between fluorescence emission and CoA concentration was derived as following: $y = 84.977 * (\text{CoA})$. This equation was used to calibrate the amount of CoA produced from the acetylation reaction.

3.3. KAT Activity at Varied Enzyme Concentrations

1. The assay cocktail was prepared with 34.4 μL 2× RB, 4 μL H4-20 stock solution, and 1.6 μL of 1 mM Ac-CoA (in 2× RB) with 32 μL H_2O. The total volume was 72 μL. The solution was incubated at 30°C for 5 min.
2. 8 μL MOF with different concentrations (0, 0.0625, 0.125, 0.25, 0.5, 0.75, 1 μM) were added. The reaction was maintained at 30°C for 10 min.
3. 8 μL of 100 μM CPM were mixed with 72 μL DMSO. The mixture was added to the above reaction mixture to quench the KAT reaction (see Note 2).
4. The solutions were kept at room temperature in darkness for 12 min, and then, fluorescence intensity at 479 nm was recorded with λ_{ex} at 390 nm. The fluorescence intensity was converted to the CoA concentration by using the above calibration equation. As shown in Fig. 3, the formation of CoA increased linearly with the enzyme concentration.

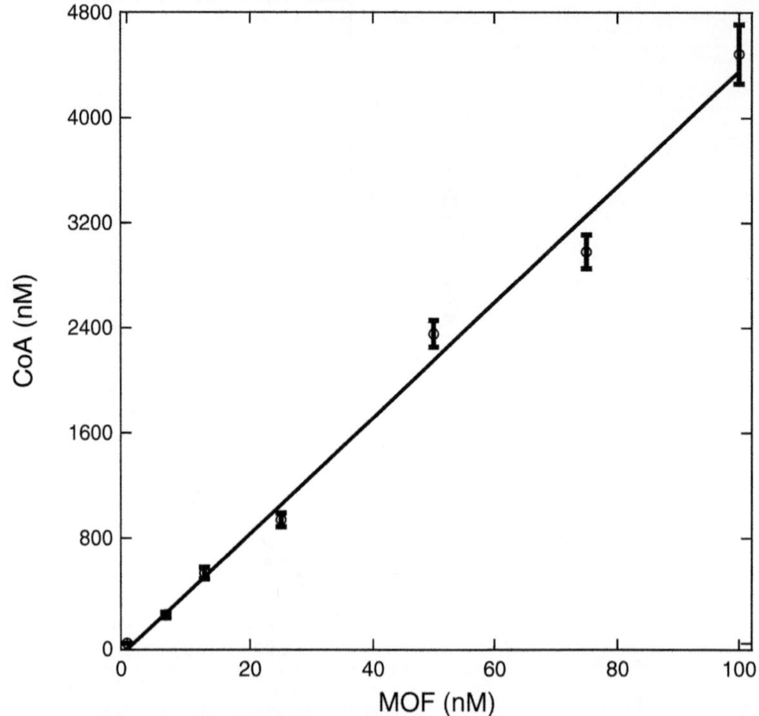

Fig. 3. Relationship of the amount of CoA generated in the acetylation reaction versus the concentration of MOF. The acetylation reaction contained 20 μM Ac-CoA, 500 μM H4-20 peptide, and varying MOF concentrations (0, 6.25, 12.5, 25, 50, 75, and 100 nM). 5 μM CPM was used for CoA detection.

3.4. Activity of MOF as a Function of Reaction Time

1. The assay mixture was prepared with 127.4 μL 2× RB, 14 μL H4-20 stock solution, 5.6 μL of 1 mM Ac-CoA (in 2× RB), and 119 μL H_2O, in a total volume of 266 μL. The solution was incubated at 30°C for 5 min.
2. 14 μL of 0.4 μM MOF were added to initiate the acetylation reaction. The solution was incubated at 30°C.
3. 40 μL reaction solution was taken out and quenched by 112 μL DMSO at different reaction times.
4. 8 μL of 100 μM CPM were added to each sample and then incubated at room temperature for 12 min in darkness.
5. Fluorescence intensity at 479 nm was recorded with λ_{ex} at 390 nm and then converted to CoA concentration by using the calibration equation. A plot of CoA production versus reaction time is shown in Fig. 4. The linear correlation suggests that MOF catalysis is in the initial kinetic range.

3.5. Measurement of the Michaelis–Menten Kinetics

1. The assay cocktail was prepared with 32 μL 2× RB, 4 μL H4-20 stock solution, 8 μL Ac-CoA with different concentrations (0, 5, 10, 20, 40, 80, 160 μM in 2× RB), and 28 μL H_2O.

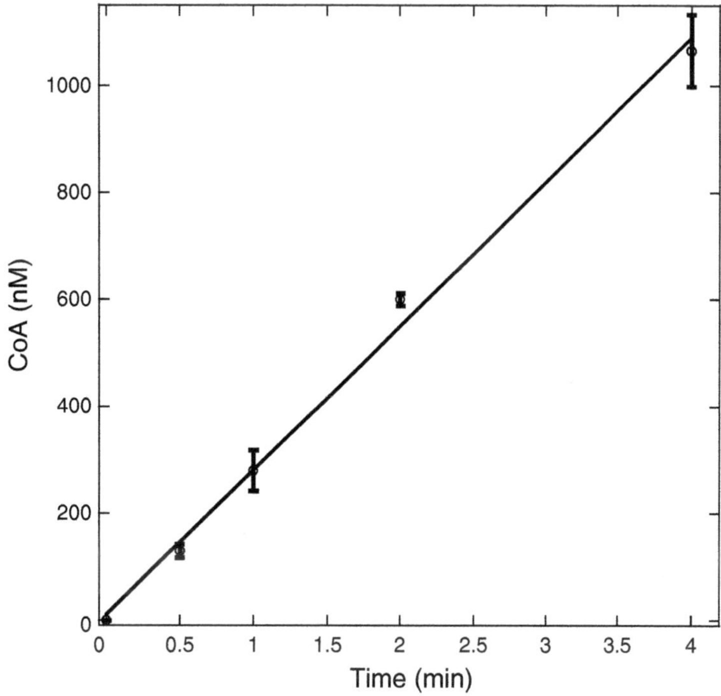

Fig. 4. Formation of CoA at different reaction times. Assay was performed in 1× RB with 20 μM Ac-CoA, 500 μM H4-20 peptide, and 20 nM MOF. 5 μM CPM was used for CoA quantitation.

The total volume was 72 μL. The solution was incubated at 30°C for 5 min.

2. 8 μL of 0.5 μM MOF were added to initiate the acetylation reaction. The solution was incubated at 30°C for 10 min.
3. 8 μL of 100 μM CPM (in DMSO) were mixed with 72 μL DMSO. The mixture was added to the above solution to quench the acetylation reaction. The solution was incubated in darkness at room temperature for 12 min.
4. Fluorescence intensities at 479 nm of each reaction mixture were recorded with λ_{ex} at 390 nm. The amount of CoA was calculated from the fluorescent signals using the calibration equation. The initial reaction velocity was obtained by dividing the amount of CoA by the reaction time and then plotted against the concentrations of Ac-CoA (Fig. 5). The data were fitted to the Michaelis–Menten equation from which K_m was calculated as 3.13 ± 0.39 μM and k_{cat} as 10.3 ± 0.45/min.

3.6. Study of KAT Inhibition with the Fluorescent Assay

1. An assay cocktail was prepared with 39.8 μL 2× RB, 2.4 μL H4-20 stock solution, 1.2 μL of 200 μM Ac-CoA in 2× RB, and 20.6 μL H_2O, with a total volume of 64 μL.

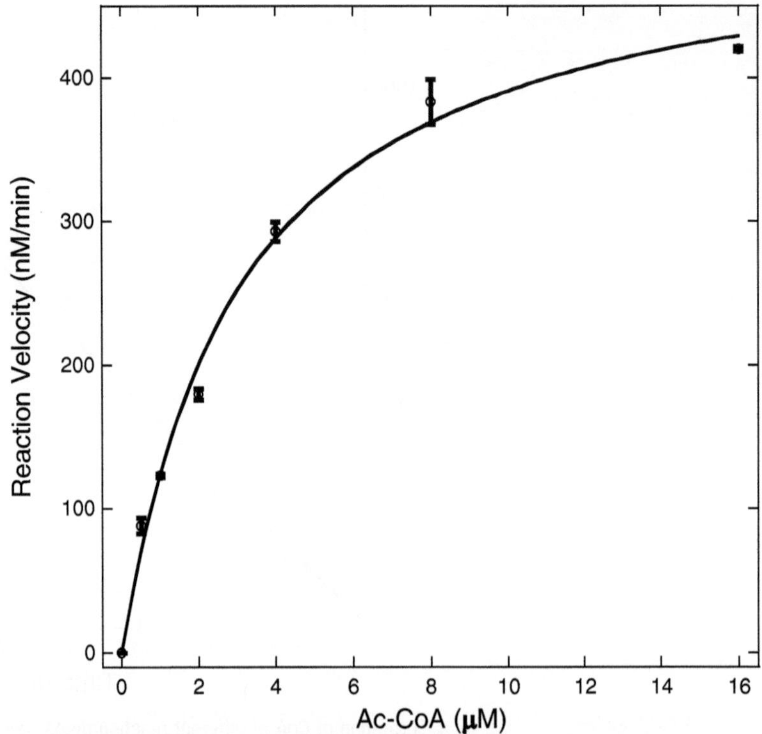

Fig. 5. Initial velocity of MOF catalysis versus concentration of Ac-CoA. The assay was performed at 30°C with 500 μM H4-20, 50 nM MOF, and varying concentrations of Ac-CoA (0, 0.5, 1, 2, 4, 8, 16 μM). 5 μM CPM was used for product quantitation.

2. Lys-CoA is a previously reported KAT inhibitor (9, 11) and used for MOF inhibition. 8 μL Lys-CoA with different concentrations (0, 1.56, 3.1, 6.25, 12.5, 25, 50, 100, and 200 μM in H_2O) were added to the above solution. The solutions were incubated at 30°C for 5 min.
3. 8 μL of 0.2 μM MOF were added to initiate the acetylation reaction. The reaction mixture was incubated at 30°C for 10 min.
4. 8 μL of 100 μM CPM (in DMSO) were mixed with 72 μL DMSO and then were added to the acetylation mixture. The solutions were incubated in darkness at room temperature for 12 min.
5. Fluorescence intensity at 479 nm was recorded with λ_{ex} at 390 nm. The result is shown in Fig. 6. The data were fitted to the equation as $y = 1/(1 + x/IC_{50})$, and the IC_{50} value of Lys-CoA for MOF inhibition was determined to be 0.79 μM.

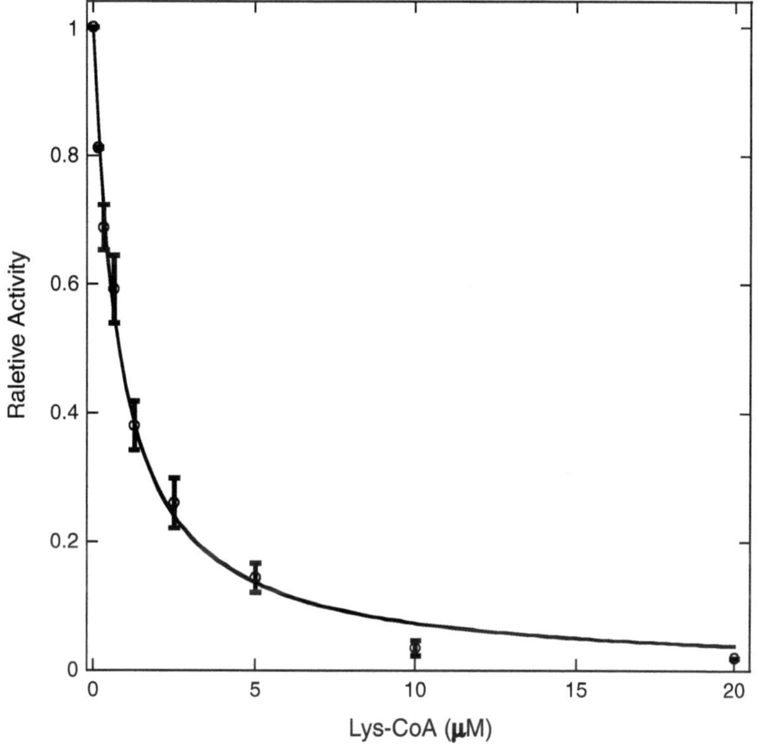

Fig. 6. Inhibition of MOF activity by Lys-CoA measured by the fluorescent assay. Assay was performed in 1× RB at 30°C with 20 nM MOF, 3 μM Ac-CoA, 300 μM H4-20, and different concentrations of Lys-CoA. 5 μM CPM was used for product quantitation.

4. Notes

1. CPM has maximum absorption at 392 nm and maximum emission at 482 nm, measured in 1× RB with 10% DMSO.
2. The test of DMSO effect on MOF catalysis showed that 50% DMSO can completely quench the acetylation reaction. As an alternative, 50% isopropanol has also been shown effective for quenching acetylation reaction (see ref. 12).

Acknowledgements

This work was supported in part by NIH grant R01GM086717 and American Heart Association grant 09BGIA2220207.

References

1. Vetting MW, S de Carvalho LP, Yu M, Hegde SS, Magnet S, Roderick SL, Blanchard JS (2005) Structure and functions of the GNAT superfamily of acetyltransferases. Arch Biochem Biophys 433:212–226
2. Lee KK, Workman JL (2007) Histone acetyltransferase complexes: one size doesn't fit all. Nat Rev 8:284–295
3. An W (2007) Histone acetylation and methylation: combinatorial players for transcriptional regulation. Subcell Biochem 41:351–369
4. Wang Q, Zhang Y, Yang C, Xiong H, Lin Y, Yao J, Li H, Xie L, Zhao W, Yao Y, Ning ZB, Zeng R, Xiong Y, Guan KL, Zhao S, Zhao GP (2010) Acetylation of metabolic enzymes coordinates carbon source utilization and metabolic flux. Science 327:1004–1007
5. Zhao S, Xu W, Jiang W, Yu W, Lin Y, Zhang T, Yao J, Zhou L, Zeng Y, Li H, Li Y, Shi J, An W, Hancock SM, He F, Qin L, Chin J, Yang P, Chen X, Lei Q, Xiong Y, Guan KL (2010) Regulation of cellular metabolism by protein lysine acetylation. Science 327:1000–1004
6. Dekker FJ, Haisma HJ (2009) Histone acetyl transferases as emerging drug targets. Drug Discov Today 14:942–948
7. Zheng YG, Wu J, Chen Z, Goodman M (2008) Chemical regulation of epigenetic modifications: Opportunities for new cancer therapy. Med Res Rev 28:645–687
8. Berndsen CE, Denu JM (2005) Assays for mechanistic investigations of protein/histone acetyltransferases. Methods 36:321–331
9. Lau OD, Kundu TK, Soccio RE, Ait-Si-Ali S, Khalil EM, Vassilev A, Wolffe AP, Nakatani Y, Roeder RG, Cole PA (2000) HATs off: selective synthetic inhibitors of the histone acetyltransferases p300 and PCAF. Mol Cell 5:589–595
10. Feng Y, Xie N, Wu J, Yang C, Zheng YG (2009) Inhibitory study of protein arginine methyltransferase 1 using a fluorescent approach. Biochem Biophys Res Commun 379:567–572
11. Wu J, Xie N, Wu Z, Zhang Y, Zheng YG (2009) Bisubstrate inhibitors of the MYST HATs Esa1 and Tip60. Bioorg Med Chem 17:1381–1386
12. Trievel RC, Li FY, Marmorstein R (2000) Application of a fluorescent histone acetyltransferase assay to probe the substrate specificity of the human p300/CBP-associated factor. Anal Biochem 287:319–328

Chapter 19

Analysis of Protein Acetyltransferase Structure–Function Relation by Surface-Enhanced Raman Scattering (SERS): A Tool to Screen and Characterize Small Molecule Modulators

Mohammed Arif, Dhanasekaran Karthigeyan, Soumik Siddhanta, G.V. Pavan Kumar, Chandrabhas Narayana, and Tapas K. Kundu

Abstract

Among the different posttranslational modifications (PTMs) that significantly regulate the protein function, lysine acetylation has become the major focus, especially to understand the epigenetic role of the acetyltransferases, in cellular physiology. Furthermore, dysfunction of these acetyltransferases is well documented under pathophysiological conditions. Therefore, it is important to understand the dynamic structure–function relationship of acetyltransferases in a relatively less complicated and faster method, which could be efficiently exploited to design and synthesis of small molecule modulators (activators/inhibitors) of these enzymes for in vivo functional analysis and therapeutic purposes. We have developed surface-enhanced Raman scattering (SERS) method, for acetyltransferases towards this goal. By employing SERS, we have not only demonstrated the autoacetylation induced structural changes of p300 enzyme but also could use this technique to characterize and design potent, specific inhibitors as well as activators of the p300. In this chapter we shall describe the methods in detail which could be highly useful for other classes of HATs and PTM enzymes.

Key words: Lysine acetylation, Surface-enhanced Raman scattering, Nanoparticles

1. Introduction

PTMs of proteins are well recognized as an important means of generating proteins functional diversity. The reversible nature of these PTMs has provided dynamic means to which various cellular functions rely upon, ranging from gene expression, DNA replication, and DNA repair to diverse signal transduction and metabolic pathways. To date,

Authors Mohammed Arif, Dhanasekaran Karthigeyan, and Soumik Siddhanta contributed equally.

Sandra B. Hake and Christian J. Janzen (eds.), *Protein Acetylation: Methods and Protocols*, Methods in Molecular Biology, vol. 981, DOI 10.1007/978-1-62703-305-3_19, © Springer Science+Business Media, LLC 2013

the most studied modifications of proteins (mainly histones) include lysine acetylation, serine, threonine and tyrosine phosphorylation, lysine methylation (mono-, di-, and trimethylation), arginine methylation (monomethylation and asymmetric and symmetric dimethylation), sumoylation, and ubiquitination. Further, these PTMs are recognized by a wide array of modular interaction domains present on various proteins, which read the state of the proteome and provide functional meaning to these coded messages. In combination, the large set of PTMs and corresponding interaction domains provide a versatile mechanism to orchestrate cellular behavior.

The acetylation of proteins by acetyltransferases is increasingly considered a biologically relevant regulatory modification like phosphorylation. Acetyltransferases transfer acetyl groups from acetyl coenzyme A (AcCoA) either to the α-amino group of the amino-terminal residue (N-acetyltransferases) or to the ε-amino group of specific lysine residues (histone/factor acetyltransferases) of substrate proteins. The reverse reaction is catalyzed by deacetylases that remove acetyl groups from specific acetyllysine residues in their substrates. The reversible lysine acetylation of histones and nonhistone proteins plays a vital role in the regulation of many cellular processes including chromatin dynamics and transcription, gene silencing, DNA repair, nuclear import, etc. To date, more than 20 acetyltransferases and 18 deacetylases have been identified, and over 40 transcription factors and several 100 other nuclear, cytoplasmic, bacterial, and viral proteins have been shown to be acetylated in vivo, and the investigation of protein acetylation continues. These modifications occur on multiple but specific residues, and the combinatorial modification profiles of histones suggest that the modification sites can also act as binding platforms for specific proteins to read those "read" as articulated by the "histone code" hypothesis. With the discovery of PTMs on nonhistone cellular proteins, e.g., tumor suppressor p53, the idea of "protein code" hypothesis has started to galvanize.

Traditionally, identification and characterization of protein PTMs has relied upon modification-specific antibodies in different immunoassay methods (e.g., western blotting and immunofluorescence). However, variable specificity of cross-reactivity presents some potential problems, or epitope occlusion by neighboring modifications presents a major problem that can lead to false signals. For these reasons, various techniques based upon mass spectrometry have become a common method for speedy identification and discovery of new PTMs on proteins.

Although Raman spectroscopy has been used in the past to get important structural information from various biological systems, like proteins, nucleotides, and peptides, including detection of proteomic analytes, it suffers from very low scattering cross sections, leading to lower sensitivity as compared to fluorescence- and ELISA (enzyme-linked immunosorbent assay)-based techniques (1, 2).

However, because of its single molecule detection sensitivity and its ability to extract valuable chemical information, SERS has become an increasingly popular tool for detection of biomolecules at extremely low concentrations (3–5).

In this chapter, we describe a detection method-based SERS which has been applied to understand the structure–function relationship of lysine acetyltransferase, p300, and its acetyltransferase catalysis domain in the context of small molecule modulator interactions (6, 7). We studied the SERS of the human transcriptional coactivator, p300, which possesses the intrinsic HAT activity (8). Interaction between small molecule HAT inhibitors and activators was also carried out to understand the nature of these molecular interactions (6, 7). A major problem encountered in identifying the specific signals during the SERS studies from the proteins and their modifications was background signals from buffer components. Thus, the buffer conditions were optimized to reduce the background signal from buffer components and other small molecules. We have therefore developed a SERS-based method to indentify the structural changes in protein (enzymes) upon acetylation and binding to specific small molecule modulators (inhibitor/activator). This is an easy, cost-effective, and a fast method, which holds great promise for the future.

1.1. Surface-Enhanced Raman Spectroscopy (SERS)

Raman and IR spectroscopy are the vibrational spectroscopy, which are used to obtain molecular fingerprints of proteins. Raman and IR spectroscopy are complementary to each other. But in the case of a protein, since there is a complicated symmetry leading to no clear selection rules for assignment of bands, Raman spectroscopy has a definitive advantage over IR for biological samples as water, a major constituent of aqueous solution, has a very weak signature in Raman scattering in comparison to the IR. The major disadvantage of Raman spectroscopy is its low scattering cross section, limiting its usage in biomolecule characterization and trace detection. In the last 30 years, SERS has emerged as an effective technique to circumvent the above-mentioned problem. In SERS, the intensity of Raman scattering is enhanced many folds when molecules are adsorbed on to certain metallic surfaces exhibiting nanoscale roughness. It was first discovered while studying pyridine on a silver electrode in an electrochemical cell, when a strong Raman scattering was observed (9). Later, it was reported that the Raman scattering from pyridine adsorbed on metal colloids was enhanced by a factor of 10^6 compared to the equivalent scattering for the same concentration of pyridine in solution (10). This enhancement phenomenon was called SERS and it depended upon the surface roughness and effective adsorption of the analyte to that surface. Surface enhancement has been mainly observed from silver, gold, and copper surfaces, although aluminum, lithium, and sodium also give enhancement to a lesser extent.

There are two fundamental enhancement mechanisms currently accepted as the underlying reasons for SERS (11, 12). First is the electromagnetic enhancement mechanism, where the local electric field created by surface plasmon resonance at the vicinity of the roughened metal surface influences the molecule such that the Raman signal intensity is enhanced. The other mechanism is the chemical enhancement, in which there is a charge transfer between the molecule and the metal surface, which further influences the excited energy states of the molecule resulting in an enhanced Raman signal. Along with this, the polarizability of the molecule is influenced during this interaction, which results in certain selection rules of SERS, different from normal Raman spectroscopy. In their simplest form, the most intense bands are predicted to be from those vibrational modes which are perpendicular to the metal surface (13, 14). This information can be applied qualitatively to obtain an approximate indication of the angle a molecule unit subtends to the surface. Here we will discuss briefly the electromagnetic and chemical enhancement mechanism of SERS.

1.2. Electromagnetic Enhancement Mechanism

A small metal sphere like a silver nanoparticle which is isolated and illuminated by time-varying electric field will induce a collective oscillation of the conduction electrons on the surface of the sphere generating surface plasmon. Since metals can be considered as periodic static positive charges with a "sea of electrons" around it, a momentary displacement of the electrons causes a dipole to be formed which will oscillate with the characteristic frequency associated with this collection of electrons, generating in turn an electric field, which dies rapidly ($1/r^{12}$) at a distance from the metal sphere. When the frequency of the electromagnetic radiation matches this frequency, you have maximum absorption of this radiation leading to surface plasmon resonance. Thus a large electric field is experienced very close to the metal surface, say a molecule adsorbed on it, due to the combination of the incident electromagnetic field and the surface plasmon generated by it. The field enhancement factor $A(\nu)$ is the ratio of the field at the position of the molecule and the incoming field. The expression for this enhancement factor is given by (15, 16)

$$A(\vartheta) = \frac{EM(\vartheta)}{E(\vartheta)} = \frac{\varepsilon - \varepsilon_0}{\varepsilon + 2\varepsilon_0}\left(\frac{r}{r+d}\right)^3,$$

where the diameter of the metal sphere is $2r$ and is small compared to the wavelength of light. The molecule is at a distance d from the surface of the metal sphere and exposed to a field EM which is a superposition of incoming field and dipole field induced in metal sphere. The sphere has a complex dielectric constant $\varepsilon(\nu)$ surrounded by a medium with dielectric constant $\varepsilon(0)$. The conditions which describe the resonant excitation of the surface plasmons of the metal sphere are a large value of $A(\nu)$ when the real part of ε is equal to $2\varepsilon_0$ and also the imaginary part of the dielectric constant is small. Metals like Ag, Au, and Cu satisfy these conditions making them good substrates for SERS (17, 18). In SERS, the Raman signals from the molecules in the vicinity of the metal nanoparticles get enhanced. The electromagnetic enhancement factor for stokes scattering is given by (15)

$$G_{EM}(\vartheta_s) = \left|\frac{\varepsilon(\vartheta_L)-\varepsilon_0)}{\varepsilon(\vartheta_L)+2\varepsilon_0}\right|^2 \left|\frac{\varepsilon(\vartheta_S)-\varepsilon_0)}{\varepsilon(\vartheta_S)+2\varepsilon_0}\right|^2 \left(\frac{r}{r+d}\right)^{12}$$

From the above equation it can be inferred that the enhancement scales is the fourth power of the local field at the vicinity of the metallic nanostructure and is strong in the case where the scattered and the plasmon field are in resonance. The electromagnetic enhancement decays as $(r/(r+d))^{12}$. Thus it is a distance-dependent phenomenon. The electromagnetic factor accounts for 10^4–10^6 of the total enhancement factor.

1.3. Chemical Enhancement Mechanism

Chemical enhancement contributes to a factor of 10–100 of the total enhancement factor and requires the molecule to be in direct contact with the surface of nanoparticle (19). The Raman cross section of the molecule increases when there is an electronic coupling between the molecule and the metal. Due to a charge transfer electronic transition in the metal–molecule system as well as broadening and shifting of electronic levels in the adsorbed molecule, a resonance Raman effect can be observed which also leads to SERS enhancement. There is also a dynamic charge transfer where a photon is absorbed by a metal which results in a hot electron state. The hot electron gets transferred to the *L*owest *U*noccupied *M*olecular *O*rbital (LUMO) of the molecule. The excited electron is then transferred back to the metal from the LUMO resulting in the emission of stokes photons. This mechanism is also known as the first layer effect as direct attachment to the surface gives higher enhancement factor (16).

1.4. Surface-Enhanced Raman Spectroscopy of p300 Histone Acetyltransferase

p300 is a 300 kDa multidomain transcriptional coactivator protein. Surface-enhanced Raman spectroscopy was used as an attempt to understand its structure in the absence of its crystal structure. Vibrational analysis of the protein can also throw light into the protein—small molecule interactions (inhibitors and activators). Therefore these studies have therapeutic significance and can be used to gather specific information about the protein–ligand systems. SERS studies of both the whole-length p300 as well as the p300 HAT domain have been done to understand the structural aspects and the structural changes that occur on binding of small molecule inhibitors or activators. The SERS spectra contain information about the secondary structure of a protein as well as the relative abundance and orientation of different aromatic amino acids present in the protein. The spectra also contain bands from carboxylic groups and other side chain vibrations including those containing sulfur. The tentative band assignments of SERS are summarized in Table 1.

1.5. Band Assignments: Important Bands in SERS of p300 Histone Acetyltransferase

SERS provides information about the secondary structure of protein. It is different from that of NMR and X-ray crystallography which gives information about the tertiary structure of the protein. Characteristic bands found in SERS spectra of proteins and polypeptides include the amide I, II, and III bands (20). These bands arise from the vibrations of the amide bonds that link the amino acids. The amide I band is mainly assigned to the C=O stretching. The amide II band arises from out of phase combination of the NH in-plane bend and the CN stretching vibrations. The amide III is the in phase combination of the NH bending and CN stretching vibrations. Since both C=O and N–H bonds take part in hydrogen bonding between different residues in the secondary structure their position and intensities can be directly correlated to the secondary structure of the protein. Typically the amide I bands for α-helix, β-sheet, and random coil lie in the regions 1,640–1,658 cm^{-1}, 1,665–1,680 cm^{-1}, and 1,660–1,665 cm^{-1}, respectively, in Raman spectra. In case of SERS the strong interaction of the –CO–NH– with the silver surface weakens the C=O bond. Therefore the amide I band shifts to lower frequencies. The amide II band lies in the region around 1,550 cm^{-1}. This mode is inactive in Raman spectra but appears in SERS because of modification of surface selection rules. The amide III band lies in the region from 1,200 to 1,400 cm^{-1}. In full-length p300 (Fig. 1a) the amide I, II, and III bands lie at 1,623, 1,540, and 1,296 cm^{-1}, respectively (Fig. 1b) (8). In the SERS of p300 HAT domain (p300HD), as shown in Fig. 1e, the amide bands lie at 1,660 cm^{-1} (amide I, α-helix), 1,632 cm^{-1} (amide I random coil), 1,518 cm^{-1} (amide II), and 1,294 cm^{-1} (amide III) (21). The positions of the amide bands were confirmed by deuteration of the proteins. Upon deuterating the protein, most of the hydrogen in the amide groups gets replaced by the heavier deuterium. This increases the reduced mass of the vibrating unit which is inversely proportional to the vibrational frequency. The amide I band in both p300 and p300HD shifts by around 7 cm^{-1} on deuteration. The amide II band also shifts on deuteration by around 5 cm^{-1} and the amide II band shifts from 1,296 to 950 cm^{-1} (22) (Table 1).

The SERS of p300 as well as p300HD was dominated by aromatic side chain vibrations of tyrosine (Tyr), tryptophan (Trp), and phenylalanine (Phe) (20). The intensities of different ring modes of these aromatic amino acid residues depend on the orientation of

Table 1
Tentative band assignment of full-length p300, p300 with Ag nanoparticles aggregated with NaCl, and p300 HAT domain (Adapted from Pavan Kumar et al. 2006 and Arif et al. 2007, with permission)

p300 (cm^{-1})	p300 + 0.3 mM NaCl (cm^{-1})	p300 HAT domain (cm^{-1})	Tentative band assignment
		1,660	Amide I (α-helix)
1,623	1,624	1,632	Amide I (random coil)
	1,605		Trp, Tyr, and/or Phe (ν_{8a})
		1,596	Trp, Tyr, and/or Phe
1,588		1,583	$\nu_{as}(COO^-)$
		1,545	Trp, Tyr, and/or Phe
1,540		1,518	Amide II and/or Trp
	1,508		His
	1,454	1,449	$\delta(CH_2)$
1,442	1,433		His and/or Trp
1,389		1,369	$\nu_s(COO^-)$
1,374			ν(CH)
	1,364		Trp
	1,345		δ(CH) and/or Trp
		1,303	$\omega(CH_2)$
1,296		1,294	Amide III (α-helix)
	1,267		ν(C=O) (α-helix)
1,180	1,184	1,184	Phe (ν_{9a})
	1,153		ν(C–N)
1,114		1,132	$\nu_{as}(C_\alpha CN)$
1,091			$\nu(C_\alpha N)$
	1,008		Phe (ν_{12})
912	913	917	$\nu(C–COO^-)$
800	803	806	Tyr and/or ν_s(C–S–C)
		782	Trp_{W18}
758	761		Trp or His
723	728		$\delta(COO^-)$
		623	Phe (ν_{6b})
	642		Tyr
580	584	579	Trp
		505	ν(S–S)

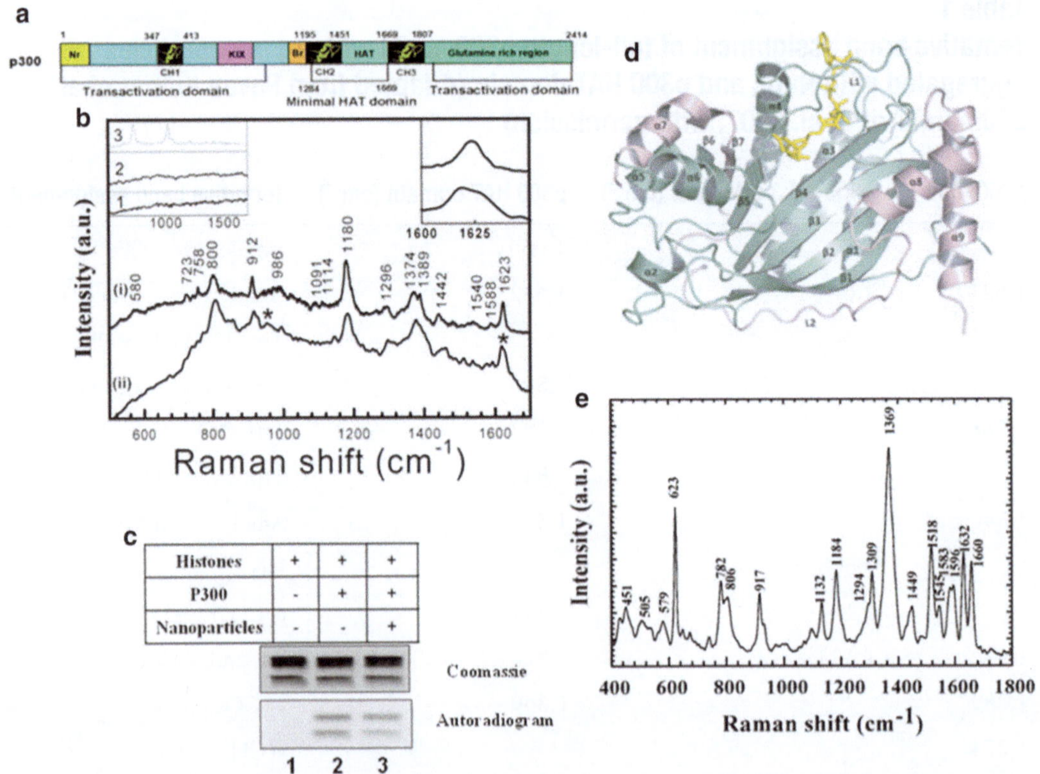

Fig. 1. (**a**) Shows the cartoon of p300 domain organization, (**b**) *inset* on the *left* shows the Raman spectra of pure p300 (curve 1), nanoparticles (curve 2), and buffer with nanoparticles (curve 3). Curve (i) shows the SERS spectra of p300 and curve (ii) shows the SERS spectra of deuterated p300. New or modified bands have been indicated by *asterisk*. The *inset* on the *right side* shows the amide I region for clarity (Adapted from ref. 8). (**c**) HAT assay was performed by using purified HeLa core histones (1.2 μM) in the absence (*lane 1*) and presence of (5 ng) p300 (*lane 2*) and the p300 enzyme preincubated with silver nanoparticle (*lane 3*) and the fluorographic images are shown in the autoradiogram with Coomassie showing the loading control for core histones. (**d**) Crystal structure of p300 HAT domain. Shown here is the ribbon model of the same. (**e**) SERS spectra of p300 HAT domain (Adapted from Arif et al. 2007, with permission).

the ring on the silver surface. The benzene ring of these amino acids can be oriented either in a tilted manner or parallel to the surface. In the former case the ring breathing vibration modes will be strong and also C–H stretching vibration, but in the later case these modes will be very weak.

Aliphatic side chain vibrations also contribute to the SERS spectra of proteins. In p300 SERS spectra the carboxylate group of the aspartic acid (Asp), glutamate (Glu), and the C-terminus group in protein interacts with the silver surface to give strong enhancements of these Raman bands. The enhancement of modes corresponding to the symmetric stretching of $C_{\alpha}N$ and the asymmetric stretching of $CC_{\alpha}N$ groups suggests that the p300 absorbs to the silver nanoparticles through the nitrogen groups.

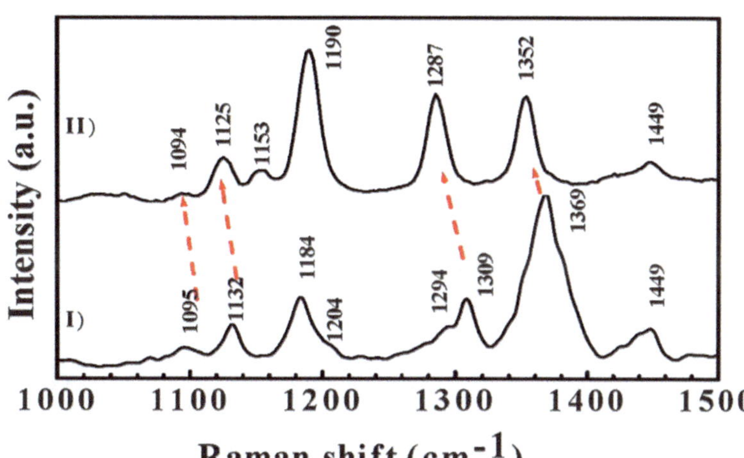

Fig. 2. SERS of (I) HAT domain compared with (II) acetylated HAT domain. The *red arrows* indicate the red shift which implies softening of some modes (Adapted from Arif et al. 2007, with permission).

1.6. SERS to Probe Acetylation Induced Specific Structural Changes in Histone Acetyltransferase Domain of p300

SERS was performed on fully autoacetylated p300HD and also on normal partially acetylated HAT domain. Unlike normal Raman spectroscopy, SERS probes the vibrational modes near to the surface of the metal nanoparticles. The intensity of the modes depends on the electromagnetic enhancement which is distance dependant and also on surface selection rules. Therefore structural modifications to the protein near the surface of the metal nanoparticles are probed. On comparing the normal HAT and the acetylated HAT domain (Fig. 2), there is a marked change in the modes corresponding to the symmetric stretching of COO^- (1,369 cm^{-1}) and amide III (1,294 cm^{-1}), ν_{9a} of Phe (1,184 cm^{-1}) and asymmetric stretching of $C_\alpha CN$ (1,132 cm^{-1}). The intensity of these modes changes and also shifts in position to lower frequencies indicating softening of modes. This softening of modes in order of ~10 cm^{-1} indicates bond weakening due to interaction of various groups and increase in hydrogen bonding due to autoacetylation (21). This change in hydrogen bonding can affect the bond strength of different groups like amides. Consequently modes corresponding to groups like CH_2 are not expected to change. This can be seen from the fact that modes like 1,449 and 1,094 cm^{-1} that correspond to $\delta(CH_2)$ and proline (Pro) do not show significant shift on acetylation. The shifts in band positions and changes in intensities of bands indicate structural reorganization of the HAT domain on complete acetylation.

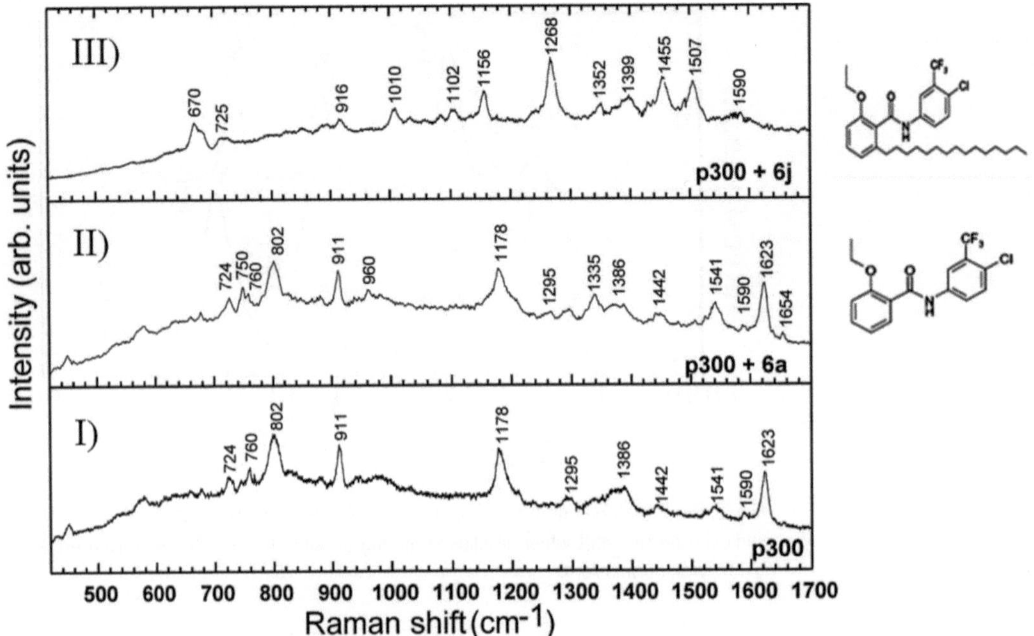

Fig. 3. SERS spectra of p300 with CTB (6a) and CTPB (6j). Shown is the SERS spectrum of p300 (I). In the presence of CTB (II), the SERS spectrum of p300 undergoes a large change. Upon addition of CTPB to p300 (III), the SERS spectrum of p300 shows the appearance of certain new modes (Adapted from Mantelingu et al. 2007, with permission). The chemical structure of the molecules 6j and 6a are shown adjacent to the respective spectra.

1.7. SERS to Probe Activation of p300 Histone Acetyltransferase by Small Molecules Altering the Enzyme Structure

SERS on p300 in complex with different functional derivatives of CTPB shows that structural changes in the protein are responsible for the activation of HAT activity. SERS was performed on p300 protein and p300 complex with CTPB and CTB (6). The binding of molecules to the protein distances the binding sites from the nanoparticle surface. The orientation of different groups with respect to nanoparticle surface also changes. These result in the change in intensity of Raman modes and sometimes even complete disappearance. This is not true in the case of normal Raman spectroscopy. The change in the spectrum of protein on binding of molecules indicates structural changes in the protein. This also gives an idea about the probable binding sites of the molecules. In case of CTPB the SERS spectrum of p300 does not show large-scale changes (Fig. 3 panel III versus I). CTPB binds to the protein predominantly to the amide groups of the α-helix and β-sheet as indicated by the increase in intensity or appearance of bands at 1,654, 1,335, and 960 cm^{-1} (6). A large-scale change in SERS spectrum can be observed when CTB binds to p300 (Fig. 3 panel II versus I). CTB also binds to the α-helix and β-sheet regions of p300 but is quite different from CTPB as both the Raman modes 1,623 and 1,654 cm^{-1} disappear. One of the major reasons for such an observation can be understood from the fact that both CTB and CTPB are insoluble in water. Secondly, both the molecules have a strong hydrophobic region dominated by the halogen groups and the hydrophilic group centers on the amide group. Hence the binding will require to be centered on a strong hydrophobic and strong hydrophilic region in the p300 and more than one molecule binds. Since CTPB has a very large pentadecyl alkane chain, there would be a strong steric hindrance while binding to the protein. This is the reason for small changes seen in the case of CTPB. This observation that the binding of these activators happens in a region of the protein dominated by α-helix and β-sheet was substantiated later on a synthetically prepared p300HD (23).

Large shifts in peaks in the region 1,150–1,550 cm^{-1} which are related to the amide II and aromatic amino acids (Trp, Tyr, Phe, and His) suggest strong binding of CTB to the binding site. These kinds of binding are brought on by strong hydrogen bonding as well as hydrophobic interactions which shifts Raman peaks by 10–35 cm^{-1}. The appearance of new modes at around 1,010 and 1,102 cm^{-1} (symmetric and asymmetric vibrations of $C_{\alpha}CN$) and disappearance of band at 802 cm^{-1} (Tyr) indicate strong binding to the protein (6). A mode at 670 cm^{-1} appears which shows that the CTB binds to the carboxyl groups of the aliphatic amino acids too (6). SERS was also used to study whether the activity of the compounds are related to the alteration of the enzyme structure. The compound CTPB and CTB were derivatized to find the structural basis of enhancement of enzyme activation. Three kinds of derivatives were synthesized: (1) with or without pentadecyl hydrocarbon chain, (2) change in relative position of $-CF_3$ and –Cl, and (3) ethyl and –Cl groups were substituted by methyl or isopropyl and $-NO_2$ or –CN groups, respectively. It was observed that only the molecules which activated p300 showed structural alteration of the enzyme as shown by SERS (6). The effect of relative position of Cl and CF_3 (Fig. 4a) on enzyme structure could also be elucidated on basis of change in SERS spectra. There was also a one to one correspondence between the degree of activation (Fig. 4b and c) and the extent of change in the SERS spectra of the enzyme–molecule complex (Fig. 4d).

1.8. SERS to Probe p300-Specific Inhibitors Inducing Specific Alteration in the Enzyme Conformation

SERS was performed on p300 in complex with garcinol, isogarcinol, and isogarcinol derivatives. These aromatic ligands form hydrogen bonds or hydrophobic interactions with the protein resulting in change in the structure of the protein. SERS spectrum of p300–garcinol complex (Fig. 5c) shows changes to the bands corresponding to the amide bands of α-helix at 1,624 and 1,296 cm^{-1} (7). There are also changes in the carboxylic group vibrations of the aliphatic amino acids (985 and 684 cm^{-1}) and Tyr modes like 803 cm^{-1} (7). Isogarcinol, which is structurally different from garcinol in having an oxygen bridge instead of a hydroxyl group in garcinol, causes large-scale changes to the structure of p300 (Fig. 5d). The modes which were affected in case of garcinol were

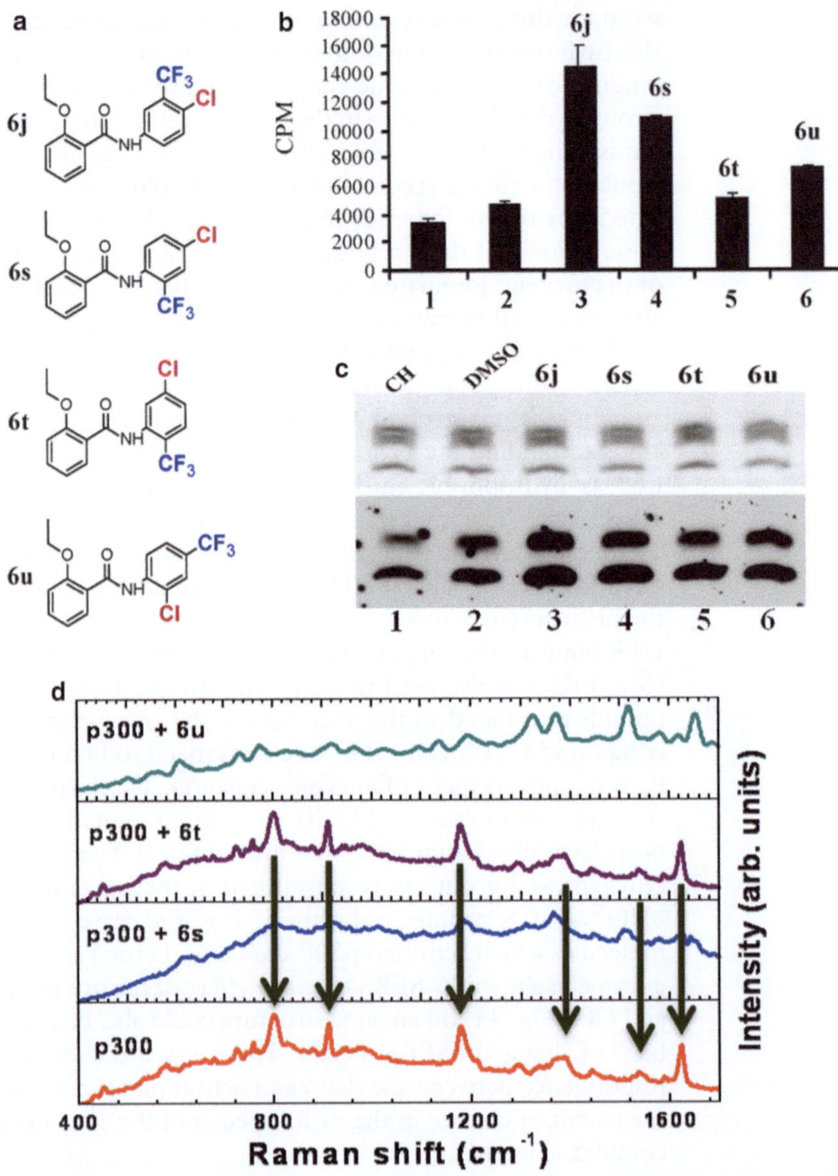

Fig. 4. The role of the hydrophobic group of CTB molecule, namely, $-CF_3$ and –Cl, in the process of HAT activation. The molecular structure is given in (**a**) and the corresponding HAT activities for these molecules are shown in both the filter-binding assay (**b**) as well as fluorographic analysis (**c**). *Lane 1* histones with p300, *lane 2* histones with p300 and DMSO and *lanes 3–6* 200 μM of 6j and 6(s-u). The SERS spectra of 6(s-u) is shown in (**d**). Note no effect on the SERS spectra in the case of the 3rd curve from bottom, concurrent with its activity on p300 (*arrows* show all the peaks exist in both the spectra).

also affected in case of IG and in addition to that it also affects the modes related to Tyr, Trp, Phe, and His. SERS of protein complex with the derivatives of isogarcinol like LTK-13, -14, -15, and -19 were also studied (Fig. 5e–h). LTK-15 is an inactive isogarcinol derivative and shows no change in SERS spectrum (7). LTK-13 changes the SERS spectrum only in the areas of amide modes and

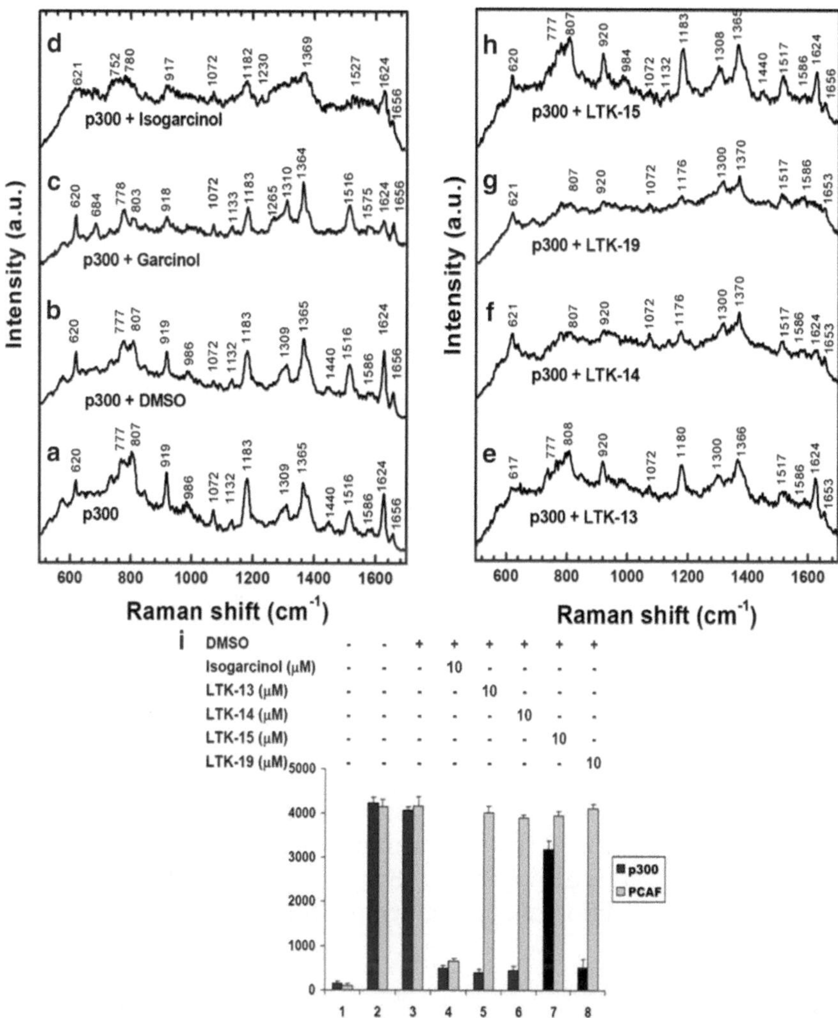

Fig. 5. Isogarcinol derivatives are specific inhibitors of p300 HAT. SERS of (**a**) p300, (**b**) p300 + DMSO, (**c**) p300 + garcinol, (**d**) p300 + isogarcinol, (**e**) p300 + LTK-13, (**f**) p300 + LTK-14, (**g**) p300 + LTK-19, and (**h**) p300 + LTK-15 (Adapted from Mantelingu et al. 2007, with permission). (**i**) HAT assays were performed either with p300 or PCAF in the presence and absence of IG or its derivatives by using highly purified HeLa core histones and processed for filter binding. *Lane 1* core histones without HAT, *lane 2* histones with HAT, *lane 3* histones with HAT and in the presence of DMSO, and *lanes 4–8* histones with HAT and in the presence of IG or its derivatives LTK-13, -14, -15, and -19.

certain carboxylic groups of the aliphatic amino acids. But in case of LTK-14 and LTK-19, there are large-scale changes in the SERS spectra, not only in amide modes but also in modes corresponding to Tyr, Trp, Phe, and His (7). Thus, through SERS it could be inferred that the specific and nonspecific inhibitors of p300 (Fig. 5i) bind to the amide groups of α-helix altering the structure of the protein differentially.

In the following section we will discuss the detailed experimental setup for carrying out SERS experiments with a custom-built Raman spectrometer converted from an epi-fluorescent

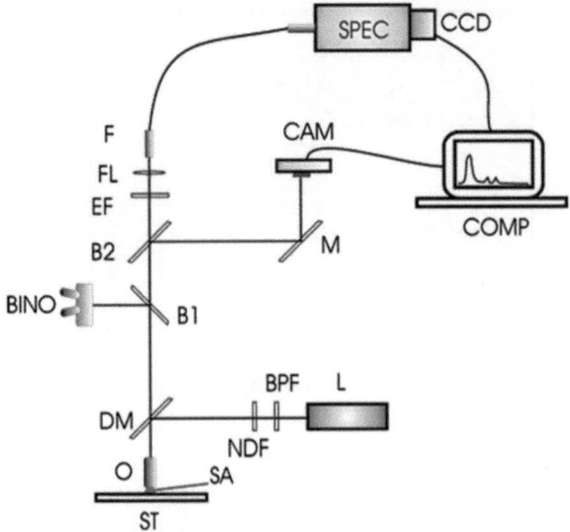

Fig. 6. Schematic of a micro-Raman system built using a simple viewing microscope with an epi-fluorescent attachment. *ST* stage, *SA* sample, *O* objective lens, *L* laser, *BPF* band-pass filter, *NDF* neutral density filter, *SM* special mirror, *B1 and B2* beam splitters, *M* mirror, *BINO* binocular, *EF* edge filter, *FL* focusing lens, *F* optical fiber, *SPEC* spectrometer, *CCD* charged coupled device, *COMP* computer, *CAM* camera.

microscope (Fig. 6). Application of SERS for studying full-length p300 histone acetyltransferase, p300 HAT domain, and protein–small molecule interactions resulting in protein structural change will be demonstrated. Band assignments of the protein in different conditions and in deuterated conditions will be presented.

2. Materials

2.1. Molecular Biology Reagents

All the reagents were from Sigma unless otherwise stated.

1. (^{3}H) acetyl-CoA (NEN-PerkinElmer).
2. Cold acetyl-CoA.
3. P-81 filter papers (Whatman).
4. Wallac 1409 liquid scintillation counter.
5. Dimethyl sulfoxide (DMSO).
6. Kodak X-OMA T film and cassettes.
7. Standard labware for sodium dodecylsulfate (SDS) polyacrylamide gel electrophoresis (PAGE).
8. Trichloroacetic acid (TCA).
9. HeLa or calf thymus histones: Core histones were purified from cultured HeLa cells as described earlier (24).

2.2. Solutions

1. 2× HAT buffer: 50 mM Tris–HCl, pH 8.0, 10% (v/v) glycerol, 1 mM dithiothreitol, 1 mM phenyl methyl sulfonyl fluoride, 0.1 mM EDTA, pH 8.0, 10 mM sodium butyrate.
2. Deuterated water.
3. All waters used were Milli-Q.
4. Amplify solution (Pharmacia-Amersham) for fluorography.
5. Coomassie Blue-R250 staining and destaining solutions.

2.3. Spectrometer

1. 0.55 m spectrograph (Jobin-Yovn 550 Triax, Instruments SA, Inc., NJ, USA) with 600, 1,200, and 1,800 grooves/mm holographic gratings.
2. Liquid nitrogen cooled CCD detector.

2.4. Laser

1. 632.8 nm He–Ne laser (model 309995, Research Electro Optics, Inc., USA).
2. Solid-state frequency-doubled 532 nm Nd-YAG laser (model No. GDLM-5015L, Photop Suwtech, China).

2.5. Components for SERS Microscopy

1. Nikon Eclipse 50i (Nikon, Japan) microscope with an epi-fluorescent attachment.
2. Band-pass filters (LL01-532-12.5 and LL01-633-12.5, Semrock, UK).
3. Edge filters (LP03-532RS-25 and LP02-633RS-25, Semrock, UK).
4. 250 μm multimode single core optical fiber with a band pass of 400–1,000 nm.
5. 60× infinity corrected water immersion objective (Nikon Fluor, NA 1.00, Nikon, Japan).
6. 50× infinity corrected SLWD (Nikon L Plan, NA 0.45, Nikon, Japan).
7. Dichroic mirror (Omega optical).

2.6. Special Component

1. Mirror consisting of 2 mm diameter Ag coating at the center of a 25.2 × 35.6 (±2) mm fused silica substrate of 1.1 mm thickness.

3. Methods

The following protocol provides guidelines for investigation of p300 protein and its interaction with small molecule activators/inhibitors. Using SERS it can be shown that there is a structural alteration on acetylation of p300 HAT domain. It also shows that there is alteration of protein structure on interaction of small molecules with p300 histone acetyltransferases which has a correspondence with its function.

3.1. HAT Assay

HAT assay is used to detect the enzymatic (functional) activity of the p300 HAT using core histone as substrate. Intact enzymatic activity also suggests the presence of a functional protein during the various assay and detection conditions. Effects of silver nanoparticle were assessed using HAT assay.

1. Take 1.8 μg of highly purified HeLa core histones, HAT enzyme and incubate in HAT assay buffer in a final reaction volume of 30 μL at 30°C for 10 min in the presence and absence of silver nanoparticle and garcinol (an HAT inhibitor; as a positive control).
2. Add 1 μL of 6.2 Ci/mmol (^{3}H)acetyl coenzyme A (acetyl-CoA) and incubate it for an additional period of 10 min at 30°C.
3. Blot the reaction mixture onto P-81 (Whatman) filter papers, and record the radioactive counts on a liquid scintillation counter (for details see ref. 6).

3.2. In Vitro Gel Assay

The effect of silver nanoparticle was also assessed using in vitro gel assay.

1. HAT assay was performed as mentioned above. To visualize the radiolabeled acetylated proteins, follow the below-mentioned steps.
2. Precipitate the protein using 30 μL of 50% (w/v) trichloroacetic acid (TCA) to get a final concentration of 25% (w/v) in the reaction mixture. Vortex slightly and incubate over ice for 30 min.
3. Centrifuge this mixture at 16,000 × g at 4°C for 30 min.
4. Remove the supernatant carefully without disturbing the pellet (white colored on the wall of the tube).
5. Add 500 μL of cold acetone (to wash the pellet for removing the residual TCA) in the tube.
6. Centrifuge it again at 16,000 × g at 4°C for 30 min.
7. Discard the supernatant and air-dry the tube for about 15 min.
8. Run the standard SDS-PAGE at room temperature. For histone and p300 HAT visualization, we run 15% and 10% SDS-PAGE, respectively.
9. Stain the gel with colloidal Coomassie Blue-R250 and scan the gel after destaining that can be used to visualize the loading and comparing the gel with the autoradiogram.

10. For fluorography, remove the destaining solution and wash with water (two times). Then dehydrate the gel using dimethyl sulfoxide (DMSO) for 30 min. Perform this step twice.
11. Now immerse the gel in Amplify solution fluorography reagent as per manufacturer's instructions. Dry the gel and expose to film for autoradiography. The exposure time required is a function of the amount of activity loaded onto the gel and the quality of the acetyl-CoA used (see Note 1) and can range from 24 h to a week's time (Fig. 1c).

3.3. p300 HAT Autoacetylation Assay

This method helps in visualizing the autoacetylation status of p300 HAT and the affect of silver nanoparticle on p300 HAT function (autoacetylation) (see Note 2).

1. Autoacetylation assay is performed similar to HAT assay.
2. Take 600 ng of p300 HAT (*E. coli* expressed recombinant HAT domain) in HAT assay buffer in a final reaction volume of 30 μL in the presence and absence of silver nanoparticle and HATi (as a positive control) (for details see ref. 21).
3. Incubate the reaction mixture at 30°C for 10 min.
4. Add 1 μL of 4.7 Ci/mmol (^3H) acetyl coenzyme A and incubate it for an additional period of 10 min at 30°C.
5. Perform in vitro gel assay to visualize the radiolabeled acetylated p300HD by autoradiography.

3.4. Autoacetylation of p300 HAT Domain for SERS Experiments

For performing the SERS we set up several autoacetylation reactions of p300 HAT using cold acetyl-CoA so that we can get good amount of protein which can be used to further repurify the p300 HAT using Ni^{2+}-NTA resins. Repurification of p300 HAT is important to get rid of the free cold acetyl-CoA which would otherwise have a strong signal during SERS experiment and can mask the protein signal.

1. Take 600 ng of p300 HAT in HAT assay buffer in final reaction volume of 30 μL.
2. Incubate the reaction mixture at 30°C for 10 min.
3. Add cold acetyl-CoA to a final concentration of 125 μM, and incubate it for an additional period of 2 h at 30°C.
4. Again add same amount of cold acetyl-CoA and incubate it further for 2 h at 30°C to get the fully autoacetylated p300 HAT.
5. Pool all the reaction mixtures together.
6. In order to remove the free acetyl-CoA and other small molecules of the reaction mixture from the autoacetylated p300 HAT that would otherwise hinder the SERS study, autoacetylated p300 HAT was affinity purified through the Ni-NTA column through standard procedure.

7. Remove imidazole and glycerol by dialysis. Give several change of buffer for complete removal of imidazole and glycerol.
8. Check the activity of p300HD by HAT assay.
9. Perform MALDI-TOF to confirm the autoacetylation status of p300 HAT.

3.5. General Setup

In general, a very advanced Raman spectrometer costs a fortune and would dissuade a biologist in investing in one. Hence, one of the authors, CN, thought of helping design a Raman spectrometer around the fluorescence microscope commonly found in a biology laboratory. The details of building one's own Raman spectrometer for a fraction of the money needed otherwise is given elsewhere (25). Hence a few vital aspects of the SERS setup will be discussed here.

A Raman excitation light from a solid-state laser (such as frequency-doubled Nd-YAG laser), or a gas laser (such as He–Ne laser), traverses through a band-pass filter and enters the input port of an upright microscope. In order to divert the laser beam towards the sample, a dichroic mirror (DM) is used. The incident light is tightly focused on to a sample using a high numerical aperture objective lens (O). The role of an objective lens is twofold: incidence of laser light and collection of the backscattered light. Depending upon the sample type, different kinds of objective lenses are used. For solid, dry samples a high numerical aperture, dry objective lens serves the purpose, whereas for aqueous samples a water immersion objective lens can be used. It is to be noted that the collection efficiency of Raman scattering significantly depends on the kind of objective lens used. Next, the backscattered light collected by the objective lens passes through an edge filter (EF) placed at the output port of the microscope. The role of an edge filter is to reject the intense Rayleigh-scattered light from the collected signal. Edge filters allow only the stokes component of the Raman spectra, whereas a notch filter allows both stokes and anti-stokes component of the scattered light. The filtered light is now focused on to an optical fiber (F). The other end of the optical fiber is f-number matched to a Raman spectrograph. The spectrograph has a computer-controlled adjustable slit and a turret, which holds three gratings for a range of measurements. Different gratings and slit widths are chosen depending upon the needs of spectral resolution. In Raman spectrographs three kinds of gratings with 600 grooves/mm or 1,200 grooves/mm or 1,800 grooves/mm can be used. The higher the number of groves, the greater will be the spectral resolution, and the lesser will be the throughput. Finally, the captured light reaches the CCD camera which converts the light signals into electronic signals for further processing. A digital camera atop the microscope allows for registration of the focused laser spot and simultaneously captures the optical image and video.

The DM can have several disadvantages. It blocks a large region of Raman spectrum (<200 cm^{-1}) close to the Rayleigh scattering. It also cuts off the high frequency Raman spectra (>3,000 cm^{-1}). In order to overcome this and to use this setup for specific applications requiring acquiring spectra over a wide spectral range (e.g., 50–4,000 cm^{-1}), the DM can be replaced by a custom-made special mirror with a 2 mm coating of Ag on the center. The diameter of the laser beam was 1 mm, which is completely reflected by the mirror of 2 mm diameter. This mirror has a reflection band between 400 and 900 nm with reflectivity greater than 95%.

3.6. Preparation of Silver Nanoparticles

1. Use Milli-Q water with a resistance of 18.2 MΩ cm at 25°C for all preparations.
2. Add 18 mg of $AgNO_3$ to 100 mL water and bring to boiling point under refluxing conditions to avoid decrease in water volume. Add 2 mL of 1% sodium citrate solution while stirring vigorously.
3. Continue boiling and stirring for 1 h and then discontinue heating and bring the solution to room temperature by stirring (see Note 3).

3.7. SERS of Aqueous Samples

1. Keep the concentration of protein at 40 ng/μL.
2. Mix the protein solution with Ag colloidal solution in the ratio of 1:4 and drop on a cavity glass slide. Allow the solution to settle for 5 min (see Notes 4 and 5). Perform SERS using a water immersion objective (see Note 6).
3. Accumulate spectra for 3–5 min (see Note 7).
4. For SERS of p300 complexed with small molecules, mix 20 μL of p300 protein with 2 μL of the compound (see Note 8). Mix this solution with 50 μL of Ag colloid and incubate for 5 min.
5. For SERS of deuterated proteins, incubate the protein in D_2O before taking spectra.

3.8. SERS on a Glass Substrate

3.8.1. Method I

1. Deposit 100 μL of Ag colloid on a glass slide and allow it to dry overnight at room temperature. A brown-colored ring formation will be seen on drying of the nanoparticles.
2. Drop 20 μL of p300 solution on the dried nanoparticles and dry the sample at room temperature for 5 h. Take SERS spectra by focusing the laser onto the upper layers of the spot.

3.8.2. Method II (Sandwich Method)

1. Deposit 50 μL solution of Ag colloid on a glass slide and dry overnight at room temperature. Deposit 20 μL of p300 solution on the dried ring followed by 50 μL of the Ag colloid. Dry the sample at room temperature for 5 h. Take SERS measurements by focusing the laser spot on the top of the sample. Control the focus to get maximum intensity of the SERS signal.

4. Notes

1. In our laboratory, we have used (3H)-acetyl-CoA exclusively; however (14C)-acetyl-CoA can also be used. The specific activity of radiolabeled acetyl-CoA can vary dramatically from production lot to production lot and typically ranges as follows: (14C)-acetyl-CoA (40–62 mCi/mmol) or (3H)-acetyl-CoA (2–15 Ci/mmol). It is recommended that material of the highest specific activity available be used in acetyltransferase assays. Acetyl-CoA is a high-energy compound (the thioester is subject to spontaneous hydrolysis) and should be immediately aliquoted upon receipt and stored per the manufacturer's instructions.
2. The nanoparticles should not bring about any change in the activity of the protein per se, as shown in Fig. 1c.
3. For characterization of the silver nanoparticles, UV-visible spectroscopy was done to obtain the absorbance maximum. Transmission electron microscopy (TEM) was performed to obtain information about particle size and morphology, and dynamic light scattering was performed to determine the zeta potential of the silver nanoparticles. The absorbance maximum was found at 425 nm. The average nanoparticle diameter was 50 nm and the average shape was spherical. Some rod-shaped nanoparticles were also formed whose numbers are negligible compared to the spherical ones. A zeta potential value of −28 meV indicates an overall negative charge of the nanoparticles.
4. It is important for the protein molecules to get attached to the nanoparticles to give good SERS enhancement. This interaction is brought about either by covalent bond formation between the nanoparticles and the protein or electrostatic interaction between the protein and the nanoparticle. It is important to consider the pI of the protein as well as the zeta potential of the nanoparticles at a particular pH. In our experimental conditions, the protein is positively charged and the nanoparticles are negatively charged facilitating strong interactions between the protein and nanoparticles. This strong interaction also brings about aggregation of the protein–nanoparticle complex creating "hotspots" and thus enhancing the Raman signal of the protein.
5. "Hotspots" are areas of strong electromagnetic field when a number of nanoparticles come together and form a cluster. Any molecule trapped into the space between the nanoparticles experience intense electromagnetic field and hence experience higher SERS enhancement. We could observe aggregation of

the protein–nanoparticle complex without addition of any external aggregating agent.

6. Water immersion objectives are used to perform SERS experiments in aqueous phase. It is essential to study the protein structure in aqueous phase in order to maintain its native structure. Water immersion objectives with high NA (numerical aperture) and high magnification like the 1.2 NA and 60× magnification help to obtain high resolution, high throughput, and tight focusing of the incident beam in SERS studies. In a typical SERS experiment the analyte molecule like the protein molecule in our case forms a composite with silver nanoparticles. This composite in the liquid is under a constant Brownian motion, and therefore, the Raman signal collected is the resultant time-averaged signal of the composite residing in the probed volume of the laser beam. There will be some aggregation of the analyte–nanoparticle complex which settles down at the glass–liquid interface. These aggregated stationary complexes will give a larger Raman signal because of the increase in scattering cross section. This behavior was consistent for all SERS experiments. Therefore, the water immersion objective has the ability to spatially resolve two regions inside the aqueous solution along the focusing direction which are separated only by a few micrometers. We also performed SERS of the dried protein sample described in ref. 8 and also protein–nanoparticle complex aggregated by adding NaCl as an external aggregating agent. We could obtain comparable results in both these conditions mentioned above.
7. The laser power at the sample for the 532 nm laser was ~8 mW and for the 632.8 nm laser was ~5 mW. The laser intensity can be reduced by neutral density filters so that no photodegradation of the samples happen. No spectral changes due to photodegradation could be seen within 5 min of irradiation of laser of both the frequencies without the use of neutral density filters. Photodegradation is generally accompanied by a change in Raman spectra with diminishing of bands of protein and appearance of D and G bands of amorphous carbon. None was observed in the SERS experiments with the p300 proteins. The accumulation time used was typically 3–5 min to obtain spectrum with high signal to noise ratio.
8. It is important to obtain the SERS spectra of the small molecules and buffer only and compare them to that of protein–molecule complexes. It is essential to prove that the spectra come from protein rather than from the small molecules or the buffer. It is also important to keep the components of the optics clean and free from contamination from other proteins to obtain good spectra as SERS is a very sensitive process.

Acknowledgment

We would like to thank JNCASR, Department of Science and Technology (DST), Govt. of India, and Chromatin and Disease Programme Support, Department of Biotechnology (DBT), Govt. of India, for funding. DK and SS would like to thank the Council for Scientific and Industrial Research for the Senior Research Fellowship. TKK is a recipient of Sir JC Bose National Fellowship, DST.

References

1. Dixit CK, Vashist SK, O'Neill FT, O'Reilly B, MacCraith BD, O'Kennedy R (2010) Development of a high sensitivity rapid sandwich ELISA procedure and its comparison with the conventional approach. Anal Chem 82:7049–7052
2. Zhang K, Song C, Li Q, Li Y, Sun Y, Yang K, Jin B (2010) The establishment of a highly sensitive ELISA for detecting bovine serum albumin (BSA) based on a specific pair of monoclonal antibodies (mAb) and its application in vaccine quality control. Hum Vaccine 6:652–658
3. Feng F, Zhi G, Jia HS, Cheng L, Tian YT, Li XJ (2009) SERS detection of low-concentration adenine by a patterned silver structure immersion plated on a silicon nanoporous pillar array. Nanotechnology 20:295501
4. Moore BD, Stevenson L, Watt A, Flitsch S, Turner NJ, Cassidy C, Graham D (2004) Rapid and ultra-sensitive determination of enzyme activities using surface-enhanced resonance Raman scattering. Nat Biotechnol 22: 1133–1138
5. Dasary SS, Singh AK, Senapati D, Yu H, Ray PC (2009) Gold nanoparticle based label-free SERS probe for ultrasensitive and selective detection of trinitrotoluene. J Am Chem Soc 131:13806–13812
6. Mantelingu K, Kishore AH, Balasubramanyam K, Kumar GV, Altaf M, Swamy SN, Selvi R, Das C, Narayana C, Rangappa KS, Kundu TK (2007) Activation of p300 histone acetyltransferase by small molecules altering enzyme structure: probed by surface-enhanced Raman spectroscopy. J Phys Chem B 111:4527–4534
7. Mantelingu K, Reddy BA, Swaminathan V, Kishore AH, Siddappa NB, Kumar GV, Nagashankar G, Natesh N, Roy S, Sadhale PP, Ranga U, Narayana C, Kundu TK (2007) Specific inhibition of p300-HAT alters global gene expression and represses HIV replication. Chem Biol 14:645–657
8. Pavan Kumar GV, Ashok Reddy BA, Arif M, Kundu TK, Narayana C (2006) Surface-enhanced Raman scattering studies of human transcriptional coactivator p300. J Phys Chem B 110:16787–16792
9. Fleischmann M, Hendra PJ, McQuillan AJ (1974) Raman spectra of pyridine adsorbed at a silver electrode. Chem Phys Lett 26:163–166
10. Hu J, Zhao B, Xu W, Li B, Fan Y (2002) Surface-enhanced Raman spectroscopy study on the structure changes of 4-mercaptopyridine adsorbed on silver substrates and silver colloids. Spectrochim Acta A Mol Biomol Spectrosc 58:2827–2834
11. Joel G, Abraham N (1980) Electromagnetic theory of enhanced Raman scattering by molecules adsorbed on rough surfaces. J Chem Phys 73:3023–3038
12. John RL, Ronald LB (2008) A unified approach to surface-enhanced Raman spectroscopy. J Phys Chem C 112:5605–5617
13. Moskovits M, Suh JS (1984) Surface selection rules for surface-enhanced Raman spectroscopy: calculations and application to the surface-enhanced Raman spectrum of phthalazine on silver. J Phys Chem 88:5526–5530
14. Gao X, Davies JP, Weaver MJ (1990) Test of surface selection rules for surface-enhanced Raman scattering: the orientation of adsorbed benzene and monosubstituted benzenes on gold. J Phys Chem 94:6858–6864
15. Baker GA, Moore DS (2005) Progress in plasmonic engineering of surface-enhanced Raman-scattering substrates toward ultra-trace analysis. Anal Bioanal Chem 382:1751–1770
16. Otto A, Mrozek I, Grabhorn H, Akemann W (1992) Surface-enhanced Raman scattering. J Phys: Condens Matter 4:1143–1212
17. Aroca RF, Alvarez-Puebla RA, Pieczonka N, Sanchez-Cortez S, Garcia-Ramos JV (2005) Surface-enhanced Raman scattering on colloidal nanostructures. Adv Colloid Interface Sci 116: 45–61

18. Maurizio M, Cristina G, Emilia G (2011) Surface-enhanced Raman scattering from copper nanoparticles obtained by laser ablation. J Phys Chem 115:5021–5027
19. Alan C, Patanjali K (1998) Surface-enhanced Raman scattering. Chem Soc Rev 27:241–250
20. Stewart S, Fredericks PM (1999) Surface-enhanced Raman spectroscopy of peptides and proteins adsorbed on an electrochemically prepared silver surface. Spectrochim Acta Part A Mol Biomol Spectrosc 55:1615–1640
21. Arif M, Kumar GV, Narayana C, Kundu TK (2007) Autoacetylation induced specific structural changes in histone acetyltransferase domain of p300: probed by surface enhanced Raman spectroscopy. J Phys Chem B 111: 11877–11879
22. Tuma R, Prevelige PE Jr, Thomas GJ Jr (1998) Mechanism of capsid maturation in a double-stranded DNA virus. Proc Natl Acad Sci U S A 95:9885–9890
23. Liu X, Wang L, Zhao K, Thompson PR, Hwang Y, Marmorstein R, Cole PA (2008) The structural basis of protein acetylation by the p300/CBP transcriptional coactivator. Nature 451:846–850
24. Kundu TK, Wang J, Roeder RG (1999) Human TFIIIC relieves chromatin-mediated repression of RNA polymerase III transcription and contains an intrinsic histone acetyltransferase activity. Mol Cell Biol 19:1605–1615
25. Kumar GV, Narayana C (2007) Adapting a fluorescence microscope to perform surface enhanced Raman spectroscopy. Curr Sci 93: 778–781

INDEX

Sandra B. Hake and Christian J. Janzen (eds.), *Protein Acetylation: Methods and Protocols*, Methods in Molecular Biology, vol. 981, DOI 10.1007/978-1-62703-305-3, © Springer Science+Business Media, LLC 2013

If you have any concerns about our products,
you can contact us on
ProductSafety@springernature.com
In case Publisher is established outside the EU,
the EU authorized representative is:
Springer Nature Customer Service Center GmbH
Europaplatz 3, 69115 Heidelberg, Germany
Printed by Libri Plureos GmbH
in Hamburg, Germany

MIX
Papier aus verantwortungsvollen Quellen
Paper from responsible sources
FSC® C105338

If you have any concerns about our products,
you can contact us on
ProductSafety@springernature.com

In case Publisher is established outside the EU,
the EU authorized representative is:
Springer Nature Customer Service Center GmbH
Europaplatz 3, 69115 Heidelberg, Germany

Printed by Libri Plureos GmbH
in Hamburg, Germany